Topics

CONTENT

✓ Planning and Researching

▷ Organizing and Drafting

✓ Style

✓ Design

✓ Graphics

✓ Revising and Editing

The document is organized into easily recognizable chunks, which match the document's title. Notice, too, that some of the subheadings echo the title. Repeated phrases such as these not only contribute to the document's organization, but also to its usability. Data doesn't help readers if they don't know why it's provided. In this example, the reader can easily see *why* this information is given.

ENLARGE

Select a thumbnail

Pages 1 - 3 of 3

Interactive Model Documents in MyTechCommLab

MyTechCommLab includes more than 80 model documents that can be used to illustrate and extend the coverage of technical document types in *The Technical Communication Handbook*. These model documents include dynamic, interactive annotations that highlight key features, design elements, and usability guidelines for document types covered in the book, including:

Abstracts	Presentations
Application and Cover Letters	Proposals
Brochures	Reports
Descriptions and Definitions	Research Reports
Emails	Resumes
Instructions and Procedures	Websites
Letters	

D0142438

The Technical Communication Handbook

Laura J. Gurak
University of Minnesota

Mary E. Hocks
Georgia State University

PEARSON
Longman

New York San Francisco Boston
London Toronto Sydney Tokyo Singapore Madrid
Mexico City Munich Paris Cape Town Hong Kong Montreal

Executive Editor: Lynn M. Huddon
Senior Development Editor: Michael Greer
Senior Supplements Editor: Donna Campion
Associate Development Editor: Erin Reilly
Marketing Manager: Colleen Kochannek
Production Manager: Bob Ginsberg
Project Coordination, Text Design, and Electronic Page Makeup: Pre-Press PMG
Cover Designer/Cover Design Manager: Wendy Ann Fredericks
Cover Illustration: © Stockbyte/Getty Images
Photo Researcher: Chris Pullo
Senior Manufacturing Buyer: Roy L. Pickering, Jr.
Printer and Binder: Worldcolor/Taunton
Cover Printer: Coral Graphic Services, Inc.

For permission to use copyrighted material, grateful acknowledgment is made to the copyright holders on p. 615, which is hereby made part of this copyright page.

Library of Congress Cataloging-in-Publication Data

Gurak, Laura J.
   The technical communication handbook/Laura Gurak, Mary Hocks.
      p. cm.
   ISBN 0-321-36507-0
   1. Communication of technical information. 2. Technical writing. I. Hocks, Mary E. II.
Title.
   T10.5.G845 2008
   808'.0666–dc22   2008030438

Visit us at www.pearsonhighered.com

ISBN-13: 978-0-321-36507-1
ISBN-10:   0-321-36507-0

2 3 4 5 6 7 8 9 10—WCT—11 10 09

Contents

Part 3 — **Visuals and Other Media**   263

Part 4 Research Strategies and Tools 387

Preface

Technical communication, essential to many businesses, industries, and technical professions, has changed focus with the rise of the Internet and the quantity of available information on almost any topic. Good technical communication, more than ever, demands *information quality*, in terms of both content and presentation. An outdated, poorly written, or badly formatted web page, report, or podcast stands little chance of making an impact on its readers in an age where technical communication abounds. In response to this rise in technical information, technical writing and communication programs as well as jobs in these fields continue to be on the rise. In addition to being critical for those who major in and enter the profession of technical communication, the skills of a good technical communicator are important for anyone in the applied sciences or engineering fields.

Effective technical communication requires key competencies and specific skill sets. More than ever, today's technical communicator needs to be able to work in environments that are defined by highly specialized, modular, digital, and global forms of communication in multiple media. Key concepts, such as audience, purpose, and clarity, must now be balanced with the challenges that are presented by new and emerging genres such as blogs and wikis. Delivery media that can be print-based, web-based, or delivered to a handheld device makes audience and design considerations more complex than ever. The *Technical Communication Handbook* is designed to provide the kind of examples, guidelines, concepts, and reference tools needed for communicators to function well in this environment. In particular, this handbook is written and organized so that readers will recognize that today's technical communication is

- **Specialized and highly technical in nature**: Technical communication workplaces are not generic sites of writing and page layout. Instead, they are highly specialized environments that require knowledge of a technical topic and attention to the particular forms of writing and communication that take place in these settings. This book offers real-world examples from a variety of high technology and science domains in order to demonstrate and provide guidelines for these settings. For example, the book features charts, maps, and reports on topics such as chemistry, geology, environmental engineering, and nutrition science.
- **Collaborative**: Technical writing is nearly always done in teams. Today, many of these teams are spread out across the globe. This book recognizes this fact, offering (in Part 1) explicit instructions on project management, content management systems (used by teams to keep track of complex documents), teamwork and collaboration, and virtual teamwork.

- **Modular and non-linear**: Very few technical documents are written in a linear fashion and then moved from draft text into production, as is sometimes portrayed in textbooks on professional writing. Technical comunicators and their teams work with various parts of a document at the same time. Depending on the project, the delivery media or the visual elements may be the first point of discussion. Or, for international audiences, issues of translation or ethics may need to be discussed first. This handbook is designed in a modular format to take advantage of whatever approach an instructor or technical writer needs to employ.
- **Digital and visual**: Almost every technical communication product is produced for more than one media format and genre; writers produce many kinds of workplace documents (reports, email, memos, web sites). Even a plain, ordinary paper report is now expected by most users to be available in PDF on a web site. Unlike some textbooks that treat digital technology as an afterthought, this handbook integrates technology concepts and guidelines throughout. Further, we live in a highly visual age where screens and images dominate our communication environments. This book addresses visual communication in all relevant entries and in addition offers an entire part (Part 3) focused on examples and guidelines for visuals and digital media, from graphs and photographs to audio, comics, and video.
- **Global and intercultural**: In the digital age, all communication is global. This book includes entries (in Part 1) on international communication, teamwork and collaboration, virtual teamwork, and writing in regulated environments. These sections take a practical, example-driven approach rather than a more generic view of writing for cross-cultural audiences.
- **Rhetorical, driven by a sense of audience and purpose**: Technical communicators need to assess complex rhetorical situations, whether in research settings, college courses, or workplaces, and produce material that is clear, concise, and accurate. This handbook offers examples from a variety of settings, from student projects to real workplace documents. It also offers extensive sections on research and on grammar and style.

In short, the design, content, and organization of the *Technical Communiation Handbook* are in synch with today's technical communication workplace environment and today's technical communication student and instructor, while also having a forward-looking approach to changes in research, writing, and designing practices. This book is comprehensive enough to stand on its own, but flexible enough to use in combination with other textbooks or as a reference. The *Technical Communication Handbook* offers a comprehensive reference guide with coverage of the major genres and strategies for creating and editing technical documents but in an easy-to-use handbook format. Its brief, well-illustrated entries are designed to be a quick reference for both students and working professionals.

Features of This Handbook

- An author team experienced and recognized in technical communication (Gurak), writing-across-the-curriculum and composition (Hocks), and digital media (both) bring this combined research and teaching experience to the handbook.
- Accessible organization with entries grouped into major topic areas. Unlike other handbooks, where all entries are alphabetized, this handbook is organized by major topic (six parts) and then arranged alphabetically within each part, for ease of use by both students and instructors.
- The handbook is designed specifically for student writers "at work" in various disciplines as well as for working professionals.
- The book offers a visually effective look and feel with real documents, not typeset ones.

Several distinct features help this handbook support teaching and learning with a focus on authentic model documents:

- Detailed annotations accompany each model document, pointing to the important features of each and highlighting key layout, design, and format concerns for each document type.

- Guidelines boxes at the end of each entry serve as a quick access checklist of central issues in the writing and design of various kinds of documents.

- Audience Considerations stress the rhetorical nature of all document types and media, as well as the complicated and ethical nature of audience analysis.

- Design Considerations emphasize the written stylistic and format decisions relevant to a specific document type or media.

- "See Also" boxes at the end of major entries (Parts 2 and 3) cross-reference sections from the handbook when appropriate, connecting entries to key topics and strategies in other parts of the book to help users understand relationships among document types and rhetorical strategies.

How to Use This Book

This book is structured to be used in a flexible, modular format. The alphabetical arrangement of entries within each part makes specific sections easy to find and use in any order. Many instructors will wish to begin with some or most of the key concepts in Part 1 but then may wish to jump to assignments based on Parts 2 and 3, using Part 4 as the guiding section for doing research, then moving back to Part 1 to add "ethics," "style sheets," or some other elements, as suited to the project. Part 6 provides examples for how to cite sources in four representative formats for humanities, social sciences, physical sciences, and engineering. Part 5 is a thorough yet accessible reference for grammar and style issues, tailored to technical communication and can be used through the term.

Part 1, "Key Concepts in Technical Communication," contains much of the introductory material (audience, purpose, stages in the writing process) but also contains critical entries (ethics, accessibility) that some instructors deal with early on but others find most useful later in the semester. Entries can be taught in any order and include core rhetorical topics such as audience, purpose, ethics, and international communication. Part 1 is unique in that it contains many cutting-edge topics that are critical in today's technical communication environment but are rarely found in other technical communication handbooks, such as accessibility, content management systems, virtual teamwork, and writing for regulated environments.

Part 2, "Technical Document Types," is the genre section, providing extensive examples of typical documents. Unique to this handbook, the entries in this section offers not only the important standard document types (resumes, proposals, reports) but also newer genres that are here to stay, such as blogs and wikis, frequently asked questions, online help, and medical communication. Each entry in Part 2 provides examples and guidelines in a range of media formats and emphasizes the textual as well as visual and multimedia elements.

Part 3, "Visuals and Other Media," provides examples, guidelines, software tips, instructions for working with and creating the vast assortment of digital media available today. Entries on the more technical forms of visual information and data display use real-world examples, with data that is publicly accessible to students through government and other sources.

Part 4, "Research Strategies and Tools," is a resource for students and other researchers at any stage of their projects. This part explains the complexities of doing research in a digital age and provides extensive examples and guidelines. Featuring sample research tools and examples that focus on wind energy research, Part 4 includes comprehensive coverage of library-based and digital research skills within an accessible, current thematic focus. Key sections include comprehensive entries on usability and user testing, surveys and questionnaires, and citing and evaluating sources.

Parts 5 and 6 offer general guidance on grammar and style as well as documentation, but in both sections, the examples are geared toward scientific and technical topics. Part 6 includes excerpts from annotated student papers, one for each of the documentation styles.

Supplemental Material

Pearson's MyTechCommLab is a comprehensive online resource-designed to work specifically with this textbook. MyTechCommLab includes a wide array of multimedia tools, over 80 interactive documents, and an instructor gradebook and course management tools, all in one place. See the inside front covers for more information.

Acknowledgments

We appreciate the comments and input we received from numerous reviewers during the planning and development of this handbook: Roger Bacon, Northern Arizona University; Lynn Dianne Beene, University of New Mexico; Stephen A. Bernhardt, University of Delaware; Philip Bernick, Arizona State University; Elizabeth Christensen, Sinclair Community College; Sherry Cisler, Arizona State University; William O. Coggin, Bowling Green State University; Cathy Corr, University of Montana; Rose Day, Albuquerque Technical Vocational Institute; Amy Earhart, Texas A&M University; Leslie Fife, Oklahoma State University; Scott Fisher, Rock Valley College; Jeffrey Grabill, Michigan State University; Lila Harper, Central Washington University; Amy C. Kimme Hea, University of Arizona; Bill Klein, University of Missouri, St. Louis; Kevin LaGrandeur, New York Institute of Technology; Thomas Long, Thomas Nelson Community College; Caroline Mains, Palo Alto College; David Martins, California State University, Chico; Chris McKitterick, University of Kansas; Tomothy Miller, Millersville University; Elizabeth A. Monske, Louisiana Tech University; Lisa Neilson, State University of New York, Ulster; Jackie Palmer, Texas A&M University; Virginia Parrish, Southeastern Oklahoma State University; Alice Philbin, James Madison University; Colleen Reilly, University of North Carolina, Wilmington; Beth Richards, University of Hartford; Natalie D. Segal, University of Hartford; Summer Taylor, Clemson University; and Craig A. Warren, Penn State University, Behrend College.

We wish to thank Lee Scholder and Elizabeth Tasker for their editorial and research assistance with this project.

Thanks also to Lindsay Mateiro at Pre-Press PMG for managing the production process and Cheryl Hauser for her insightful copyediting.

Thank you to Lynn Huddon, executive editor at Pearson Longman for her editorial support of this book. A special and most heartfelt thanks to Michael Greer, our development editor extraordinaire, for his creativity, guidance, and input throughout the entire process. Finally, thank you to our friends and families for their ongoing support.

LAURA J. GURAK
MARY E. HOCKS

PART 1

Key Concepts
in Technical
Communication

Accessibility

To say that a technical document is *accessible* means that it can be read or used by the widest possible range of people. Accessible documents can be read and used by readers with visual impairments or other disabilities that may require them to use screen-reading web browsers or other assistive technologies. Technical communicators working in many industries and government agencies need to understand and apply the principles of accessibility to most, if not all, of the documents they compose and deliver to public audiences.

The Americans with Disabilities Act, known as the ADA, "prohibits discrimination and ensures equal opportunity for persons with disabilities in employment, state and local government services, public accommodations, commercial facilities, and transportation." Since being signed into law in 1990, the ADA has often come to be associated with building entrances and other facilities-related issues. Yet ADA compliance can be required for spoken and written documents and for web sites, too. If you are working on a form of technical communication where you need to consider ADA compliance, you should follow established accessibility guidelines.

For web sites, follow the guidelines from the World Wide Web Consortium (W3C). Their *Web Content Accessibility Guidelines* (WCAG) offer principles for page authoring and site design. Their guidelines begin with a list of four principles:

1. Content must be perceivable.
2. Interface elements in the content must be operable.
3. Content and controls must be understandable.
4. Content must be robust enough to work with current and future technologies. (W3C consortium 2004)

These principles each come with a set of technical guidelines that are detailed on the W3C web site (see the Works Cited section of this handbook). Because web sites contain information in a variety of media formats (text, images, sound, color, and so on), people with different visual, auditory, and cognitive abilities often need to use special software or devices to navigate and read or listen to a web document. The goal of designing for accessibility is to make web pages that are as user friendly as possible to the widest set of users under the widest set of circumstances.

For example, you can test to make sure your site will be accessible to a visually impaired person who accesses it using a screen-reading program. Often, if you are doing business with the state or federal government, you must prove ADA compliance. Many of the major software companies do this by running usability tests.

One of the benefits of including people with disabilities is that web developers can learn how people with disabilities interact with the Web and with assistive technologies. For example, consider a developer who does not know what it is like to use a screen reader. To meet the web accessibility guideline of providing text alternatives for all nontext content, the developer might provide the alternative text: "This image is a line art drawing of a dark green magnifying glass. If you click on it, it will take you to the Search page." However, observing a person who uses the site with a screen reader will clearly show the developer that this alternative text is ineffective and overly descriptive: "search" is all the user needs.

Another consideration for accessibility is the fact that computer users are getting older, an important fact when designing web sites and online information. A report by the National Institute on Aging and the National Library of Medicine offers guidelines for making web sites more accessible to older people. These guidelines include suggestions about

- What size type to use (larger sizes, such as 14 point), what fonts are best (sans serif, such as Helevetica), and other typographic ideas
- How to write and organize text so that it is clear, easy to follow, and broken up into short sections
- Visuals and animation, such as avoiding large video or image files that might take too long to download on older computers
- Navigation, such as using a standard, consistent page design and designing large navigation buttons

(National Institute on Aging 2002)

See also

Document Design (Part 3)

Web Page Design (Part 3)

Usability and User Testing (Part 4)

Audience

To whom are you writing? Audiences for technical communications are often categorized as *internal* or *external*. An internal audience consists of people who are in the same company or organization as the writers or designers of the communication. For example, content managers in the IT (information technology) group who create and maintain information on an employee web site for their company would have an internal audience. An external audience is one that is outside of the organization producing the communication. Customers, prospective customers, people from other organizations, or even the general public are all different categories of external audiences.

Other important considerations of audience are location, expertise, physical limitations, and other pertinent demographics. When thinking about your audience, consider the following:

- **Experience**—an audience's experience with the subject matter could range from novice to intermediate or expert.
- **Location**—audiences may be geographically dispersed around the country or around the world; this is often the case with a product's customer base.
- **Physical limitations**—some audience members may have physical limitations or unusual working conditions that require special considerations. This type of information should influence your design decisions, as it most certainly impacts the way the audience will interact with your documents.

Communicating with Multiple Audiences

Many technical subjects need to address multiples audiences, and each audience includes busy people who may not want to sit through tutorials and computerized training. They want information on demand. The challenge for technical communicators is to design fewer sources of information but make them suitable for a variety of audiences and scenarios. This strategy can be accomplished by providing one information repository with different views of topics and multiple navigational aids, such as tables of contents, indexes, and search functions designed for different levels of users.

Readers with Differing Levels
of Technical Expertise

A common challenge for technical communicators is to design documents for multiple audiences with different levels of subject matter expertise. Often these audiences will need different types of documents. For example, in the case of computer software or hardware documentation, the new user might need a tutorial; the experienced user might need context-sensitive help and an online index of tasks; and the administrator might need a reference guide, a compatibility guide, and a quick reference card of administrative procedures. In this case, each audience needs a different type of document.

Variability in the audience's subject matter expertise is made more complex by variability in audience members' levels of computer experience. Take, for example, the wide variety of users of tax preparation software. What kind of technical documentation do they need? It depends on their knowledge. Users' financial expertise may range from novice to expert, as might their computer skills. Thus at one extreme is the user with expert financial background and expert computer skills and at the other extreme is the user with novice financial background and novice computer skills. Also, what about the intermediate-level, infrequent user, such as the person who does taxes only once a year and who may have forgotten how to do certain tasks? The technical communicator needs to work with software developers to design a product, documentation, and online help that will help the novice and infrequent users as well as power users.

Before you begin any technical communication project, you need to understand who your audience is and what tasks or purposes need to be accomplished. For most projects, audience needs can be analyzed by completing an audience and purpose assessment.

See also

Accessibility (Part 1)

Audience and Purpose Assessment (Part 1)

Purpose (Part 1)

Audience and Purpose Assessment

Technical communication and information design are acts of communication prompted by rhetorical situations. What this means simply is that all technical communication is driven by the writer's or designer's purpose, which is to fulfill an audience's real or perceived need for information. Sometimes the audience's need for information is urgent and overt. For example, diagnostic and troubleshooting information often are needed immediately when equipment fails, and instructions may be needed to assemble a product or file taxes. Other times the audience's need may be less urgent, such as the information in a policy manual, which may go unread for a long period but likely will be needed by employees at some point. In some situations, the writer or designer of technical information may be trying to create feelings of need in the audience, such as in the case of sales proposal. Whatever the case, to begin any technical communication project, the designers and writers must first consider the purpose of the communication, the needs of the audience, and the situation in which the audience will receive the information.

Performing a Needs Assessment

To best achieve your purposes for writing, you must anticipate the needs and situations of your audiences. These can be identified through an audience and purpose analysis, sometimes referred to as a *needs assessment*.

By conducting a needs assessment, you can identify precisely the purpose, audience, and special requirements involved in a proposed technical communication project. Ideally, you should conduct a needs assessment before you start designing and writing your document. The needs assessment template shown here is a tool for determining audience needs and gathering project requirements.

**QUESTIONS TO CONSIDER WHEN YOU START
A TECHNICAL COMMUNICATION PROJECT**

- Why are you writing?
- To whom are you writing?
- How will your readers receive this information?
- Will they know what to do with it?

Needs Assessment Template

Use this form to gather information for your technical communication project. Record answers beneath the questions in the right column.	
Statement of purpose	Why create this technical communication?
Business need	What is the business need for the communication?
Audience	1. Who is the audience? Internal or external? Multiple audiences? What are their job titles and responsibilities?
	2. How large is the audience?
	3. What is their expertise related to the subject matter, their technical skills, the industry, etc.?
Situation	1. What is the context or environment in which the audience will receive the communication?
	2. Geographic locations? One or multiple languages?
	3. Other physical limitations or special needs?
	4. Will the audience be multi-tasking or distracted?
	5. Is the audience likely to be resistant?
	6. Will the audience require special equipment or software to receive the message?
Goals of the communication	1. What results do you expect the technical communication to deliver?
	2. What new skills, actions, or behaviors are you trying to elicit in the audience?
	3. What other quantifiable changes should the technical communication bring about? How might these be measured? Should they be measured?

(continued)

Needs Assessment Template *(continued)*

Requirements	1. Are there any special requirements for the technical communication?
	2. What type of medium is most suitable? (paper vs. screen, web based, etc.)
	3. What types of materials would be best? (help system, instructor's guide, slides, quick reference, video presentation, etc.)
Scope	Would the project be best served with one document or a collection of related documents?
Topics	List of topics that you would like covered:
Audience prerequisites	What are the required prerequisites for the audience?
Expertise of writers/designers	1. Who will design and develop the communication?
	2. What is their level of expertise?

Needs Assessment Template

The needs assessment template is a form for gathering data about your audience and the project as a whole. You might be able to fill out the template yourself, but it is best to talk to the proposed audience and subject matter experts (SMEs) to help you fill in the blanks with the most realistic responses. If several different audiences will use the technical communication, you may want to perform a separate needs assessment for each audience and compare their responses.

Depending on the nature of your technical writing project, you may need to modify, add, or delete some of the questions posed in the template. One way to determine whether the questions are appropriate is to try the template first as is, then modify as needed.

The sample document shows how a fluvial geomorphologist (an environmental scientist who specializes in the study of watersheds and river formations) used the needs assessment template in determining the requirements for an environmental engineering project report. (See Reports, Part 2, for a complete copy of this report.)

Sample Needs Assessment

Needs Assessment for Seeley Creek Geomorphology Report Completed by Steve Gough _December 2003_	
Statement of purpose	Why create this technical communication? _Government regulators at the local and state levels need to understand how changes to the flow in a suburban creek might affect its morphology and ecosystem. The report will summarize a field study using language and graphics to help laypeople understand complex stream processes._
Business need	What is the business need for the communication? _A municipal wastewater treatment plant seeks to increase the flow in a creek, and regulators need to decide whether to allow it._
Audience	1. Who is the audience? Internal or external? Multiple audiences? What are their job titles and responsibilities? _Multiple audiences including civil engineers, state government regulators who are biologists, fisheries biologists, and urban planners._ 2. How large is the audience? _Probably about 20–30 people._ 3. What is their expertise related to the subject matter, their technical skills, the industry, etc.? _See above—none are specifically trained in fluvial geomorphology, the main subject of the report (fluvial geomorphology is the study of river morphology and the processes that produce it)._
Situation	1. What is the context or environment in which the audience will receive the communication? _Written report._ 2. Geographic locations? One or multiple languages? _A single location, English only._ 3. Other physical limitations or special needs? _None._ 4. Will the audience be multitasking or distracted? _Not sure how to answer._ 5. Is the audience likely to be resistant? _Yes, regulators are required to read the report skeptically._ 6. Will the audience require special equipment or software to receive the message? _The report will be delivered in Adobe PDF form only._
Goals of the communication	1. What results do you expect the technical communication to deliver? _Understanding of the situation as stated above and persuasion to allow increase in flow of Seeley Creek._ 2. What new skills, actions, or behaviors are you trying to elicit in the audience? _Understanding and buy-in to project._ 3. What other quantifiable changes should the technical communication bring about? How might these be measured? Should they be measured? _The regulators will understand Seeley Creek's geomorphology as it would be affected by the wastewater treatment plant, and allow or deny a permit to it._

(continued)

Sample Needs Assessment *(continued)*

Requirements	Are there any special requirements for the technical communication? *No.* What type of medium is most suitable? (paper vs. screen, web-based, etc.) *The best medium would probably be verbal presentation by the author, but this is not practical; second best is the PDF document.* What types of materials would be best? (help system, instructor's guide, slides, quick reference, video presentation, etc.) *Engineering report backed by site surveys.*
Scope	Would the project be best served with one document or a collection of related documents? *In this case, regulators and others require a concise document with all relevant data well summarized.*
Topics	List of topics that you would like covered: *Description of current geomorphology. Prediction of impact of increased flow. Data to support prediction. Site surveys.*
Audience prerequisites	What are the required prerequisites for the audience? *Basic knowledge of environmental science.*
Expertise of writers/designers	1. Who will design and develop the communication? *Steve Gough of Little River Research and Design.* 2. What is their level of expertise? *Senior fluvial geomorphologist; MS in hydrology with extensive postgraduate experience and training.*

As the Seeley Creek example shows, the needs assessment template captures the following important facts about the rhetorical situation:

- The information will be read by multiple audiences (civil engineers, state regulators, biologists, and urban planners) for a total of 20 to 30 readers.
- The document will need to use scientific data to persuade the readers whether or not the project is environmentally friendly.
- Readers are likely to be resistant, which suggests that the document must have a persuasive element. The writer must decide how to be persuasive. In this case, clearly organized scientific evidence will provide persuasion.
- The writer feels that face-to-face verbal delivery of the information would be best but would be impractical. Also the need for scientific data to be referenced and studied by readers makes a printed report the best choice.

Thus, a needs assessment is useful in gathering not only information about the audience and purpose of a technical document but also the context or situation in which it will be used.

Content Management Systems

Content management systems (CMS's) are software programs (or web sites) developed to help project teams to keep track of documents going through multiple revision and comment processes. When multiple authors work together on a single document, as is often the case in technical communication, keeping track of the most recent version of a document file can become a task in itself. Content management systems are designed to respond to this need by organizing and tracking information that teams working across multiple work sites can access.

Content management software allows teams of technical writers to work on large, complex documents and sets of information by storing text and visuals in a central database, keeping track of the different versions and changes to information, and maintaining information in chunks, or units of information, that can be put together to form a variety of kinds of documents (reports, web pages, brochures) without requiring new text to be written and reformatted every time. Content management systems are also called a form of "database publishing," since text modules are often stored, tracked, and maintained in a database. Content management systems provide a systematic way to ensure that information written for regulated environments can be reviewed by legal and other compliance experts within the organization, and that the text the writers use is text approved for that particular setting.

Content management systems use different forms of *version control*, to keep track of each new version of a manual, set of instructions, web site, or other document. Version control is a process that keeps the older versions of information and visuals in a database and maintains associated information such as date changed, items changed, person or teams who made the change, authorization of the changes, and so on, for each update or revision. For many organizations, it is critical, or even required, to keep track of such changes for legal and compliance reasons. Content management systems often include *structured authoring* tools, which allow writers to use tags [also called markup languages, such as Standard Generalized Markup Language (SGML) or Extensible Markup Language (XML), and other *single-sourcing* tags] to indicate structure and content. Single-sourcing tags ensure that the company, organization, or writers can control appearance and structure. For example, a tag might be used to indicate that a chunk of text is a "Headline," and this tag leaves the interpretation to the particular style sheet and format desired. In this way, actual formatting (bold, font size) is not embedded into the text but is based on how this tag is interpreted. Text will thus appear in a consistent manner and will be compliant with, for example, the approved appearance of the company name or the FDA-approved way of describing a medical device.

Because almost all technical writing is done as a team activity, most technical writers will work in an environment where they must use a content management system. Desktop programs such as FrameMaker have some of this functionality built in, but most large organizations use database-driven content management systems that are either built for that company or purchased and customized.

■ Drupal is a popular open-source content management software that allows groups and organizations to publish and manage web site content

Drupal (www.drupal.org) is a free content management system that can be used for content management on web sites. Given how frequently web site information gets changed, it is important to use a system that keeps track of changes and helps the web administrator or site manager to update new content without losing the older information. Drupal and other content management systems also allow you to categorize content by using tags that users don't see but that help keep track of the type of information.

See also

Teamwork and Collaboration (Part 1)

Writing for Regulated Environments (Part 1)

Virtual Teamwork (Part 1)

Delivery Medium

Writers cannot control many things about audiences and their situations, but writers can control the design and delivery medium of the message. The final stage for any technical communication project includes distribution of the documents and getting feedback whenever possible before distributing the next version. Users or readers and their needs will help you determine the best delivery methods for your project.

What type of materials are you trying to deliver? The most common delivery media are screen and paper. Screen-based documentation (often called electronic documentation) is generally more interactive than paper-based documentation. Many types of electronic documentation, such as help systems and online tutorials, can be tightly integrated into a product to provide context-sensitive help, just-in-time training, and other effective collaborations of information and technology. For example, sophisticated database technology, with its powerful indexing and search features, can turn collections of technical documents into vast *knowledge bases*. Advances in digital technology have effectively blurred the line between product and documentation. The ability to link to any web page on the Internet and to attach all types of media files and programs to a web page allows designers to put diverse and highly interactive information at the reader's or user's fingertips.

Electronic delivery often includes converting files to a PDF (portable document format) file, which retains all formatting in a print-quality document.

Advantages of PDF files are that they can be password protected and can be read by anyone with the free version of the program Adobe Acrobat. However, users can't copy or edit these files without a full version of the software and permission to access the file.

Some technical information, however, does not lend itself to electronic devices and does not require interactivity. Specifications, reports containing long narrative, and other chronological and linear types of information may be cumbersome to view on screen and unusable on pocket technologies or hand-held displays. Also, large-format displays of information, such as posters or signage, may be more quickly produced and look best when created with paint, ink, and paper products. Digital graphics files can also be sent straight to production in the full-service copy shops and graphic design houses that own sophisticated output devices. Professional production equipment for on-demand printing, such as a wide-format, high-resolution printer, produces your large-format documents quickly and efficiently, saving you both time and money.

In addition to situational factors, budget is always a consideration in delivery medium. With limited resources to develop technical communication, organizations want to keep information development efforts streamlined. Even with multiple audiences, the overall trend in technical communication is to move toward a paperless work environment.

See also

Document Design (Part 3)

Style Sheets and Templates (Part 1)

Tools and Technologies (Part 1)

Descriptive Documents

A common purpose of technical communication is to describe or explain something. Descriptive documents can consist of anything from program overviews to product designs, site descriptions, or any other type of descriptive information that an audience might need. Descriptive information can appear in all kinds of technical communications, from printed reports to web sites, for a wide range of audiences, including technical experts inside an organization to external lay audiences such as prospective customers. The following illustration

■ Descriptive product information on the Web

shows an example of descriptive product information found on an environmental scientist's web site.

This figure shows a mechanical model that illustrates the same processes described by the Seeley Creek report included in Reports, Part 2. Water pumped through the model simulates river hydraulics and geomorphologic change of the water channel. The text on the left describes the Emriver model's purpose, while the photograph on the right gives the human scale of the model stream. Together, they create a better description than either one could alone.

Typical Descriptive Documents

Descriptive documents include:

- Reports
- Product descriptions

- Memos
- White papers
- Specifications
- Lab and field notebooks

Other documents, such as letters, brochures, web sites, and emails, often combine descriptive and persuasive information. See Part 2 for detailed entries on all of these types of technical documents.

Modes for Description

All descriptions—whether used alone or within task-based and persuasive documents—need details and organization to be effective. Description can benefit from using the classical modes of development:

- Classification (breaks into labeled parts)
- Cause and effect (if ... then statements)
- Compare and contrast (common features + differences)
- Definition (denotes meaning)
- Process (step-by-step organization)
- Narrative (chronological organization)

These modes are rarely used alone, but instead provide ways to develop well-organized, detailed descriptions and evidence for any type of communication.

> **See also**
>
> Document Design (Part 3)
>
> Persuasion (Part 1)
>
> Photographs (Part 3)
>
> Task-based Documents (Part 1)

Ethics

Technical communication is not a neutral activity. Rather, all aspects of technical communication can present situations where your sense of right or wrong is in conflict with decisions that may be more profitable or easier for a company to justify. Yet because technical communication involves important issues about technology, science, medicine, and other areas that can impact people, an ethical perspective in technical communication is essential. Should a pharmaceutical company leave known negative health effects out of its patient brochures or information for doctors? Is it acceptable for visual displays, such as charts, to be designed so that they deemphasize or skew the actual statistical data? Should photographs be altered to meet a public relations or company demand, even if this means changing the image so much that it barely resembles the original?

Ethical decisions should not be confused with legal ones. Many times, the law is quite clear about what one should and should not do legally. For instance, if you write an overtly untrue statement about a person, one that causes harm to this person's reputation, you may be charged with *libel*. But if your manager asks you to write a statement that leaves out critical information, on purpose, in order to skew reader understanding, you may be on solid ground legally but instead find yourself faced with an ethical decision.

Other areas where ethical issues can be important include the following:

- **Content**—In technical communication, you must work with large amounts of information, from a variety of sources, and synthesize this information into the right content for your intended audience. In so doing, you will make important decisions about what information can be left out, what information needs to be made extremely clear and placed up front, and what the balance is between providing just enough and too much information.
- **Structure and style**—Document structure and writing style offer important ethical choices. For example, sentence structure and usage creates emphasis in technical communication. If you hide the most important sentence in the middle paragraph of a long email or memo, using passive voice, you may be able to claim that you didn't leave the information out. But you also didn't make the information very clear or visible to your readers.
- **Charts and graphs**—Any graphical display of quantitative data, such as a pie chart or line graph, is susceptible to misleading presentation. For instance, in a pie chart, a color or 3D graphic could be used to make one part of the pie appear to be a larger or smaller proportion than it really is.
- **Photographs and visuals**—Photo editing and graphics software packages make it very simple to alter an image to suit a new purpose. Adding a person who was not in the original scene, altering someone's appearance, or modifying a visual so it differs from the original purpose are all occasions where ethical standards may be breached.
- **Design and page layout**—Technical communication is as much about design as it is about content. Using small fonts that make text hard to read may make important legal or financial information inaccessible to readers. Using a page layout that is confusing or hard to follow could mean that readers miss important instructions or procedures.

- **Research**—Many federal, state, university, and professional standards must be followed when doing research that involves people. These standards, often enforced by human subjects review boards (also called institutional review boards, or IRBs) are important ways to ensure that the right legal and ethical guidelines are in place for any study.

Technical Communication Situations Involving Ethics

Over the past few years, several high-profile ethical cases involved technical communication. If you research these cases, you will quickly discover that an important aspect of each case involved communication: memos, Microsoft PowerPoint slides, or altered photographs.

- **Tobacco trials**—A series of highly publicized trials determined that some tobacco companies knew about the serious addictive and health effects of smoking for many years, but that they purposely did not take adequate steps, based on available scientific evidence, to inform the public. Most of the evidence to support these findings came from careful analysis of the memos that were written by the lawyers, scientists, and managers at the companies involved. The wording of these memos raises ethical concerns when read today.
- *Columbia* **disaster**—On a January morning in 2003, NASA launched the space shuttle *Columbia*. Some time during launch, foam fell from an external fuel tank and struck the underside of the left wing. As a result, the shuttle broke up during reentry and all seven astronauts were killed. Before the space shuttle returned, scientists and engineers at NASA analyzed the potential impact of the foam that had struck the wing, using PowerPoint to present their findings to the team. Edward Tufte, an expert on visuals and graphics, has suggested that PowerPoint, which encourages the use of templates, bulleted lists, and "canned" style formats might have contributed to the disaster. Tufte (2005) argues that the format and templates of PowerPoint, when used to the extreme, allow people to substitute the tool—PowerPoint—for good, solid scientific logic and reasoning.
- **Photo altering**—In several cases photographs were altered to make them appear that a university campus was more diverse than the original photo suggested. Faces of ethnic minority students were added to a group photo that, in its original and true form, did not present a diverse student body. Here we see a double ethical problem: the new, altered group photograph was not a true representation, and the person whose face was added was never asked permission.

- **Medical information**—In several recent cases a new drug or medical device was released to the public and later, after health problems arose, it became clear that the companies knew about these problems but chose not to provide this information to patients and their doctors. In all cases of this sort, people along the way made the decision to keep this information out of publicly available documents.

Important Questions about Ethics and Technical Communication

You can ask yourself the following questions in situations where you have ethical concerns. You will need to decide when and how to approach a manager or other member of the organization if you feel that a decision could have serious consequences.

- Are you being asked to write content that is misleading, leaves out or buries important information, or exaggerates or confuses key ideas?
- Would the use of editing software to change an image for a brochure or web site go beyond what you feel comfortable doing?
- Is a chart or graph technically accurate but visually misleading to readers?
- If you are working on a team project, do you feel pressure from other team members to "lighten up" and not worry about something that you consider ethically unacceptable?
- What are the consequences of a using small font and very crowded page layout for important medical or financial information?
- Is the material respectful of privacy and confidentiality?

See also

Audience (Part 1)

Charts (Part 3)

Data Display (Part 3)

International Communication (Part 1)

Presentations (Part 2)

Purpose (Part 1)

International Communication

Many technical documents today are distributed internationally. Technical communicators often need to design and compose documents that may be translated into multiple languages and disseminated in many different cultures. Even documents written in English may be distributed to audiences in Australia, the United Kingdom, Canada, and Asia who may expect a different set of conventions for representing and understanding technical information. Technical communicators thus need to be prepared to develop documents that will travel across national, cultural, and linguistic boundaries.

Translation and Localization

Translation involves the conversion of text from one language to another. *Internationalization* means preparing information and materials so they are less culturally specific and thus easier to localize. *Localization* involves not only translating text to another language but also making sure that the information is optimized for the culture that will be using the material. Internationalization and localization ensure that the translation uses the correct terminology as well as appropriate idioms and expressions for a particular region and audience. In addition to accurate, regionalized translation, localization also involves

- Considering local standards, including laws and customs
- Converting currency for accurate rates and pricing
- Applying regional symbols and signs
- Adjusting visuals and removing potentially offensive presentation elements
- Accommodating regional formats, such as different paper size
- Using examples and images that are culturally relevant and appropriate

As businesses, customers, research communities, and web-based virtual teams are increasingly global, the need for translation increases. Translations are in demand for every conceivable type of document, from customer brochures and employee handbooks to entire web sites. Here is a sample of the types of documents that are translated today:

- Software, help systems, documentation, and web sites
- Brochures and marketing materials

- Employee handbooks, training materials, and benefits manuals
- Financial reports and presentations
- Project proposals and legal contracts
- Video transcripts

Translation Services

Most companies do not employ full-time translators. Rather, they have designated individuals, departments, or contractors that work with professional translations services or with globalization companies that offer both translation and localization as one branch of service. Translation services use a variety of specialists, including professional linguists, native speakers employed by the translation companies, and in-country freelance translators. Charges for translation and localization done by translations services are usually (but not always) calculated by the number of words translated, but cost depends on volume of material, type of translation being done, level of specialty (technical or legal), and other factors. Translation services offer the following assistance to other companies:

- Translating all types of documents and online information, including software, help, web sites, scripts
- Creating a list of usable terms
- Reviewing, editing, and proofreading
- Linguistic testing
- Localized formatting (media formats, currencies, and number and date formats)
- Compiling a style guide
- Reviewing cultural context and providing in-country research
- Globalizing web site design to account for the screen real estate requirements of various languages
- Revising legal contracts or customized reports based on national or regional regulations.

Electronic Translation Tools

Some translation and localization companies offer electronic translation tools that automatically translate web-based materials for selected languages as well as language conversion programs that translate a text file from one language to another. Globalization companies also offer other automated localization tools, such as automatic currency conversion calculators and multinational financial reporting tools. Translation companies often use software that relies on "translation memory," a comparison tool where segments (sentences or phrases) in a source language are compared with those in another language, and a form of logic is used to suggest the appropriate translation. Human translators are always needed to review the material and fine-tune or edit the translation.

Preparing a Manuscript for Translation

Companies take a variety of approaches when it comes to preparing documents for translation. The overall idea is to write in English in such a way that reduces as much ambiguity as possible, so that whether a document is translated by a computer or by a person (or by a combination of both), the outcome will be accurate.

One approach, suggested by expert John Kohl (1999), is called "syntactic cues." These cues are "elements or aspects of language that help readers correctly analyze sentence structure and/or to identify parts of speech." Using these cues appropriately can help reduce the ambiguity of the original sentence, such that the translation is more accurate. For instance, Kohl notes that English speakers often omit the relative pronoun and the verb to be. He gives us these two examples:

> Programs *that are* currently running in the system are indicated by icons in the lower part of the screen.

> Programs currently running in the system are indicated by icons in the lower part of the screen.

Removing "that are" might make the first sentence shorter, in English, but it could cause confusion because the second sentence is less clear.

Kohl offers a list of guidelines, with examples and explanations, for using syntactic cues when writing in English that will be translated into other languages. Often, these guidelines result in English sentences that contain more words than they might. But the extra words allow a translator or machine translation package to do a better job. Basic strategies for using syntactic cues to prepare manuscripts in English for translation to other languages include the following:

1. Do not omit direct or indirect articles (a, an, the).

   BEFORE   Reports will be placed in mailboxes.
   AFTER   The reports will be placed in the mailboxes.

2. Look for past participles and consider adding "that."

   BEFORE   Final copies made by the agencies will be delivered tomorrow.
   AFTER   Final copies that were made by the agencies will be delivered tomorrow.

3. Break up noun phrases and hyphenate compound adjectives.

BEFORE   The initial file mode creation mask value is zero.

AFTER   The initial value of the file-mode-creation mask is zero.

See also

Accessibility (Part 1)

Audience and Purpose Assessment (Part 1)

Usability and User Testing (Part 4)

Persuasion

Persuasion, or the activity of convincing an audience, is one of the chief aims of any communication, including technical communication. Persuasion is often associated with *rhetoric*, a term defined by the classical philosopher Aristotle as the art of finding and using the best available means of persuasion to move an audience to action. Persuasion involves making *claims*, then backing your claims with *evidence*, *reasoning*, *authority*, and emotional *appeals* in an effort to sway your audience to a particular choice, point of view, or course of action.

In the broadest sense, all technical communication is rhetorical. Any type of document that is written for an audience requires some level of persuasiveness. Even if the primary purpose of a document is to report factual information rather than argue a point, the writer has the responsibility of persuading the reader that the information is credible and accurate. And while the chief purpose for many technical documents is to describe complex technical information, for many others persuasion is the most important goal.

To better understand persuasive appeals, think about proposals, project plans, technical specifications, and web sites that a technical communicator might create. In addition to summarizing or providing details about a product, service, project, or company, these documents must also "sell" ideas and make the objects of discussion appealing to the target audience. Here are some examples.

If you were...	Then you would want...
A small business consultant who needed to write and submit a technical proposal to a potential client...	Your proposal to stand out and be chosen over the competition's.
A project coordinator charged with drafting a plan for your team at work or at school...	To generate enthusiasm and get "buy-in" from those who will participate in the project and those who must approve it.
A designer who needed to deliver specifications for a product so that others in your company could build it...	Your specifications not only to communicate your vision but to promote it in a positive and convincing way.
The web designer for a web-based provider of vitamins and health supplements...	A unique web site that appealed to your client base and persuaded them that your company's products and services were superior to alternative web-based vendors in your field.

Even documents as technical as engineering reports can have a strong persuasive element.

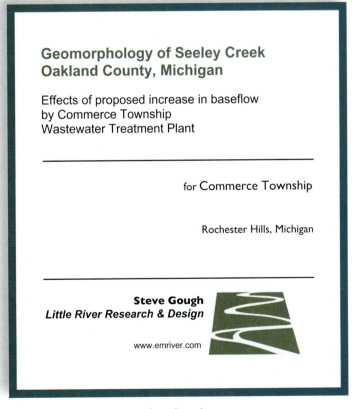

■ Cover of an environmental engineering report

As the cover indicates, this report uses scientific research to predict the effects of a proposed increase in flow from a water treatment plant into Seeley Creek. Nothing in the cover suggests persuasion, but the reader does expect to see plenty of scientific evidence. The following example shows the executive summary of the same report.

Executive Summary

The Township of Commerce proposes to increase output at the Commerce WWTP in Seeley Creek's watershed. The increased discharge, which may eventually reach 8.5 mgd, will increase baseflow from nearly zero to about 13 cfs.

Stakeholders have raised concerns about channel stability and tree mortality under the higher flow regime. To address these concerns I carried out a geomorphological survey of Seeley Creek and its watershed that included observation of most of Creek's channel, surveys of channel cross sections, bank materials tests, and hydraulic analysis. We also tested bank and bed materials for particle size and strength characteristics. This report also addresses specific concerns about erosion of cohesive soils and bankfull discharge theory.

Seeley Creek's channel was largely excavated by farmers decades prior to urbanization. Its channel has changed little since then because hydraulic forces are generally low and bank and bed materials, especially when vegetated, are highly resistant to erosion. Numerous culverts and other structures have prevented channel downcutting. The channel now runs in plant root-reinforced sediments that it is unable to rework, even under flood conditions.

Increases in current velocity and hydraulic shear from the higher baseflow are small, and not high enough to initiate channel instability of any kind. Seeley Creek appears to be little affected by even high return period (e.g., 10-year) floods, which strongly indicates that it is not a self-formed channel and cannot adjust to such a small change in its flow regime. I found only two instances of significant lateral movement of the channel. In both cases, the lateral movement in the past 45 years has been less than 18 feet.

We did locate areas of high tree mortality just south of 13 Mile Road. The poor condition of trees and shrubs in this reach, however, is caused by standing water on the floodplain and, more importantly, strong competition from grasses. The increase in baseflow will have no effect on tree mortality in Seeley Creek.

Seeley Creek's bed is generally composed of gravel and sand-dominated materials. These sediments are underlain, however, by cohesive strata that are resistant to erosion. Coupled with control of vertical erosion by built structures and generally low hydraulic energy, the chances for downcutting in this system are very slight.

Seeley Creek's morphology and vegetation will be unaffected by the increase in baseflow, and the chances of significant instability or woody plant mortality are very small.

■ Executive summary of an environmental engineering report

The highlighted sentences show that the document will provide evidence supporting the argument to allow the water treatment plant to increase its baseline output into Seeley Creek:

1. The first paragraph states the proposition under consideration: to increase the baseline flow from the water treatment plant into Seeley Creek.
2. The second paragraph raises the claim made by opponents of the plan that such a plan would make the channel unstable and cause an increase in tree mortality.
3. The rest of the executive summary counters the opponents' objections by stating that the proposed hydraulic increase would not create channel instability or increase tree mortality.
4. The last sentence summarizes an explicitly persuasive recommendation to allow the baseline flow of Seeley Creek to be increased.
5. Following this executive summary, the body of the report provides detailed evidence, discussion, and illustrations to support the argument presented here. (For a complete discussion of this report, see Reports, Part 2.)

Compelling evidence is one element of persuasion. But persuasion involves other factors as well. How do you go about creating persuasive technical communication? Customers and others in your target audiences do not usually want to debate or argue, nor do they want to feel manipulated. To be persuasive without being obnoxious requires understanding the types of persuasion that are effective in technical and professional contexts. An effective document should employ the best available means of persuasion, constructed with appeals that are appropriate to the document's purpose, situation, and audience.

Classic Methods of Persuasion

Traditionally, Western forms of communication have emphasized logical argument, consisting of claims, evidence, and reasoning.

A speaker or writer who is attempting to persuade an audience can use three common types of persuasion, which are often referred to as the three persuasive appeals (or *proofs*) of classical rhetoric:

- **Ethical appeals** (*ethos*)—refer to the credibility or character of the writer or speaker. Ethical appeals demonstrate the speaker's expertise or authority.
- **Logical appeals** (*logos*)—describe the reasonableness of the argument in itself. Logical appeals consist of factual evidence and logical reasoning based on facts, common knowledge, and shared beliefs.
- **Emotional appeals** (*pathos*)—describe the impact of the material on the audience's passions, feelings, or physical senses.

What gives a document...

...ethos (credibility)	...logos (logic)	...pathos (emotional appeal)
Appropriate tone	Clear subject and purpose	Emotional impact: sympathy, empathy, identification, etc.
Well organized and smoothly written	Evidence (facts and reasons)	Nostalgia
Good, thorough examples	Well-organized details: analysis (deductive), inference (inductive), probability (inductive)	Humor and other strategies to engage readers' emotions
Sophisticated analysis and insight	Modes of development: classification, cause/effect, compare/contrast, process, narrative, definition	Appeals to positive emotions: happiness, love, calmness, innocence, patriotism, spirituality, sexuality
Visually appealing Appropriate medium	Sound conclusions	Appeals to negative emotions: sadness, anger, guilt, fear
Research: external expertise		
Speaker's own expertise		

The three types of persuasive appeals are often pictured as in the form of a "rhetorical triangle" to show how they work together in any given context or rhetorical situation.

The triangle shows the relationships among the writer's ethos (credibility and authority), the audience's pathos (attitudes and emotions), and the message or document's logos (evidence, facts, and logical reasoning). Surrounding the writer, audience, and speaker is the context in which the communication occurs, which includes not only the communication medium but also other contextual variables:

- Geographic proximity
- Time frames
- Physical factors like space limitations
- Cultural background
- Attitudes
- Emotions

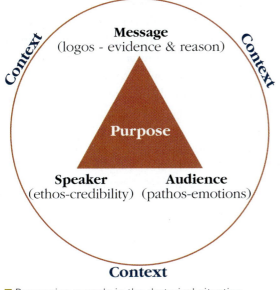

■ Persuasive appeals in the rhetorical situation

Still, whomever the audience, whatever the medium, and whatever contextual constraints may exist, the components of the rhetorical situation are always present, and they are critical considerations in designing persuasive appeals.

Logical persuasion is often the dominant appeal in technical communication, but ethical and emotional appeals are also important. For example, appeals to fear are often used in proposals and project plans—if the proposed plan is not adopted, negative consequences may result. Ethical appeals are also important in documents such as usability studies, market research reports, and product specifications. These types of documents must communicate that researchers of a study or the developers of a product understand the audience's needs, that they stand behind what they are purporting or selling, and that they have data to prove that their offering is solid.

Modern Methods of Persuasion

When designing persuasive communication in digital environments, in addition to the three traditional appeals of ethics, logic, and emotion, technical communicators need to consider multiple modes of communication. Text, visuals, sound, music, spoken word, gestures, and designed spaces can work together as subtle forms of persuasion in terms of how they encourage a user to interact with and interpret information in a multimedia document. The available means of persuasion have changed to include all forms of media.

Cultural connotation is also very important in persuasive technical communication, especially in the contexts of global business and worldwide distribution of documents over the Web. The aggressive style of a formal debate seems ordinary in U.S., Canadian, and Western European communication because those systems of law and government use these kinds of persuasion. Forms of inductive argument based on inference from data and mathematical probability are also common in scientific studies. But many other forms of inductive argument exist in non-Western cultures. Using story-telling techniques and parables for persuasion is common in many cultures. If you are designing and writing technical communication for a global audience, try to balance these different forms of persuasive appeals, examples, and styles of argument by learning as much as you can about your audience.

See also

Audience (Part 1)

Reports (Part 2)

Situation (Part 1)

Visual Communication (Part 3)

Project Management

Teamwork and group projects require clear and realistic goals. Project management involves setting goals, organizing tasks and team roles, and maintaining systems and schedules to keep a project moving smoothly. Specific project management strategies and tools can help you accomplish tasks by better supporting the work of the group. A work process method and a project schedule usually determine how well a group works together to complete a complex task.

A project schedule may include much of the following:

- Scope of the work
- Time to complete work and progress report dates
- Individual task assignments
- Detailed calendar

- Equipment list
- Deliverables

Project management visuals using tools like a *Gantt chart* can help manage your projects as it schedules all the benchmarks backward from when the work must be completed. With complex projects or large teams, a chart that schedules work based on other work and benchmarks is absolutely essential. Take, for instance, a renovation of a bathroom. Tile cannot be completed until the drywall is finished. Wiring should be done before the drywall begins but after the demolition. Plumbing must be roughed in before the appliances are placed. Fixtures must be selected before installation can be completed. A chart for this remodel shows all these interdependent activities and can be printed and displayed in the bathroom itself for common access. The chart looks like this:

■ Sample schedule with benchmarks

Project management tools are used for a variety of situations in the technical communication workplace. The development of a complex software program or the research, writing, design, and production for a set of user instructions are two other examples where project management tools would play a critical role in keeping the team on track and making sure key benchmarks and due dates are not missed.

A wide range of work styles, technologies, equipment, and other factors may complicate the process. If a programmer works all night authoring code in a lab, for example, while the graphic designer works at her home office and the software development team leader attends her meetings in various locations, how will they exchange their work and communicate about their progress? Scheduling becomes the most critical issue. A software tool like Microsoft Project focuses on the scheduling of work for a project and resides on the Internet so that all team members can access and modify the schedule.

■ Sample Microsoft Project schedule

Here are some general tips for managing large projects:

- **Display** your project management documents where all members have access (often online) and have each member check off tasks when they are completed.
- **List** a preferred method of communication for each member (email, phone call, memo, text message)
- **Identify** the file types and other acceptable formats for exchanging work.
- **Determine** the normal amount of response time or turnaround (i.e., 24 hours).

- **Coordinate** schedules in advance for holding meetings, whether virtual or face to face.
- **Schedule** progress reports and stick to the schedule for presenting them.

Having a centralized document management system can help keep from confusing versions of a collaborative project (see content management systems). Knowing your collaborators' access to email on weekends or their cell phone use will allow planning your communication strategies. Regular phone conference calls with documents and computers in front of each participant makes a very effective team strategy for managing a long and complex project (as with this book). In general, staying with the schedule and using communications via email or other communication technologies allows those who work on different schedules to collaborate efficiently and effectively.

> **See also**
>
> Content Management Systems (Part 1)
>
> Teamwork and Collaboration (Part 1)
>
> Project Management Visuals (Part 3)

Purpose

Why are you writing or creating this document? The primary purpose of all technical communication is to provide information to an audience or multiple audiences. When beginning a technical communication project, think about your goals.

Some technical documents are designed to serve a dual purpose. For example, a training manual could provide an overview of the various features available in a product or system and then provide the steps for using that product or system to perform particular tasks. Web sites often describe a company and its products, and services, while also attempting to sell those products or generate qualified customer leads.

Purposes for technical documents

IF your goal is to...	THEN the document's primary purpose is...	For example:
Help readers perform a task or group of tasks.	**Instructional** Emphasizes task-based information	You might need to help readers build, install, or use some type of equipment. Examples of instructional documents are user guides and help screens.
Explain how something works or to explain its parts or features.	**Descriptive** Emphasizes modes of development	You might need to give a general overview of a project or the detailed inner workings of a program or system. Examples of descriptive documents are product overviews, detailed specifications, and reference guides.
Persuade readers to make decisions or take actions.	**Persuasive** Emphasizes audience-based appeals	You might need to persuade users to start new projects, adopt a technology, or purchase a product. Examples of persuasive documents are brochures, proposals, and recommendations.

See also

Audience (Part 1)

Audience and Purpose Assessment (Part 1)

Task-based Documents (Part 1)

Descriptive Documents (Part 1)

Persuasion (Part 1)

Situation (Part 1)

Situation

The situation (or context) of any technical communication includes not only where and how the audience will receive the information but also their general attitudes, reading styles, and habits. For example, will an audience be in a hurry or have time when viewing the information? Hurried readers, especially of screen-based media, including computer monitors and hand-held electronic devices, are often impatient and easily distracted when it comes to reading technical information. Readers of paper-based technical communication are usually not reading for pleasure; they are often in a hurry to find a specific answer or quickly gain general knowledge on a real-world subject. Most readers of technical communication are scanning text as quickly as possible with a particular goal in mind.

When designing technical communication, here are some useful questions to reflect on about the situations of potential readers:

- Is the audience likely to be multitasking, or will they be focusing their undivided attention on reading the technical communication?
- Is the technical communication expected and invited by the audience, or did it present itself unexpectedly?
- Will the audience see the information on a computer screen, in a brochure, on a large wall poster, or elsewhere?
- Will the information be delivered on a disk with software or in a box with equipment?
- Will the information be heard as an announcement or recorded training session, either in addition to or rather than being seen?

See also

Audience and Purpose Assessment (Part 1)

Audience (Part 1)

Delivery Medium (Part 1)

Persuasion (Part 1)

Purpose (Part 1)

Stages in the Writing Process

Technical writers will have a better chance of producing quality communications if they understand and agree to follow a general design and writing process. Technical communication projects should begin with thorough planning and research to determine the scope, effort, and resources needed, as well as a predictable time frame for completing the project. When planning a technical document or project, think carefully about your audience in light of the following:

- Purpose: to persuade, to inform, to explore, to teach, to decide on a course of action.
- Situation: is the communication related to the past, present, or future? What should be done?
- Required or hoped for outcomes.

- Sources for research and background information.
- Points of contact and communication between the audience and writer or speaker.
- Constraints, such as resources and time.
- Risks that could prevent or delay the successful completion of the project.

Most organizations and groups of professional technical writers have an established planning and writing process. The process needs to be flexible, but some basic steps are always important, as shown in the following diagram.

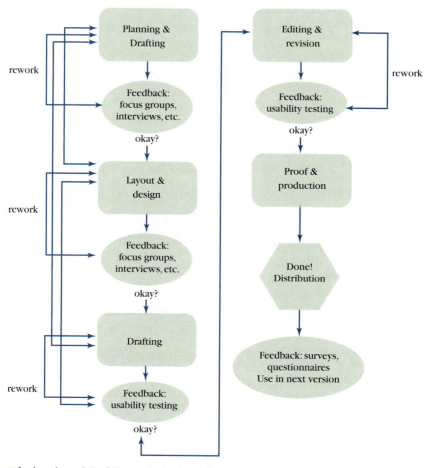

■ A visual model of the technical writing process
- Squares represent steps by writers and designers.
- Circles represent possible points of audience feedback.
- Questions "okay?" and "rework" represent how writers and designers evaluate along the way in light of the audience.
- Arcs and arrows represent the nonlinear looping back and forth in this process.

This graphic shows typical steps in the writing process, where the squares indicate what writers and designers do: planning and research, layout and design, drafting, editing and revising, proof and production, and finally, when done, distribution to the audience. Because these processes are nonlinear, writers and designers need to reevaluate everything along the way and ask themselves questions in light of their purpose and audience. Thus, the circles in the diagram show possible steps for audience feedback, which helps prompt these questions and ultimately improve the document.

Designing and writing technical communication is an *iterative* process, which means that looping back and forth between steps is a normal part of writing. Feedback loops should be built into the writing process at multiple stages to allow document designers and writers to validate their ideas with potential readers and to make revisions based on reader feedback. These feedback loops help ensure that a document serves its intended purpose and meets the needs of its audience. By soliciting reader feedback, writers can

- Learn more about readers' attitudes and expectations.
- Measure the effectiveness of a document.
- Revise parts of the document that are unclear or are not meeting the readers' needs.

Feedback loops can occur at any point during the design and writing processes, but they are most useful during the design and research phases because readers can help writers envision what is most important in the proposed communication. Technical communicators can then build reader requests into the design. However, writers need to review feedback during the design phase carefully because readers do not always know what they need.

Another good time to gather feedback is early in the drafting process when a sample section of a paper document or a prototype of the user interface for an electronic document is available. This technique, called *rapid prototyping*, lets readers react to a sample and provide valuable early feedback. Many times, feedback is not solicited until a document is complete but before it is finalized and distributed. The danger with waiting for the finished project to solicit user feedback is that missing information, design flaws, and other problems are more difficult and costly to fix at the end of a project and may delay delivery.

Unfortunately, sometimes delivery schedules are so tight that writers have no time to conduct feedback loops at all. If this is the case, feedback can still be gathered after the document is delivered to its audience. The feedback can then be used when the document is modified in the future. Ideally, feedback loops should be built into more than one phase of the project so that you continuously revise and improve throughout the writing and designing process.

For more on	See also
Planning and Drafting ▶	Part 1, Audience, Purpose, Situation
Starting Research ▶	Part 4, Digital Research Strategies, Image Searches on the Web, Primary and Secondary Sources
Gathering Requirements ▶	Part 4, Focus Groups, Surveys and Questionnaires
Creating Project Plans ▶	Part 1, Project Management, Teamwork and Collaboration
Audience Feedback ▶	Part 4, Usability and User Testing
Layout and Design ▶	Part 3, Document Design, Visual Communication
Planning Data Visuals ▶	Part 3, Charts, Graphs, Data Display, Technical Illustrations
Planning Web Sites ▶	Part 3, Web Page Design, Part 2, Web Sites
Editing and Revising ▶	Part 5
Proof and Production ▶	Part 5, Part 6
Delivery and Distribution ▶	Part 1, Delivery Medium, Tools and Technology

Style Sheets and Templates

Before you deliver your final materials, you need to consider layout and what software platform you will use for delivery to the audience. Style sheets and templates both provide shortcuts for formatting many standard parts of a document and can be used at any point in the writing and designing process.

Style sheets refer to a set of specific layout features that are predefined for documents. Traditionally called "style guides," these requirements usually contain company guidelines and specifications for document design, and the details of organization and writing style for print texts. Style sheets in word processing and other software programs let you define these elements in a list of named styles, such as "Heading 1," "bulleted text," and so on, to apply repeatedly throughout a text. In Microsoft Word, for example, all styles are defined based on a "normal" style, which includes the font name and size, the paragraph alignment, and text features such as underlining and color. Styles can be created and modified using any kind of formatting that the program allows. That list is called a style sheet and it applies to the specific document that you are writing, as pictured next:

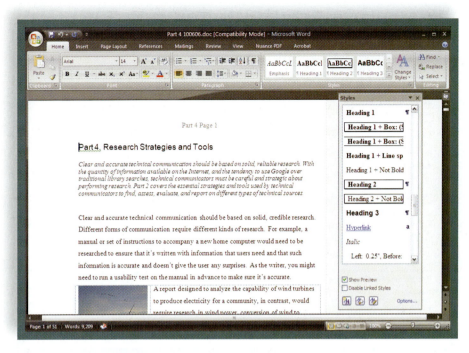

These styles offer the advantage of automatic layout features so that you don't have to format constantly while writing. Cascading style sheets (CSS) are similarly used in hypertext markup language (HTML)–coded documents for the Web so that you automatically set layout features for web pages instead of entering every element in the code. Styles and style sheets can also be saved and transferred between documents.

Templates are generic documents with predefined page layouts and graphical elements that a writer uses to generate a professional looking document. A master style sheet is also a template. Most word processing, page layout, web design, and presentation software applications provide a variety of templates to help users build documents quickly. This sample document shows the trifold brochures template available in Microsoft Word.

HOW TO CREATE A BROCHURE

Using this template, you can create a professional brochure. Here's how:

1. Insert your words in place of these words, using and/or re-arranging the preset paragraph Styles.

2. Print pages 1 and 2 back-to-back onto sturdy, letter size paper.

3. Fold the paper like a letter to create a three-fold brochure (positioning the panel with the large picture on the front).

WHAT ELSE SHOULD I KNOW?

To change the Style of any paragraph, select the text by positioning your cursor anywhere in the paragraph. With your cursor blinking in the paragraph, choose Styles and Formatting from the Format window. Finally, from the Styles and Formatting work pane, choose a new Style.

To change the picture, first click on it. Then, point to Picture on the Insert menu and choose From File. Select a new picture, and click Insert.

BITS, BYTES & CHIPS

CUSTOMIZED TURNKEY TRAINING COURSEWARE

Future Solutions Now

&

Bits, Bytes & Chips
123 Main Street, Suite 100
Any City, State 12345-6789
Phone (123) 456-7890
Fax (123) 456-7890

HOW TO CUSTOMIZE THIS BROCHURE

You'll probably want to customize all your templates when you discover how editing and re-saving your templates would make creating future documents easier.

1. In this document, insert your company information in place of the sample text.

2. Choose Save As from the File menu. Choose Document Template in the Save as Type: box (the filename extensions should change from .doc to .dot) and save the updated template.

3. To create a document with your new template, on the File menu, click New. In the New Document task pane, under Templates, click On my computer. In the Templates dialog, your updated template will appear on the General tab.

ABOUT THE "PICTURES"

The "pictures" in this brochure are Wingdings typeface symbols. To insert a new symbol, highlight the symbol character and choose Symbol from the Insert menu—select a new symbol from the map, click Insert, and Close.

HOW TO WORK WITH BREAKS

Breaks in a Word document appear as labeled dotted lines on the screen. Using the Break command, you can insert manual page breaks, column breaks, and section breaks.

To insert a break, choose Break from the Insert menu. Select one option. Click on OK to accept your choice.

HOW TO WORK WITH SPACING

To reduce the spacing between, for example, body text paragraphs, click your cursor in *this* paragraph, and choose Paragraph from the Format menu. Reduce the Spacing After to 6 points, making additional adjustments as needed.

To save your Style changes, (assuming your cursor is blinking in the changed paragraph), click on the down arrow for the Style in the Styles and Formatting work pane. Select Update to match selection to save the changes, and update all similar Styles.

To adjust character spacing, select the text to be modified, and choose Font from the Format menu. Click Character Spacing and enter a new value.

OTHER BROCHURE TIPS

To change a font size, choose Font from the Format menu. Adjust the size as needed, and click OK or Cancel to exit.

To change the shading of shaded paragraphs, choose Borders and Shading from the Format menu. Select a new shade or pattern, and choose OK. Experiment to achieve the best shade for your printer.

To remove a character style, select the text and press Ctrl-Spacebar. You can also choose Default Paragraph Font from the Styles and Formatting work pane (accessible from the Format menu).

BROCHURE IDEAS

"Picture" fonts, like Wingdings, are gaining popularity. Consider using other symbol fonts to create highly customized "Icons."

Consider printing your brochure on colorful, preprinted brochure paper—available from many paper suppliers.

AT FEES YOU CAN AFFORD

We can often save you more than the cost of our service alone. So why not subscribe today?

Call 555-0000

■ Template for a trifold brochure

This template provides a predefined layout of a trifold brochure along with instructions for replacing and customizing the text and graphics to the writer's own subject matter. The advantage of templates is that they have been created by professional graphic designers and can give a professional look to a document. The disadvantage is that templates impose a layout, which can be restrictive, and they can be difficult for inexperienced users to modify. You can create and save your own templates in many of these same programs once you identify the specific styles and formats that work best for your project.

See also

Teamwork and Collaboration (Part 1)

Document Design (Part 3)

Web Page Design (Part 3)

Task-based Documents

A common purpose of technical communication is to provide instructions for completing a task or a series of tasks. Task-based technical documents most commonly consist of instructions, which, if followed correctly, should allow a reader to achieve a certain result. The set of instructions for completing a single task is called a *procedure*.

Documenting Procedures

Procedures are used in technical writing to document a wide variety of tasks: everything from installing a software application to putting together a piece of furniture. The sample shown on page 42 shows a procedure for adding a new contact to an online address book of an email program.

There are many types of procedures common to technical communication. These include the following:

- Standard operating procedures: Often referred to by the abbreviation "SOP," these procedures are common in factories and other facilities where it can be required by law that the company maintain a set of approved safety and operating procedures.
- Medical and health care procedures and protocols: Medical and pharmaceutical companies as well as hospitals normally must maintain an up-to-date set of procedures.

- Procedures that accompany policy documents: Many organizations keep high-level concepts ("we value a diverse workforce") in a policy document but then design procedure documents to discuss the details on how such a policy gets implemented.

To add a contact to your address book:

1. Select **File, New, Contact,** or press **Ctl+Shift+C**. An untitled Contacts window appears.
   TIP: If you already have the Contacts folder open, you can use one of these methods to display a untitled Contacts window:

   - Click the **New Contact** icon in the toolbar.
   - Select **Contacts, New Contact**.
   - Press **Ctrl+N**.
   - Double-click in a blank area of the Contacts folder workspace.

2. Type a name for the contact in the Full Name field.
3. If desired, type a job title and company for the contact.
4. Select an option for the File As field. The File As field gives you several choices for filing the contact, such as by first name, last name, or company.
5. After completing all contact information, select **Update**. The address book saves the new contact information.

■ Procedure for adding a contact to an online address book

When writing procedures, observe these guidelines:

- Use a descriptive heading or short introductory phrase to introduce the procedure.
- If the order of steps is important, be sure to number the steps. If your procedure is longer than ten steps, consider breaking it into two procedures if possible.
- Begin each step with the action that the reader is to perform. Then describe the result of the action. What happens?
- Avoid passive voice because it makes the doer of each step unclear. Instead use the active voice or the imperative or command form of a verb.
- If the organization has a set format for the procedure, follow that. Often this format is dictated by a regulatory agency or the company's legal department.

Many types of technical communication require procedures. For example, user manuals and help systems contain large collections of procedures that describe how to use the many features of a technically complex product, such as a software application or an electronic device.

Documenting Processes

A *process diagram* is a type of chart used to represent interdependent procedures in a task-based process. When a process becomes complex or involves multiple steps, a process diagram, also called a *flow chart*, can help readers grasp the whole series of related steps. This example shows how the US Federal Government decides when federal officers and officials can use Government aircraft:

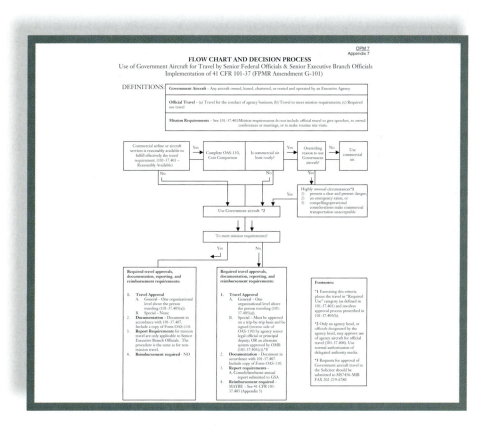

■ Flow chart for managing government aircraft usage by officials

The document this flow chart represents, called an Operational Procedures Memorandum (OPM), outlines complex policies and procedures for improving the management of Government aircraft. This chart provides a clear visual explanation of the complex, multi-step policies and procedures, as well as how the steps are related in the decision-making process. Each step requires a "yes" or "no" answer in the decision making process. Readers can then look in the OPM text and find the specific procedures related to each step in the process.

Process diagrams and other charts work best when they provide your audience with an accurate visual outline of the whole process.

See also

Charts (Part 3)

Graphs (Part 3)

Instructions (Part 2)

Purpose (Part 1)

Teamwork and Collaboration

Technical communication is rarely a solitary task. Teamwork is common in technical workplaces and the actual coauthoring of documents is also not unusual. Writing and other forms of communication will incorporate many ideas and suggestions from others before they are complete. Even when you work on projects independently for periods of time, you will already spend countless hours talking, listening, and gathering ideas—collaborating with other people. Strategies presented in this section for both leading and working within teams will make you a better technical communicator. Likewise, new communication technologies can help you in the collaborative process, especially when you are managing large documents, complex teams, or long-term projects.

Leading a Team

Effective teams require organized capable leaders. Good leadership requires that you recognize and can use specific skills and behaviors, such as

- Setting clear and specific goals and outcomes for the team
- Having realistic expectations about meeting those goals

- Holding productive team meetings
- Outlining action steps and deliverables for the group
- Providing regular, periodic reviews
- Using good time management and organizational skills
- Delegating work regularly to qualified team members
- Avoiding micromanagement of team members and their projects
- Building community and common purpose among the team members by rewarding positive behavior and encouraging ideas that support a team approach

Leadership abilities, though harder to achieve, can be developed over time with effort, research, and practice. Some of the capabilities seen in strong leaders include these:

- Vision and enthusiasm for ideas and projects
- Knowledge about the subject, market, users, and the competition
- Trust in qualified team members
- Ability to build community and find consensus among people and differing ideas
- Supervisory abilities
- Open communication to and from members about challenges and mistakes
- Flexibility toward others

Being a Good Team Member

Many of the abilities listed for a strong leader require interpersonal skills and the ability to communicate and get along with other people in a group or on an individual basis. But these same abilities are important when you are a team member, too. All members of a team must communicate well in order to accomplish common goals. This communication is particularly challenging when people with different specialties and personalities comprise the technical team. Some of the most important strategies in a team include these:

- Willingness to ask for help
- Willingness to review work when asked
- Recognition of another's success
- Offering constructive criticism
- Providing clear channels of communication
- Timely replies to inquiries
- Giving honest answers, including "no"
- Sensitivity to and respect for differences
- Using the right communication technology (instant messaging, phone, email, virtual team meetings) to accomplish the task

Some of these strategies will be more important to you than others depending on your leadership or support role in the group. With so many teams spread out over vast distances clear communication is key. Some general tips for team communication include the following:

- Speak in a calm voice whenever possible with your team.
- When arguments arise, find someone who can act as mediator. List the pros and cons, get input from all sides of the argument, and reach a compromise solution.
- Agree to disagree when necessary and move on.
- Avoid the use of email for overly complicated problem solving and never engage in email arguments, which escalate quickly without nonverbal cues to soften the language. The telephone, or, if possible, a quick face-to-face meeting can solve many problems more easily than email can.
- Use project management software so that all team members can stay up to date on the project timeline and common documents.
- Listen patiently to a colleague's point of view without interrupting.
- Reflect back specifics of what you hear in a conversation, as in "what I hear you saying is that…" or "your concerns seem to be…."
- Correct another's errors firmly but with specific suggestions for improvement.

If you frequently act as the dominant member of a group, you may need to hold back and allow others the opportunity to contribute. Likewise, shy or nervous group members may need more time to feel comfortable and offer their contributions. For the quiet members, offering suggestions in writing can be a good way to break the ice and get vocal in your group.

See also

Content Management Systems (Part 1)

Project Management (Part 1)

Virtual Teamwork (Part 1)

Tools and Technologies

Technical communicators must use new technologies regularly to design and produce their documents. With so many tools available to support research, document design, job management, and production, it is more important than ever to choose the right one for the job. Before you begin a new project, ask yourself what technologies and tools you need to use.

Will you need...

Company-specific technologies?	Email and communication systems	Project management and scheduling tools	Mandatory templates or style guides
Authoring tools?	For page layout	To create media elements	For web page creation
Research technologies?	Proprietary databases	Specific Internet research sites	Intranet or content management systems
Pocket technologies?	To gather primary data	As a virtual office	For webcasting

Once you begin a project and have determined your audience, purpose, and situation, you can select the technologies and tools that will best enable you to complete the project.

Choosing Appropriate Technologies

Most projects you create as a technical communicator will be portable, except those projects that become built installations or displays. Your delivery platform for documents will help you determine what technologies you need and how much storage space your project will require. Will other people access your project over the Internet, in networked folders, on high-storage media, or on pocket technologies or other drives? Those decisions about delivery will affect how complex the technologies will be to design your project because speed, bandwidth, processing power, and data storage space still vary widely with each delivery platform. Here are some tips:

- **Use the simplest tool for the job**—Often you will need to use whatever tool is available at the company or organization where you work. If you can choose a software tool, try choosing the simplest one so that you don't have to work with complicated features that you don't need. Many personal computers come with pre-installed programs for video editing, word processing, and audio editing, and most of these are easy to use and cross-platform compatible.

- **Download trial versions first**—Many 30-day versions are available to ensure the software does what you need.
- **Consider creating PDF files to deliver print-friendly files over the Web**—If all you need is simple layout, use a word processor and save your document as a PDF file instead of using elaborate page layout software like Quark XPress or Adobe InDesign.

Learning to Use Technology and Software

Learning to use a new program or a new technology can be intimidating for many people. At the same time, the ability to grasp and learn to use new technology and software is an essential skill for technical communicators in the workplace. Unfortunately, many companies and organizations do not adequately support training or instruction when it comes to new equipment and software. You may find yourself more or less on your own, with a looming deadline and a project that requires you to use a new program for its completion. You may, for example, need to learn a new database program and develop a procedure for using it to track inventory. Or you may be asked to "take the lead" on setting up a process that enables off-site team members to collaborate on a project in real time using web-based virtual collaboration tools. Beyond the immediate task of learning to use new technologies, you may also be asked to train others or write instructions so that they can use the new tools as well.

Where do you start? What are the most effective ways to learn new technologies?

- Discover how you learn best and then develop a plan or process that fits with your own style of learning.
- Practice using the new technology in the context of a real project.
- Find a partner or a team to collaborate with and reinforce the new technologies you are learning.
- Seek out formal training outside the company (seminars, short courses, online instruction) for more complex tools and skills.

It is important to recognize that the ability to learn is a vital skill in itself. Many people find they need to use a tool in order to know what questions to ask. Fortunately, many features of software functions and interfaces are similar, so that once you learn one type of tool, learning others may come more easily. Most technical communicators find that their troubleshooting and problem-solving skills improve with time and practice. Take the time to find "low stakes" projects or other opportunities where you can learn and apply your new skills and tools.

See also

Audience (Part 1)

Audience and Purpose Assessment (Part 1)

Purpose (Part 1)

Situation (Part 1)

Virtual Teamwork

Technical industries have set the trends for telecommuting, contract work, and working from home. Much collaborative work is now completed using remote communication technologies and sometimes without participants ever meeting face to face. The growing popularity and use of weblogs, or blogs, and instant messaging (IM) offer workers many options for online communication. Chat tools also offer an immediate and informal means of communicating within a group. Pocket technologies like BlackBerries or the iPhone can combine complex communication functions and for all practical purposes can function as a portable home office. All of these technologies and applications have changed how technical communicators can manage projects and complete their work with colleagues.

Meetings and Collaborative Work

While instant messaging and text messaging are good tools for brief discussions, brainstorming, and problem solving, blogs and wikis offer file sharing and archival storage that make them more appropriate for exchanging documents for review and intense collaborative work. Free blog hosting sites give collaborators an easy-to-use, preformatted space to communicate and save all their "discussions." Many companies also offer tools that allow teams to share ideas and keep track of a discussion. Blog software supports links, pictures, and document attachments, so they can support focused group discussions, although with large group discussions it becomes hard to keep up with the posts. Blogs work particularly well for small groups where members can create their individual blogs and then check in and communicate with another smaller group there. Wikis use a form of blogging software that allows all users to have full editing privileges. Wikis tend to be used less for conversation and discussion and more for the collection of ideas and information (much like an encyclopedia). The wiki immediately becomes a collaborative document in itself and offers another possibility for collaborative projects, though it takes time to learning its structure and how to modify a wiki web site. Plans and processes for projects will determine how group members will use the online space to complete and discuss their work.

Video Conferencing

Desktop cameras and video conferencing software have become inexpensive enough to make video a good choice for holding virtual face-to-face meetings over an Internet connection. The obvious benefits of video over other forms of conferencing are that vocal nuances and nonverbal expressions become part of your virtual communication. These features may reduce the time it takes to hold a conversation because of fewer misunderstandings and less verbal explanation. Video conferencing offers a solution for holding meetings in multiple locations simultaneously and recordings can be reproduced and viewed again. The downside of Internet-based video is its quality and speed, which still do not equal the full-motion broadcast video of television.

See also

Teamwork and Collaboration (Part 1)

Blogs and Wikis (Part 2)

Pocket Technologies (Part 3)

Video and Animation (Part 3)

Writing for Regulated Environments

Many technical communicators work in organizations where products, procedures, and written documentation are regulated by governmental organizations. For example, the medical device and the pharmaceutical industries are regulated in the United States by the Food and Drug Administration (FDA). Many manufacturing firms (such as production plants) are regulated by the Occupational Health and Safety Administration (OSHA). Organizations working with energy or the environment may be regulated by the Environmental Protection Agency (EPA); groups working in agricultural situations may be regulated by the United States Department of Agriculture (USDA).

In these organizations, there is a level of review and compliance associated with writing and designing information that may not exist in other situations. You not only need to follow all of the important guidelines for technical communication (know your audience, understand the purpose of the communication, design effectively, and so on) but you will also need to follow the procedures and conventions specified by the particular regulatory agency and approved by your organization's legal and management teams. For instance, the FDA regulates not just the content of documents but also the design, delivery medium (print, DVD, web), and structure.

If your products are complicated, with a volume of information that needs careful tracking, you may want to use a content management system that allows you to keep track of each new version of a manual, procedure, or documentation, so that there is a record of all changes and updates. Such a record provides important information in the event of an audit or lawsuit. Content management systems also allow you to keep track of the review process and to control that any regulatory standards for writing style and formatting.

Examples of Regulated Information

Industry or Area	Regulated Information	Guidelines	Regulating Agencies and Related Links
Pharmaceutical (prescription drugs)	• Drug inserts and patient information • Prescription labels • Advertising • Web sites	Language use must be accessible to a wide range of reading levels. Labels must use standard, approved icons and symbols.	FDA www.fda.gov
Medical (devices, such as pace-makers)	• Instruction manuals • Documentation that accompanies the product or device • Standard operating procedures	Know your audience. Information written for a physician is different from that written for a patient.	FDA www.fda.gov OHSA www.osha.gov
Manufacturing (facilities, such as automobile manufacturing plants)	• Instructions • Safety procedures • Standard operating procedures	Understand how the information will be used. Are instructions designed to train new equipment operators or to document the companys compliance with federal regulations?	OHSA www.ohsa.gov

See also

Accessibility (Part 1)

Content Management Systems (Part 1)

Document Design (Part 3)

Tools and Technologies (Part 1)

Technical Document Types

Blogs and Wikis

Blogs and wikis are web sites designed to allow people from across different locations to collaborate and share ideas and links to other sites. But unlike traditional web pages, which anyone can read but which may have limited features for updating or making comments, blogs and wikis are designed to be sites for social and workplace collaboration and networking. Most blogs and wikis have an open access policy, where anyone can add or update information, make comments, and track back to other related discussions.

Blogs, a shortened term for web logs, are web pages that publish individual entries organized by topic. Items usually appear with the most recent entry first and can range from technical news notes to more personal journal entries. Blogs allow one person to post ("blog") ideas regularly and communicate quickly to a community of readers. The author who creates and "owns" a blog invites other participants to respond by setting options for the entries, including whether users can add topics and or post comments. Depending on the software used, entries can include links between words, illustrations, and other multimedia. Blogs allow people with similar interests in business, news, media, education, and other communities to share information and ideas easily.

Wikis are web pages that allow multiple users, from any location, to add, edit, archive, and update information. Although Wikipedia is the most familiar site, wikis have in fact become quite popular for other purposes, including in the workplace. With many organizations spread across the globe, and with almost all workplace writing being collaborative, wikis are a powerful and easy-to-use way for teams to store, share, update, and revise information.

One feature that distinguishes wikis from blogs is the writing style. Unlike blogs or some web sites, where personal opinion is the norm, wikis generally are not used to express personal points of view. Instead, the writing style of most wikis is factual and encyclopedic. Neutral point of view (NPOV) entries are encouraged and often enforced by the editorial policies of a wiki.

Below is a sample of a blog, created in WordPress software, which offers news for technical communicators with a special interest in information design and architecture. Notice that a wiki may be more appropriate if the only purpose were to provide a factual, encyclopedic set of ideas.

Design Matters

News and information from the ID–IA SIG, Society for Technical Communication

Using Design Visuals To Communicate Ideas
March 7th, 2008

I just saw this post over at Boxes and Arrows, looks very interesting:

A Podcast from Vizthink '08 by Jeff Parks on 2008/03/04
http://www.boxesandarrows.com/view/using-design-visuals

Posted in News: ID and IA, Design | No Comments »

Review: GUI Bloopers 2.0-Essential Updated
November 27th, 2007

Reviewed by David Dick

Once upon a time, graphical user interfaces (GUIs) were found only in operating systems for PCs. Today, we confront GUIs when using self-service checkout counters, when paying bills online, and when using mobile phones, to name a few examples.

Whether we can complete our transactions or accomplish our tasks depends on GUIs that are easy to use and understand. No doubt you have seen people confused with the touch-screen menu at the self-service checkout, or abandon an online shopping cart because the form is confusing. You may well have chosen a competitor's brand of income tax preparation software that is easier to use. Frustrated users mean lost profit and products that fail in the marketplace. When GUIs fail, companies call a UI designer such as Jeff Johnson to change poor design to great design.

The first edition of GUI Bloopers heralded Johnson's first work as a book author. *GUI Bloopers* was intended for software developers who often double as UI designers, development managers, and new UI designers. But *GUI Bloopers* also gained popularity among teachers and technical writers who wanted to understand the rules of good user interface design. Readers' feedback, new software products and Web applications on the market inspired Johnson to write an updated version—*GUI Bloopers 2.0*.

Search

Participate
» Register
» Login
» Entries RSS
» Comments RSS
» WordPress.com

Pages
» About the Information Design and Architecture SIG
» Leadership Team
» Newsletter Archive
» Volunteer Opportunities

Categories
» DEPARTMENTS (1)
  » Definitions (13)
  » Education (14)
  » Employment (6)
  » Events (29)
  » Features (7)
  » News: ID and IA (24)
  » News: ID–IA SIG (22)
  » News: STC (3)
  » Related organizations (9)
  » Resources (22)
» TOPICS (1)
  » Design (13)
  » History of ID/IA (18)
  » Information Architecture (29)
  » Information Design (40)

New on STC.org
» STC President to Lead Delegation to China

■ Example of a blog used to share information for technical communicators

Features of the Blog

1 The banner across the top displays the sponsoring organization, the Society for Technical Communication (STC), and the blog title.

2 Links back to the main blog and to other topics appear on the right.

3 Main news entries appear by date, in reverse chronological order.

4 The entry includes links to sites or other files.

5 Each entry includes its topic and a place to comment on this post.

6 The postings on this blog are designed to provide up-to-date information and related discussions.

Audience Considerations

You will usually create a blog for the following users:

■ A particular group to read and communicate regularly—whether for a class, a work project, or just a hobby

- A team in an organization where people are spread out but need a common site to discuss project ideas and share information
- A person who wants a writing space to record regular ideas on certain topics and to seek response from others that you invite to your blog

Online conversations can create community but can also incite strong feelings and can spill into offline situations. As with other forms of digital communication, users might feel inclined to share personal information on a blog. Remember that an Internet document is always a public document. Although you can preserve anonymity and mask personal information from other users, be aware that information can always be obtained in digital systems. The hosting site will also have policies about appropriate content.

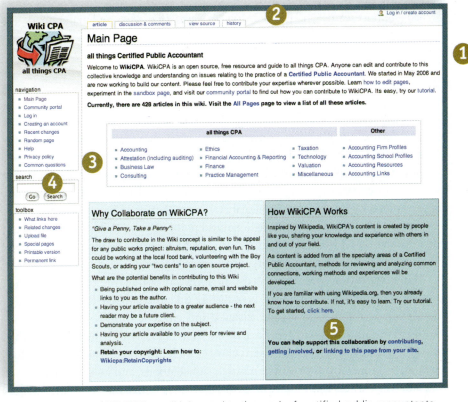

The main page of WikiCPA, a wiki devoted to the work of certified public accountants

Features of the Wiki

1 Written in a factual, neutral style.

2 Designed using the same template as the popular Wikipedia.

3 Entries are set up alphabetically, similar to an encyclopedia or dictionary.

4 Search box makes it easy for readers to find what they are looking for.

5 Allows users to post, edit, and revise entries, thus encouraging a peer review process from accounting experts around the world.

Audience Considerations

You will usually create a wiki for the following users:

- Customers on web sites who can obtain the latest technical or user information about a product. Customers can also contribute information via a wiki.
- Employees or team members on internal sites who can contribute to a knowledge database about a project or product. Wikis not only allow for immediate updates from anyone on the team, but they also provide a place where information can be archived.
- Professionals from across many different companies who can input and locate up-to-date information on the regulatory, research, or marketing issues in their field. In the sample above, certified public accountants from any company can use this wiki.
- Tech support teams who use sites to track customer issues and maintain a tech support database of information.

Design Considerations

Blog entries vary in length and range from the informal style of one writer to highly focused collaborative conversations—even ones with almost no text and all illustrations. Active blogs include frequent (daily or weekly) postings. Journal-like blogs (for example, those on *LiveJournal.com*) may use a highly personal voice to express interests, observations, or strong opinions. Even the most personal blog assumes a hypothetical or an actual reading audience. Wiki entries, however, are factual and to the point, seeking a neutral point of view. Wikis normally provide information on when the entry was updated, places where the facts still need to be checked, or places where more information is needed.

Many institutions and commercial hosting sites offer free blog spaces and software. The blog is saved on the hosting site (like the popular *Blogger.com*) that offers choices about page layout, visuals, and content. Both blogs and wikis use software templates with choices about page layout and content set by the hosting site and the software. Design choices include screen sections and color schemes, text fonts, icons or emoticons that signal moods, and more. Experienced blog and wiki owners can modify most or all of these items, but complex changes often require a paid account rather than a free one. Commercially sponsored blogs and wikis can also dictate visual features as tie-ins to their products and corporate identities.

FOR CREATING BLOGS AND WIKIS

- Determine your audience and purpose for the information and decide whether a blog or a wiki makes more sense.

- Name your blog or wiki site title clearly.

- Write entries that are consistent and relevant to your purpose.

- Invite users who are part of the conversation.

- Post entries regularly to keep your content up to date (at least once a week).

- Link information in the blog or wiki to other internal entries or to other web sites.

- Test links to other sites before posting your entries to make sure they're still working.

- Make sure to refresh your browser window to check the layout of each page in your blog or wiki.

- Go back over your post and add links to earlier posts, if appropriate and when possible. This approach helps remind readers about the contexts for the current post.

- Check color, pictures, or load issues on other computers with slower connections.

See also

Document Design (Part 3)

Multimedia (Part 3)

Style Sheets and Templates (Part 1)

Web Page Design (Part 3)

Brochures

A brochure uses a pamphlet format to present information about a product, event, or service for a targeted audience and distribute this information easily. Brochures in technical communication are used for sales and marketing, product descriptions, procedures and other technical information, or for informing the public about important services and resources.

This brochure from the Contractors State License Board (CSLB), part of the Department of Consumer Affairs, circulated widely in urban neighborhoods after a tornado hit downtown Atlanta. The CSLB professional organization offers the affected residents tips and resources about how to hire a reputable contractor after a storm. The brochure is written and designed for a general audience of homeowners, whose experience with home repair may vary widely.

⑤ **⑥** **①** **②** **③** **④**

AFTER A DISASTER DON'T GET SCAMMED!

It is an unfortunate fact that unscrupulous individuals will try and make a profit on another's misfortune. Don't let these individuals take advantage of you during this vulnerable time.

Before you hire a contractor or sign any documents for repairs, the California Contractors State License Board offers the following basic advice:

- Don't rush into repairs, no matter how badly they are needed;
- Get at least three bids. Don't hire the first contractor who comes along;
- Watch out for door-to-door offers of repair services and flyers or business cards that are left on your doorstep;
- Ask friends, family and associates for recommendations about contractors they have hired;
- Never hand over a cash deposit;
- Even for the smallest job, get proof that the person you are dealing with has a contractor's license for the type of work that needs to be done; and
- Get a written contract that details every aspect of the work to be done.

Additional Resources

Some other CSLB publications to ask for that may help you avoid problems before they occur:

"What You Should Know Before You Hire a Contractor"
  Provides information about hiring and working with contractors.

"Terms of Agreement – A Consumer Guide to Home Improvement Contracts"
  Information about what to look for in a home improvement contract.

"Preventing Mechanic's Liens"
  A guide to make sure subcontractors and suppliers are fully paid.

For a free copy of these publications, log onto the Internet at: www.cslb.ca.gov or call 1-800-321-CSLB (2752).

Additional Resources

CSLB Hotline for Disaster victims only: 1-800-962-1125

California Governor's Office of Emergency Services
P.O. Box 419047
Rancho Cordova, CA 95741-9047
(916) 845-8510 or www.oes.ca.gov

Federal Emergency Management Agency (FEMA)
www.fema.gov or 1-800-621-FEMA

Regional Office (Region IX)
1111 Broadway, Suite 1200
Oakland, CA 94607
Main Office (510) 627-7100

CONTRACTORS STATE LICENSE BOARD
P.O. Box 26000
Sacramento, CA 95826-0026
1-800-321-CSLB (2752)
www.cslb.ca.gov

DEPARTMENT OF CONSUMER AFFAIRS

05/06

AFTER A DISASTER
Don't Get Scammed!

CONTRACTORS STATE LICENSE BOARD

Department of Consumer Affairs

■ Front and back panels of a brochure for home owners

Features of the Brochure

① The three-panel design displays three "front" panels: one inner panel on the left; the back panel in the center; and the front panel or cover on the right.

② The cover captures interest with the main title "Don't Get Scammed!" because unethical contractors can take advantage of desperate homeowners after a big storm.

③ A color photograph of a destroyed house and yellow caution tape matches the negative title and helps appeal to emotions in the audience.

④ The CSLB name, affiliation, and logo on the front establish credibility thorough government authority.

⑤ The inner panel explains the brochure's purpose and offers general tips in reversed text (white) on a lilac background with dark red bullets for emphasis.

⑥ The back panel, visible when folded, offers lists of additional resources, their locations and web address, and provides complete CSLB contact information.

7

If your home or property has been damaged by fire, flood, earthquakes or other disasters, here are some resources to help you from being victimized a second time.

8

Other Considerations

With "Service and Repair Contracts," a consumer's 3-day right to cancel expires when the work begins. Check the paperwork before you sign, to see if it is a regular contract or the "Service and Repair" kind.

Renters should check with their landlords and their rental agreements about damages and repairs. Major repairs are almost always the responsibility of the landlord. Renter's insurance policies may cover personal property damage.

If you're a homeowner, contact your insurance company to find out what's covered and how to proceed.

Hire a Licensed Contractor

Deal only with licensed contractors — Ask to see the contractor's "Pocket License", along with other identification. If the person claims to represent a contractor, but can't show you a "Salespersons Registration Card", call the contractor to find out if the person is authorized to act on their behalf.

Contractors working on a job — from debris removal to rebuilding — totaling $500 or more for labor and materials must be licensed by the CSLB. To become licensed, a contractor must pass a licensing examination, verify at least four years of journey-level experience, and carry a license bond.

Some out-of-state contractors and unlicensed California contractors want to help with rebuilding in disaster areas. However, it is illegal and punishable as a felony to perform contracting work in a declared disaster area without a California contractor's license. Punishment may include a fine of up to $10,000 or up to 16 months in state prison.

Get the contractor's license number and check it out on the Internet at: www.cslb.ca.gov or call the CSLB's toll-free automated telephone number at 1-800-321-CSLB (2752) to verify that the license is valid.

Get it in Writing

Don't sign the contract until you fully understand the terms.

Make sure everything you have asked for is in writing and clearly described. A verbal promise may not give you the results you wished for.

Avoid Payment Pitfalls

- By law, a down payment on a home improvement contract cannot exceed 10 percent of the contract price or $1,000 whichever is less;
- Don't let payments get ahead of the work;
- Keep receipts and records of payments;
- Don't pay cash;
- Make sure you have the names of subcontractors, material suppliers and confirmation that they have been paid;
- Don't make the final payment until you are satisfied with the job and the building department has signed off on it.

Where to Complain

If you do have problems with a licensed or unlicensed contractor there are some places to turn.

- File complaints against contractors at: www.cslb.ca.gov or call 1-800-321-CSLB (2752);
- Small Claims Court – for disputes and losses under $7,500;
- The Consumer Division of your local District Attorney's Office; and
- Your local sheriff or police department if a crime is in progress.

■ Inside panels of a brochure for homeowners

Features of the Brochure *(continued)*

7 Inside panels show an effective use of colors, callouts, and a grid layout that aligns all the text and graphic elements and organizes information.

8 Color photo of hand hammering a nail appears on first inside panel and continues the emotional story from the front by offering a solution in the photo, state seal and text.

9 Color scheme balances the beiges and yellows inside with lilac panel backgrounds. The graphical bands alongside each picture help unify all these document colors.

10 Body text uses black, sans serif font with red used for highlights.

11 Headings organize the tips into meaningful categories and also stand out as contrasting dark red text.

Audience Considerations

All the different visual components of a brochure must work together with the text to inform and also persuade readers. But readers of this brochure will have different levels of experience with contractors, while other readers may be learning about these issues for the first time. The panels balance the general information under headings or in bullets for those new to the topic, and also display immediate resources on the back for those who are already familiar with the topic. Most brochures must account for multiple levels of experience and expertise in targeted readers.

Design Considerations

This brochure organizes information with checklists, visuals and contacts for the person who needs immediate help with a damaged home. Panels on a folded brochure add another level of organization, where each panel can be organized as a separate layout and emphasis area, like the front and back panels above that contrast enough to keep each panel distinct. Also, your back cover panel can be used as a mailing label, a tear-off form that users can send back to you, or an attention-getting piece of information. This brochure is printed on high quality glossy paper using full-color printing to obtain a professional product.

Color scheme and photos tie all elements of the brochure together and offer the same message as the text: here is the best solution to your problems after a serious storm.

Guidelines

FOR CREATING BROCHURES

- Identify the purpose of your brochure—to describe and sell a product or service, to educate readers, to provide specifications, or some combination.
- Identify all the technical and consumer audiences for the brochure, what they need to know about the topic, and their characteristics.
- Organize the information in a way that new readers can use quickly, but provide details and technical illustrations for those seeking more in-depth information.
- Make the most of your front cover—as a persuasive appeal and to draw attention—and your back cover as a mailer, a tear-off, or an attention-getting list.
- Select high-quality paper and color printing for the best results.
- Consult a graphic designer and printing company, if possible, to ensure a professional design and appealing visuals in your brochure.

See also

Email and Attachments

Email is the most common form of business communication. Designed after its paper predecessor, the memo, email uses the standard From, To, Date, and Subject fields in its header. An email usually includes the header, a message, a signature line, and attached files. Almost all email systems allow the same functions: compose the email, print, send, reply, forward, and add attachments. Email as a written form seems to encourage an informal tone that resembles spoken conversation. Often, there is no salutation ("Dear Joan") or closing ("Sincerely"). Yet, as with all correspondence, email should be written in a way that is appropriate for the situation. If you are writing to someone you have never met, you should err on the side of being formal, polite, and respectful, just as you would in a business letter.

The sample email on page 64 is shared among people who work together and are on friendly terms with each other.

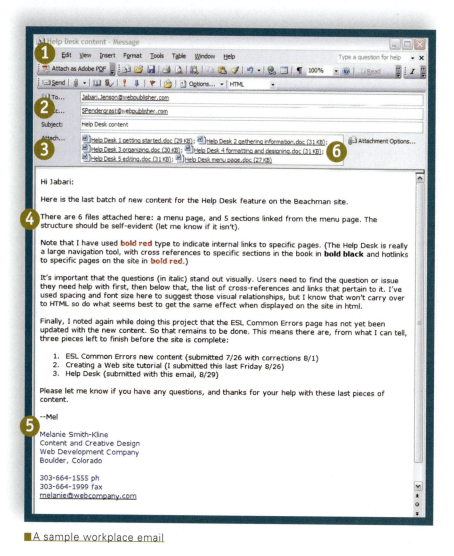

■ A sample workplace email

Features of Email

1 Commands (send now, link, options) are included in toolbars across the top of the message window.

2 Header at the beginning of the email provides recipient (to), cc, and subject of the email.

3 Subject line informs readers of the message's content.

4 Content is a transmittal (cover) letter referring to the attachments.

5 Signature line provides contact and other information about the writer.

6 Attachments are files that are sent with an email.

Audience Considerations

Every email you write at work, to a professional contact, or at school needs to be:

Professional—Email is often forwarded to readers not addressed by the original sender. You may write a casual note to a colleague or supervisor that gets forwarded later, along with a project status report, to a manager or executive. Write professional emails with the assumption that *anyone* in the organization may read it. Avoid complaining or "venting" about coworkers or policies. Strive for messages that are informative and useful, not emotionally charged.

Considerate—Readers in the workplace are busy and sometimes stressed. Be considerate of their needs by crafting a clear and useful subject line. A subject header that reads "Re: re: re: re: update" does not help a reader filter and sort on the receiving end. A subject header like "Presentation draft 2 feedback" allows readers to see what your message is about without having to open and read the message first. When you respond or forward emails sent to you, make sure your response does not require your audience to reread an entire string of messages. Delete forwarded text like "Me too" or "I agree," which are not helpful responses. Write responses that are self-explanatory and self-contained to save your readers' time and energy. Include those parts of the original message you are responding to so all recipients will be informed and aware of the issues being decided.

Concise—Most professionals today complain of "email fatigue" and spend at least an hour a day reading and responding to email. Respect your busy and overworked readers by writing emails that are clear, concise, specific, and detailed. Don't write as you might in an instant message, "mtg th 2, J's ofc" when you could write "product development working group meeting: Thursday, October 11, 2 pm, Jordyn's office." Vague or jumbled information requires readers to write back in order to ask for clarification, sometimes more than once, generating more email rather than conveying information. Avoid this problem by taking the time to write a clear, complete, yet concise message the first time.

Careful—Issues of privacy and personality come up when an email is used for workplace communication. Most organizations have the legal authority to monitor the email of their workers as well as any attachments sent with email. Email systems are not perfect; it is possible for email to be delivered to someone other than the intended audience. Because email users are not communicating face to face, they sometimes use email to let off steam. Avoid "flaming," or email that is unnecessarily angry or makes personal attacks. A good principle is not to write anything in an email that you would not say aloud in a formal meeting.

Ethical—The ability to easily "forward" a message received to another person brings up ethical issues. The person forwarding the email needs to consider whether the person who originally sent the message would want it forwarded to others. As someone writing email you need to be aware that any email you send can be forwarded to others without your taking any action. Before forwarding any email consider any ethical complications it may cause. Blind carbon copying is another complex ethical situation. By using the "bcc" function some recipients of an email are not aware of others who also received the message. Email writers also have no control over when and where their messages are read. Since users can email at all hours of the day and night, consider timing when reading and sending email. For instance, when emailing someone late at night, the sender must consider that the email might not be read until the next morning, no matter how urgent the message may be. All of these ethical situations must be resolved and handled appropriately.

Design Considerations

Email can take on many styles depending on the context in which it is being sent. Email offers a blank screen and no direction to the author about format or length. Because of its fluidity, the message of an email may mimic other forms like letters and memos or even a report. Take context into account when determining the style of each email you send.

- Avoid ALL CAPS. All capital letters are hard to read. And, in electronic communication, ALL CAPS is considered the same as shouting.
- Use emoticons with caution. Emoticons are combinations of keyboard symbols that are used to indicate a facial expression or gesture. The most common of these is the smiley face :-). Do not use emoticons in workplace or formal communiciation.
- Include signatures when you use email for professional purposes (software can be set to do this automatically). Professional signature lines consist of your name, title, work email, office phone, fax, web page and mailing address.
- Do not go overboard on the use of fonts, pictures, background images, and other visual features within your email. Not every email program will display these features correctly.

You can read email with a variety of software tools or a web browser, but not every tool can accommodate special formatting or fonts. You cannot predict how your message will show up on users' screens, so try to keep formatting to a minimum. Instead, send formatted attachments, such as documents, photographs, or other types of files. The attachment function gets files to individuals you are working with easily and quickly. Be aware, though, of how much space attachments take up and that they can be very slow to download. Not all email systems accept all kinds of attachments or they may only allow attachments under a certain size to be included with the email. Always check with your recipients first about file size and format before sending attachments.

Guidelines

FOR CREATING EMAIL

- Be professional and assume your email may be read by other people beyond your intended recipients.

- Be concise, using clear subject lines and complete sentences. Readers are impatient and don't want to scroll through long screens full of information. Large batches of information should be placed within a document and then attached to the email.

- Use an appropriate tone. Be aware that misunderstandings often result when writers don't take the time to take their tone and rhetorical situation into consideration.

- Pay attention to spelling and formality. Email users tend toward an informal writing style. Even though many email programs contain spell checkers, people rarely use them. A correctly spelled message will be more credible than a sloppy one.

- Break up the information. Instead of giving your reader long sections of ongoing text, break up the text by subject into paragraphs and use numbered lists or headings.

- Don't send large attachments without asking first. Or, put the attachment on an intranet or common file server.

- Be specific when you expect recipients to act or respond. If your email asks recipients to respond or follow up with action, be specific about what it is you are asking them to do, and suggest a date by which you need their response.

See also

Letters (Part 2)

Memos (Part 2)

Tools and Technologies (Part 1)

Forms

Forms are designed to capture specific information for an organization. For instance, a U.S. tax return form is designed to collect federal income tax information; a motor vehicle registration form is used to collect information for your county and state about a new or used car or truck. In the past, forms were available only in printed format, but today, most government and many commercial forms are available online. Forms offered in PDF (portable document format) are usually intended to be printed, but increasingly, many forms (such as registration forms for new software) can be filled out and submitted entirely via the Web.

Forms should be structured according to the conventions that your users expect. Tax forms and job applications, for example, always begin with fields that collect information about the taxpayer or applicant, such as name, address, and social security number. A good form is clear, organized, easy to read, and intuitive to fill out.

The following example is a general application form used to apply for a job with the federal government. Instructions for completing and mailing the form appear on a separate page (not shown). This PDF form is completed in digital format and then printed out for submission by mail.

OPTIONAL APPLICATION FOR FEDERAL EMPLOYMENT - OF 612

Form Approved
OMB No. 3206-0219

Section A - Applicant Information

Use Standard State Postal Codes (abbreviations). If outside the United States of America, and you do not have a military address, type or print "OV" in the State field (Block 6c) and fill in the Country field (Block 6e) below, leaving the Zip Code field (Block 6d) blank.

1. Job title in announcement	2. Grade(s) applying for	3. Announcement number

4a. Last name	4b. First and middle names	5. Social Security Number

6a. Mailing address	7. Phone numbers (include area code if within the United States of America)
	7a. Daytime

6b. City	6c. State	6d. Zip Code	7b. Evening

6e. Country (If not within the United States of America)

8. Email address (if available)

Section B - Work Experience

Describe your paid and non-paid work experience related to the job for which you are applying. Do not attach job description.

1. Job title (If Federal, include series and grade)

2. From (mm/yyyy)	3. To (mm/yyyy)	4. Salary $ per	5. Hours per week

6. Employer's name and address	7. Supervisor's name and phone number
	7a. Name
	7b. Phone

8. May we contact your current supervisor?   Yes ☐   No ☐
If we need to contact your current supervisor before making an offer, we will contact you first.

9. Describe your duties, accomplishments and related skills (If you need to attach additional pages, include your name, address, and job announcement number)

Section C - Additional Work Experience

1. Job title (If Federal, include series and grade)

2. From (mm/yyyy)	3. To (mm/yyyy)	4. Salary $ per	5. Hours per week

6. Employer's name and address	7. Supervisor's name and phone number
	7a. Name
	7b. Phone

8. May we contact your current supervisor?   Yes ☐   No ☐
If we need to contact your current supervisor before making an offer, we will contact you first.

9. Describe your duties, accomplishments and related skills (If you need to attach additional pages, include your name, address, and job announcement number)

U.S. Office of Personnel Management
Previous edition usable

NSN 7540-01-351-9178
50612-10
Page 3 of 4

OF 612
Revised June 2006

First page of a form used to apply for a job with the federal government

Section D - Education

Upon request from the employing Federal agency, you must provide documentation or proof that your degree(s) is from a school accredited by an accrediting body recognized by the Secretary, U.S. Department of Education, or that your education meets the other provisions outlined in the OPM Operating Manual. It will be your responsibility to secure the documentation that verifies that you attended and earned your degree(s) from this accredited institution(s) (e.g., official transcript). Federal agencies will verify your documentation.

For a list of postsecondary educational institutions and programs accredited by accrediting agencies and state approval agencies recognized by the U.S. Secretary of Education, refer to the U.S. Department of Education Office of Postsecondary Education website at http://www.ope.ed.gov/accreditation/

For information on Educational and Training Provisions or Requirements, refer to the OPM Operating Manual available at http://www.opm.gov/qualifications/SEC-II/s2-e4.asp.

Do not list degrees received based solely on life experience or obtained from schools with little or no academic standards.

1. Last High School (HS)/GED school. Give the school's name, city, state, ZIP Code (if known), and year diploma or GED received:

2. Mark highest level completed: Some HS ☐ HS/GED ☐ Associate ☐ Bachelor ☐ Master ☐ Doctoral ☐

3. Colleges and universities attended.
Do not attach a copy of your transcript unless requested.

| 3a. Name | | | Total Credits Earned | | Major(s) | Degree (if any) |
			Semester	Quarter		Year Received
City	State	Zip Code				
3b. Name						
City	State	Zip Code				
3c. Name						
City	State	Zip Code				

Section E - Other Education Completed

Do not list degrees received based solely on life experience or obtained from schools with little or no academic standards.

Section F - Other Qualifications

License or Certificate	Date of Latest License or Certificate	State or Other Licensing Agency
1f.		
2f.		

Section G - Other Qualifications

Job-related training courses (give title and year). Job-related skills (other languages, computer software/hardware, tools, machinery, typing speed, etc.). Job-related honors, awards, and special accomplishments (publications, memberships in professional/honor societies, leadership activities, public speaking, and performance awards). Give dates, but do not send documents unless requested.

Section H - General

1a. Are you a U.S. citizen? Yes ☐ No ☐ → 1b. If no, give the Country of your citizenship

2a. Do you claim veterans' preference? Yes ☐ No ☐ → If yes, mark your claim of 5 or 10 points below:

2b. 5 points ☐ → Attach your Report of Separation from Active Duty (DD 214) or other proof.

2c. 10 points ☐ → Attach an Application for 10-Point Veterans' Preference (SF 15) and proof required.

3. Check this box if you are an adult male born on or after January 1st 1960, and you registered for Selective Service between the ages of 18 through 25 → ☐

4. Were you ever a Federal civilian employee? Yes ☐ No ☐ → If yes, list highest civilian grade for the following:

4a. Series	4b. Grade	4c. From (mm/yyyy)	4d. To (mm/yyyy)

5a. Are you eligible for reinstatement based on career or career-conditional Federal status? Yes ☐ No ☐
If requested in the vacancy announcement, attach Notification of Personnel Action (SF 50), as proof.

5b. Are you eligible under the ICTAP? Yes ☐ No ☐
*ICTAP (Interagency Career Transition Assistance Plan). A participant in this plan is a current or former federal employee displaced from a Federal agency. To be eligible, you must have received a formal notice of separation such as a RIF separation notice. If you are an ICTAP eligible, normally you will be provided priority consideration for vacancies within your commuting area for which you apply and are well qualified.

Section I - Applicant Certification

I certify that, to the best of my knowledge and belief, all of the information on and attached to this application is true, correct, complete, and made in good faith. I understand that false or fraudulent information on or attached to this application may be grounds for not hiring me or for firing me after I begin work, and may be punishable by fine or imprisonment. I understand that any information I give may be investigated.

1a. Signature | 1b. Date (mm/dd/yyyy)

Previous edition usable
U.S. Office of Personnel Management

NSN 7540-01-351-9178
50612-10
Page 4 of 4

OF 612
Revised June 2006

[Print Form] [Save Form] [Clear Form]

■ Second page of a federal job application form

Features of the Form

1 Each section of the form has a title and letter within gray separation lines, providing clear visual cues about where each section begins and ends.

2 Areas to be completed have labels within the fields and sample date formats.

3 Columns within each section are aligned (e.g., Education).

4 General section uses arrows to indicate follow-up actions.

5 Application Certification section explains legality and requires a signature.

Audience Considerations

Remember that most people do not enjoy filling out forms and may even find them to be depersonalizing. Bureaucratic language and technical jargon can make readers feel unintelligent, which in turn causes them frustration, makes them impatient, and may cause them to make mistakes in filling out forms. If you are creating a form to collect medical or financial information, be aware that you are also asking your audience to divulge personal information. Ask for as little information as necessary, and keep forms as concise and useful as possible. If an international audience will use the form, you may need to create alternate versions in different languages. Privacy and disclosure laws and customs vary across national boundaries, so be sure to research how a form will be used in an international setting before designing it.

Design Considerations

Forms should be easy to read, with plenty of white space. Language should be brief, concise, and clear. Use language that is appropriate for the particular audience. If the audience potentially contains users whose native language is not English, consider creating a translated form and give users a choice. Forms should use words and technical terms that are understood by the specific audience. Forms do not include long paragraphs or unwieldy explanations. Such information, if necessary, belongs on a separate instructions page or screen.

Effective page layout is essential if a form is to be useful to anyone other than the agency or individual who created it. Many forms can be created using templates or wizards. Many web sites offer free-form templates for print forms. Forms specialist Caroline Jarrett (2004) recommends the following:

- Limit instructions to 150 words
- Organize questions into groups that users understand
- Use groupings or a progress indicator to show progress
- Indicate required fields
- Make the privacy policy easy to find
- Explain or justify any requests for private information
- Align form elements to a vertical and horizontal grid

When designing forms intended to be filled in and submitted on a web page, work with a skilled web designer. Users do not want to type in long numbers to log in or access information, so use a short login form when designing forms for the Web. Web forms design expert Brian Crescimanno (2005) suggests these items for web forms:

- Use the right field for the particular task or item
- Give users enough room to type

- Mark mandatory fields clearly
- Shorten forms by asking yourself what is really needed
- Provide descriptive labels for all fields
- Let the computer handle information formatting
- Use informative and helpful error messages

Guidelines

FOR CREATING FORMS

- Ask only for the information you really need. Be aware of local, national, and international laws about information privacy and disclosure. U.S. government forms are regulated by the Paperwork Reduction Act and other federal regulations.

- Use terminology and definitions that will be familiar to your users, and create a form that makes sense from their point of view. Avoid corporate or technical jargon, and opt for plain language throughout your form.

- Create clear visual boundaries that help users recognize where they need to provide information. Make "boxes" large enough to accommodate handwriting when necessary.

- Use templates, if available, rather than redesign an entirely new form. Design individual fields so they are long enough to accommodate a typical entry, but not too long.

- Work with a skilled web designer for web-based forms.

- Don't ask for the same information more than once. If you are collecting shipping and billing addresses, for example, make it easy for users to indicate if these are the same.

- Test your form with real users, and revise any sections that users find confusing, ambiguous, or unnecessary.

See also

Delivery Medium (Part 1)

Document Design (Part 3)

Style Sheets and Templates (Part 1)

Web Page Design (Part 3)

Frequently Asked Questions (FAQs)

Frequently asked questions, often called FAQs, present key information on a topic in the form of a list of questions and answers. FAQs originated on newsgroup (email) lists in the 1980s when list administrators grew tired of repeatedly answering the same basic questions and created FAQ pages as a starting point for new readers. Frequently asked questions pages have become common on organizational and commercial web sites, and sometimes appear in printed documentation as well. FAQ pages often serve as an introduction to an organization or product and help to provide essential information to first-time visitors in a convenient, reader-centered format.

■ A government FAQ page on organ donation

Features of Frequently Asked Questions

1 A summary of the major question categories allows users to find an answer to their specific questions.

2 A search box provides users with an alternative way to find answers.

3 The logo and other identifying information is easy to find in the banner across the top of the web site.

4 Question categories are written in short, clear language.

Audience Considerations

Frequently asked questions are usually written for audiences who are new to a topic or organization. FAQs may be the first place new readers go when visiting a web site or looking for information on a product or software tool. Such readers need information that is clear, pertinent, and helpful. Audiences viewing FAQ pages may often be intimidated, uninformed, confused, and possibly frustrated by an inability to find what they need. These readers turn to the FAQ page for guidance. An effective FAQ page is built around questions that everyday users of the product or service are likely to have. Ideally an FAQ page is developed as a result of user testing or surveys, in order to be sure that it addresses the right questions and is worded in a way that readers will understand quickly. Any specialized terms or concepts should be defined for general audiences.

FAQ questions are commonly written in the first person, to match the style in which a reader would pose a question in conversational speech: "What organs and tissues can I donate?" The appropriate style and voice on an FAQ page is informative, helpful, supportive, and patient. Answers should be worded with sensitivity to the needs of readers who may be confused or fearful. For example, the Organ Donor FAQ page answers sensitive questions with clarity and brevity:

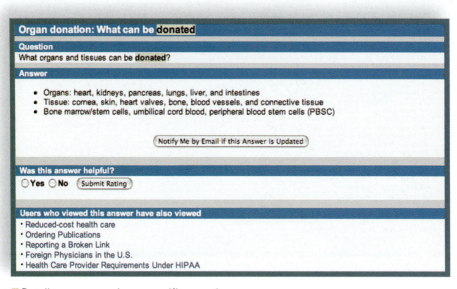

Detail page answering a specific question

Design Considerations

Frequently asked questions emerged out of a new technology, the email discussion list, and the need to provide basic information to audiences unfamiliar with both the content and the etiquette governing behavior in discussion lists. The concept of the "FAQ" has broadened to include many applications, so that many things labeled as FAQs today, on the Web or in printed documents, are not designed as questions at all but as guidelines or fact sheets. Readers who find something called "Frequently Asked Questions" expect a document designed as a series of questions, with clear and easily accessible answers. Perhaps more than anywhere else on a web site or in printed documentation, a FAQ page is the place for clean and minimalist design, concise writing, and an uncluttered visual look. Readers need to find their questions answered clearly and quickly.

Guidelines

FOR CREATING FREQUENTLY ASKED QUESTIONS

- Organize FAQ pages around a focused list of specific questions.

- Make sure the list of questions includes the questions your readers are likely to raise about the topic.

- Write questions in the interrogative voice and use first person address ("How do I report a problem?")

- Links from questions to answers should be visually clear and self-evident.

- Provide answers that are honest, clear, and helpful to readers.

- Don't overwhelm readers; create FAQs that fit easily on one page or screen.

- Design FAQs so the list of questions is the visual focus of the page.

- Make sure your FAQ page presents your organization or company in a positive, informative light.

- Resist the urge to put promotional materials or marketing pitches on an FAQ page.

See also

Audience (Part 1)

Document Design (Part 3)

Product Descriptions (Part 2)

Usability and User Testing (Part 4)

Web Page Design (Part 3)

Instructions

Most technical information contains some form of instructions. Often called task-oriented communication because people come to the document needing to perform a task, instructions are part of every product or technical service you encounter. Consumers do not want to read a long story about how the computer works or how the cell phone was engineered; rather, they want to know how to use the computer or how to program phone numbers on their phone. Instructions can be delivered in printed documentation or product manuals or posted on a product support web site. Software instructions are often embedded in the software or included on an install CD.

This example provides users with instructions on how to install drivers for a Lexar card reader. The instructions were published on the Web for ease of access.

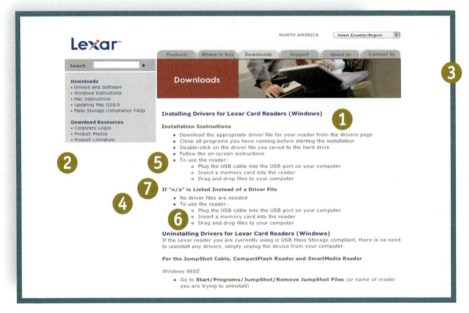

■ Online instructions for a consumer product

Features of Instructions

1 The main heading identifies the specific task described in the instructions.

2 Ample white space in the left margin and short lines of text combine to allow for easy reading.

3 The web page design is clear and uncluttered, to draw visual attention to the steps in the process being described.

4 Instruction sequences are presented as short, bulleted lists.

5 Each step starts with an action verb ("download," "close").

6 Indented bullets are used to indicate substeps within a specific process.

7 The condition statement ("If 'n/a' is listed instead of a driver file") provides users with information on how to handle an alternate condition.

Audience Considerations

Most people who read instructions do not want to be reading; they want to be using their new iPod or their new computer. As a writer of instructions, you need to put yourself in the position of being a user of instructions. Keep the words to a minimum, and be clear and concise. Write short instruction sequences in lists using either numbers or bullets. Test your instructions on others to see whether there are places where users fail to understand your message. Provide good quality illustrations for all parts and processes. Instructions are often used by audiences across the globe. While it is possible to translate instructions, it is increasingly common to see instructions created entirely out of pictures.

Design Considerations

Instructions are almost always written in imperative voice (starting with a verb and speaking directly to the reader):

Turn on the computer.
Insert the disk.

Instructions should be brief and concise. If the user needs to know about several items, chunk your instructions into logical sections. For example, if you purchase a new DVD player, the first set of instructions should be about setting the player up, while the next set should be on using it to play movies.

Instructions are sometimes available only in print format. Increasingly, instructions are packaged with the product on a CD or DVD in order to save paper and shipping costs. While products still come with a manual or short instruction

Guidelines

FOR CREATING INSTRUCTIONS

- Instructions should be self-explanatory and deliver what they promise.
- Keep the words to a minimum, and be clear and concise.
- Chunk your instructions into logical sections.
- Help your reader get started right away.
- Test your instructions on others to find places where users fail to understand your message.

- Number the steps involved in the instructions if the process is complex or involves more than three to five steps.
- Use pictures and illustrations that correspond to the numbered steps, especially if your audience is global or multilingual.
- Deliver the instructions in whatever is the most appropriate medium—print, Web, or PDF file, or in multiple media.

set, most companies realize that they can keep instructions up to date if they direct readers to a web site. Most users eventually lose the paper manual or throw it out, so give users a logical online location for finding instructions.

See also

Document Design (Part 3)

Online Help (Part 2)

Symbols (Part 3)

Task-based Documents (Part 1)

Technical Illustrations (Part 3)

Job Search Documents

When looking for a new position, job seekers use a conventional set of documents. Job advertisements and job descriptions help you locate the job you want. A cover letter introduces you to a prospective employer, while the thank you letter follows an interview. A resume allows you to summarize and list your qualifications. A longer version, called a curriculum vitae (or c.v.) is more common in academic and scientific settings (although in many European countries, "c.v." is used interchangeably with "resume").

Advertisements and Job Descriptions

Job advertisements specify the duties and requirements for a particular job and how to apply. Used when applying for a job, you may also find yourself writing a job advertisement or description for your organization. Job descriptions also provide the necessary information to create fair salary scales, help management understand the duties of their employees, and set expectations for prospective as well as current employees.

In most organizations, a job description is a detailed description of the duties of a position as well as the relationship of the position to the management and organizational structure. When searching for a job, it is not always possible to see the full job description. Instead, many job postings provide a shorter form of the description, usually by listing, as is the case in the next example, "principle responsibilities," as well as both required and prefered qualifications. In the following example, a posting for a Fire Protection Technician at Fermilab, principle responsibilities help create the de facto job description. The "essential functions" section lists additional aspects of the job.

① Fire Protection Technician (Technician II)
Job Code: 070168
Department: Facilities Engineering Services Section (FESS)

Open Date: 03/03/2008
Close Date: 04/03/2008

Principal Responsibilities

③

- Fabricate, assemble, install, and remove a variety of electronics parts, sub-assemblies, controls, and devices for Laboratory projects/equipment under limited supervision;

- Assist in design of equipment and systems; construction and assembly of experimental and support equipment;

- Prepare and test equipment for operation;

- Perform preventive and standard maintenance of equipment, troubleshooting of problems, and repairs;

- Conduct safety inspection and assist with identification, packaging, and removal of waste;

- Compose procedures for finished work; enter information into database; produce reports; prepare and compile technical data and safety records for equipment;

- Procure and maintain inventories of small parts, supplies, materials, and equipment;

- Direct and assist the work of lower level technicians;

- Abide by all environmental, safety, and health regulations.

② *Required Qualifications*

- 4-6 years of Electronics or technical experience.

  and
- Associates Degree or Technical Certificate <u>or</u> equivalent experience with High School diploma or equivalent.

Perform repair maintenance, modifications and routine testing of fire detection, sprinkler, and suppression systems. Troubleshoot system malfunctions and implement corrective measures. Respond to off-hours calls for system malfunctions. Solve field problems in conjunction with other technicians and engineers. Interface with trade workers, customers and subcontractors.

⑤ <u>In addition to the above, candidates must possess the following to be considered **highly qualified** for the position. Applicants must describe how they meet the following essential functions and preferred qualifications in their application materials.</u>

Essential Functions and Preferred Qualifications

- Ability to climb ladders.

- Ability to access confined spaces.

- Ability to lift 50 pound.

- Availability for occasional scheduled and unscheduled overtime work.

- Ability to participate in on-call rotation schedule.

- Ability to operate light trucks and passenger vehicles.

■ Job description for a position at a science laboratory

Features of the Job Description

This job description has four major parts:

① The organization's logo is clear and visible.

② Required qualifications specify the education and work experience necessary.

③ Responsibilities and accountability are outlined in the bullets.

④ Job code, department, and other descriptors follow the conventions of the agency or company.

⑤ Preferred qualifications provide additional information.

These sections will vary widely depending on the organization and the type of job. Most detailed job postings contain both "required" and "preferred" qualifications. The next example provides less detail, which is typical of some types of job postings. The responsibilities are written in very general language, and the requirements are brief. Interested applicants are asked to visit the web site for more information.

■ Advertisement for a research fellowship position at a laboratory

Features of the Job Advertisement

1 Job title is listed as Postdoctoral Research Fellow

2 Location of job can include choices, but here specifies San Francisco.

3 Date of posting keeps information current for prospective applicants.

4 Job responsibilities are described briefly.

5 Requirements for education and experience appear next.

6 Information about the nature of the company includes links to further information.

7 Instructions on how to apply include a request for references.

Job advertisements are written and designed to help the employer find and fill a job with a capable worker. Thus, the employer must give some general information about the vacant position. Most positions are advertised both in print and online or exclusively online. The job shown in the previous figure for a scientific research position represents a typical online job listing.

An ad lists primary responsibilities and the kind of training and education that the hiring company requires. Some ads will include salary ranges. The company will typically give prospective employees one preferred way for contacting the company and submitting an application.

Audience Considerations

A job description has a variety of audiences including prospective employees, human resources personnel, new hires, and supervisors who perform reviews or calculate salaries. New hires need to know the basics of their position and should be aware of what their supervisors expect, and the job description gives such information. Therefore, the job description needs to be accessible to both management and workers. In addition, each job description needs to be understood by those who are assessing salaries, because an accurate description of duties helps ensure that person has a salary congruent with others working above and below them.

Design Considerations

Job descriptions and advertisements require short, clear prose that gives relevant information to the audience. While job descriptions can be more detailed, their focus is on management structure and pay scales. Most job descriptions don't need to include a listing of every responsibility of the employee. Advertisements need to be short because printed space costs money in newspapers and other publications.

The advent of the Internet recruiting agency has changed the size and scope of job advertisements. Internet space does not cost the same as space per words or lines in a newspaper, so ads on the Internet can be longer, more detailed, and include color images. Web sites like Monster.com and CareerBuilder.com work to amass large quantities of job ads so that they attract job searchers. Also, the use of these Internet sites makes job searching international. Internet job sites normally require you to submit scannable or completely digital resumes and cover letters, usually using a template or form on their web site.

Cover Letters

The cover letter accompanies your resume in response to an advertised position or a position you may have learned about from a colleague or a professional network. Except when requested by the hiring manager or HR to only send a resume, you should not send a resume without a cover letter. Because a cover letter is a representation of your writing skills, you should make sure it is well written, with no typographical errors or grammatical mistakes. Your cover letter is the place where you can point readers toward specific skills on your resume and highlight a few areas that are especially important to that employer. Indicate how and where you learned about the position. Use a confident, professional, polite tone. Your goal is to obtain an interview.

15 March 2007

711 Oak Park Rd.
Ames, IA 50010
515-293-9155
aluraj@emailadd.com

Dr. Beverly Taylor
California University
Department of Information Technology
Los Angeles, CA xxxxx

Dear Dr. Taylor:

I wish to apply for the position of network supervisor in the Information Technology department, as posted on the California University job information web site. I have been employed for the past five years in a similar position with a staff of 15 network technicians. My bachelor's degree in business and network administration makes me especially suitable for this job. I will be moving to the Los Angeles area later this month.

Although my current position is challenging, I seek a position in a university or college setting so that I can apply my education and experience in an environment that involves the more complex challenges of a college environment. In my resume, you will note that I have continued my training by becoming certified in numerous operating systems and Internet protocols. In a college environment, file sharing is especially important—my current project in managing a large peer-to-peer network matches quite well with the job requirements listed. I have uploaded my resume, this letter, and my network certifications for your review.

I would appreciate the opportunity to speak with you to learn more about this position and the hiring plans. I will email you, as stated in the job posting, within the next two weeks. You are welcome to contact me at any time. Please let me know if there is any additional information that I can provide.

Thank you for your consideration.

Sincerely,

Max O'Hara

Max O'Hara

■ Cover letter for a job that was advertised in a newspaper, professional publication, or online site

Features of the Cover Letter for an Advertised Position

1 Uses formal, professional salutation.

2 Return address and all contact information is easy to find.

3 First paragraph states clearly how the applicant learned about this position.

4 Second paragraph points out specific items from the resume that the applicant wants highlighted.

5 Final paragraph indicates how and when the applicant will be in contact.

15 March 2007

711 Oak Park Rd.
Ames, IA 50010
515-293-9155
aluraj@emailadd.com

Dr. Beverly Taylor
California University
Department of Information Technology
Los Angeles, CA xxxxx

Dear Dr. Taylor:

I enjoyed meeting you at the Information Technology Academic Summit and appreciated your thoughts on your department's position of network supervisor. As I mentioned, I have been employed for the past five years in a similar position with a staff of 15 network technicians. My bachelor's degree in business and network administration makes me especially suitable for this job. I will be moving to the Los Angeles area later this month.

Although my current position is challenging, I seek a position in a university or college setting so that I can apply my education and experience in an environment that involves the more complex challenges of a college environment. In my resume, you will note that I have continued my training by becoming certified in numerous operating systems and Internet protocols. In a college environment, file sharing is especially important—my current project in managing a large peer-to-peer network matches quite well with the job requirements listed. I have uploaded my resume, this letter, and my network certifications for your review.

I will email you, as you suggested, within the next two weeks. Please let me know if there is any additional information that I can provide.

Sincerely,

Max O'Hara

Max O'Hara

■ Cover letter for a job that the applicant learned about through a professional connection or meeting

Features of the Cover Letter for a Position Learned about through a Professional Network

1 Uses formal, professional salutation.

2 Return address and all contact information is easy to find.

3 First paragraph reminds the reader of how he or she met the applicant.

4 Second paragraph points out specific items from the resume that the applicant wants highlighted.

5 Final paragraph indicates how and when the applicant will be in contact.

6 Tone is slightly less formal than the previous example.

Job Interviews and Follow-Up

The goal of your cover letter and resume in most cases is to obtain an interview for the job. You can also ask for an informational interview simply to inquire about potential positions. In either case, your initial interview offers your best chance to make a strong impression and get the job. Once granted an interview, you will have opportunities to continue your correspondence.

Interviews

There are many kinds of job interviews, including informational interviews, first interviews, phone interviews, and on-site interviews. These interviews can be highly structured, open-ended, or somewhere in between. In many major corporations, the interviews follow a pre-determined format, with candidates taking tests, meeting with managers, and so on. You should learn as much as you can in advance about how the job interview will be conducted at this company: search the Internet, look at job search web sites, and ask your primary contact (usually someone in Human Resources) for more information. In general, the goals are to look professional, demonstrate that you are interested, demonstrate that you have done research about the position and the organization, and show that you are qualified for the job. Interviews take practice, so use every opportunity to hone your interviewing skills. Write out questions and answers, practice talking to others, or take advantage of mock interviews available from school and community career centers. For more information about interviews, work with the career center at your college or university.

Thank You Letters

Immediately following an interview, write the interviewer a letter expressing appreciation for the interview. The purpose of this letter is to show appreciation for the employer's interest in you, reiterate your interest in the position and organization, remind the employer about your qualifications for the position, and follow up with any information the employer may have asked for during the interview.

22 April 2007

711 Oak Park Rd.
Ames, IA 50010
515-293-9155
aluraj@emailadd.com

Dr. Beverly Taylor
California University
Department of Information Technology
Los Angeles, CA xxxxx

Dear Dr. Taylor:

(1) Thank you for inviting me to campus for an interview and for your time last week. I appreciated the opportunity to learn more about the position. I am especially grateful that you and your staff were able to give up an entire day for the interview process. I learned a great deal and am even more enthusiastic about the position.

(2) As noted during the interview, my current position include s supervision and team building with a staff of 15. The position at California University, twice this size, offers the kind of challenge I am ready for. Your Information Technology strategic plan lists management of the campus peer-to-peer network as a top priority, and I was pleased to see how far along you have already gotten. My background and experience make me the perfect fit to lead this effort.

(3) You asked for more information about my networking certifications. I have enclosed the certificates attesting to my most recent training and certification levels. Please do not hesitate to contact me if you require any additional information.

Thank you again for a most informative interview. I look forward to hearing from you soon.

Sincerely,

Max O'Hara

Max O'Hara

■ Thank you letter

Features of Thank You Letter

(1) First paragraph is clear and direct, thanking the potential employer and reminding him or her of the specific meeting date.

(2) Second paragraph points out one feature that was learned during the interview and makes note of how the applicant is well suited for the position.

(3) Third paragraph provides additional information that was requested during the interview.

Resumes

The resume provides a summary of your work experience, qualifications, and educational background for a job. Accompanied by a cover letter (see above), the resume lists your qualifications for the position to a prospective employer in a succinct, quickly readable format. Professional job searching services offer resume templates that prompt you for the proper information to generate your resume online or on paper. Many companies require a scannable resume—a simply formatted document with keywords that can be scanned electronically. Most schools and public libraries have career resources available in-house or online. Whether you're looking for an internship during school, your first professional job, or a more attractive position mid-career, a nicely designed, complete, and up-to-date resume is an essential part of your application.

Chronological Resume

A typical resume for a student or recent graduate follows, with education listed first and experience listed in reverse chronological order. The resume is limited to one page. Note that related experience and skills appear on the first page.

Skills Resume

The example on pp. 89–90 shows a mid-career professional searching for a new position. Note that expertise and experience appear first, giving analytical details that relate directly to the job.

Sarah M. Barbour

(1)

Current Address:
100 Houston Street
Blacksburg, VA 24060
(540) 555-6666
smbarbour@vt.edu

Permanent Address:
22141 Cabin Road
Square, VA 23456
(703) 555-1234

OBJECTIVE   To obtain a governmental affairs position utilizing language skills **(6)**

EDUCATION   **BACHELOR OF ARTS, INTERNATIONAL STUDIES AND POLITICAL SCIENCE** May 2003 **(7)**
Virginia Polytechnic Institute and State University (Virginia Tech), Blacksburg, VA
Minor – Spanish   Overall GPA: 2.9/4.0; Dean's List last 3 semesters

(2)

VIRGINIA TECH'S "WASHINGTON SEMESTER," Alexandria, VA Summer 2002 **(8)**
- In conjunction with internship with U.S. Agency for International Development: senior seminar in U.S. public policy and political institutions.
- Site visits at the Environmental Protection Agency, Senator John Warner's office, the Campaign Center, and the Library of Congress.

(4) **CENTER FOR EUROPEAN STUDIES AND ARCHITECTURE,** Riva San Vitale, Switzerland Fall 2001
- Studied Italian, Roman history, humanities, and art.
- Traveled to Spain, Germany, Austria, France, Italy, and England studying culture, art, history, politics, and languages.

(3)

LANGUAGE SKILLS
- Written and oral fluency in Spanish.
- Basic writing skills and conversational proficiency in German, French, and Italian.

RELATED EXPERIENCE   **INTERN, U.S. Agency for International Development, Washington, D.C.** Summer 2002
- Assisted in the creation of an agency-wide database.
- Performed technical analysis of various agency programs and communicated their status to USAID missions throughout the world.
- Attended USAID and State Department meetings concerning global environmental issues.

(5)

OTHER EXPERIENCE   **Receptionist, George Mason University School of Law,** Arlington, VA Summer 2001
- Processed and filed incoming student applications and sent brochures to prospective students.

(9) **Receptionist, Chesapeake Materials, Inc.,** Dumfries, VA Summer 2000
- Organized the filing system for a branch office, performed general office work, and made bank deposits.

Office Assistant, Cedar Systems, Inc., Woodbridge, VA Summer 1999
- Awarded August "Temp of the Month."
- Assisted in the organization of the company's computer classes and performed general office work.

ACTIVITIES   Phi Beta Delta International Honor Society
International Studies Organization
Spanish Club of Virginia Tech
(10) Dance Company of Virginia Tech, Stage Manager 2002–03

■ **Chronological resume:** Student example

Features of the Chronological Resume

1 Name and contact information (school and permanent) appear across the top.

2 Indents and white space surround headings and body of resume.

3 First-level headings list categories of information.

4 Second-level headings distinguish individual positions.

5 Formatting (**bold**, CAPS, bullets) sets off the details.

6 Objective (optional) gives a statement tailored to the position.

7 Education comes first with experience items in reverse chronological order.

8 Study abroad and Washington semester are included in "Education" section (not buried

in activities where they may be overlooked).

9 "Related experience" and "other experience" are separated with inclusive dates for each.

10 Other sections highlight distinctive skills. "Languages" relates to objective and is placed higher on the page.

NORISHA M. PRICE

(1) 1223 Maple Lane • Springtown, MN 55666 • (656) 222-1234 • nprice@notasite.org

PROFILE (3) Fifteen years experience in human factors and user interface design in the software industry, including experience with medical devices and related Web sites. Academic background in science and art, and a focus on visualization of multivariate quantitative data. Strength in combining creative and analytical approaches to conceptualize complex interfaces.

Expertise in the following areas:

Human Factors
- Implementing knowledge acquisition methods including user requirements identification, task analysis, user profiles, and usage scenarios
- Imp lementing usability evaluation methods including formal usability testing in a lab environment, contextual inquiry, heuristic evaluation, cognitive walkthrough, and field studies

User interface Design
- Designing user interfaces for enterprise server/client software applications
- Designing page layout and navigation structure for Web pages and Web applications
- Researching technologies that support Web-based UI architecture and display
- Writing detailed GUI specifications for user interface designs

Concept Generation
- Collaborated with a team to generate a method, a basis for implementation, and a conceptual graphical user interface to visualize large amounts of data for an expert decision support system
- Filed (with team) several patent applications in 2004 based on the method; one patent applications received the 2004 Intellectual Property Award
- Extensive experience with brainstorming activities, including experience with electronic brainstorming tools

Academic Research
- Cognitive processes in visualization
- Visual representation of multivariate quantitative data
- Statistical analyses and usability research methods

PROFESSIONAL EXPERIENCE

STARTUP SOFTWARE *Springtown, MN*
Enterprise Data Management Group

Senior Staff Usability Engineer 2000–Present

 Conduct a wide range of usability and system analysis activities for cross-platform enterprise backup and disaster recovery software. Activities include user and task analysis, close collaboration with customers and partners, product definition, GUI design, heuristic evaluation, and usability testing. Led development of a program for disabled accessibility compliance that meets federal government standards. Designed and supervised construction of a digital, state-of-the-art usability lab. Accessibility testing of software for compliance with Section 508 of the American Disabilities Act Developed innovative graphic display for large amounts of data in an operations monitoring system

IN BANIKER CONSULTING INC. *Wintertown, MN* (4)
President and Lead Consultant 1998–2000
Consulted for several companies, resulting in five major usability projects. Project reports and related materials available on request.
- Implantable Device Usability Test: Tested interface for an implanted device that supports data transmission via phone line with an at-home transmission hardware unit. Conducted usability test of the data transmission process with heart failure patients at a local hospital.
- Web Site Usability Evaluation: Evaluated Web interface to a database of clinics worldwide that are qualified to provide care to patients with MedDevice implanted devices. Conducted user analysis and usability evaluation of the navigation structure and page layouts and reported findings and recommendations based on severity.

■ **Skills resume:** Professional example

- Field Study: Wrote detailed project plan for observation and interviews with clinicians using the web site during the clinical trial release.
- Documentation Usability Evaluation: Evaluated organization, layout, and content of manuals, with a focus on the effectiveness of navigation components.

PARENTHETICAL INFORMATION CORPORATION *Wintertown, MN*
Director, Usability 1997–1998
Principal Engineer, Usability 1995–1997
Launched a new usability group and managed all aspects of usability. Developed user interface style guidelines, strategy, and guidelines for incremental changes to the GUI of acquired products for consistency with parent product. Performed on-and off-site usability evaluation using a variety of methods. Created a structured GUI design process and developed training oftware engineers. Led a cross-functional UI architecture team that defines long-term UI strategy and short- and midterm tactical solutions. Launched a new usability group. Managed the design and usability evaluation of a task-focused user interface for a Java-based, Web-enabled software application. Collaborated on strategic plan for development and release of the application.

POLYTECHNIC INSTITUTE *Upstate, NY*
Usability Project Lead, 1993
As a consultant, directed usability component of a joint project between a major international medical instruments manufacturer and a university research team on medical instrument documentation. Collaborated with graphic artists and technical writers in evaluating and redesigning documentation for biomedical equipment. Designed usability study, ran statistical analysis and supervised graduate students in conducting usability tests.

JCN CORPORATION *Klangston, NY*
Graphics Systems Laboratory

Information Developer, 1990–1995
Participated in an international corporate UI design team to design and prototype a user interface for an object-oriented hypertext information retrieval system. Collaborated on the design and implementation of an online information authoring environment which received the 1993 Corporate Innovation Excellence Award. Participated in the User Interface Design Work Group, Information Architecture Project, and Corporate Information Development Council.

RONDO CORPORATION *Iron, NY*
Intern, Online Tutorial Design, 1989
Collaborated with a team of online tutorial designers and software engineers to design a multimedia online help system for an integrated software package.

EDUCATION

UPSTATE POLYTECHNIC INSTITUTE
Ph.D., Cognitive Psychology
Research area: Cognitive processes in data visualization and visual representation of quantitative data
Coursework: cognitive psychology, experimental design, human factors, user interface design, statistics

UPSTATE POLYTECHNIC INSTITUTE
MS, Technical Communication
Concentration in Human Factors

UNIVERSITY OF NEW YORK
BA, Biology
Concentration in Neurophysiology

■ **Skills resume:** Professional example

Features of the Skills Resume: Professional Example

1 Name and contact information appear at top.

2 Indents and white space surround body of resume.

3 First-level headings list general categories of information.

4 Second-level headings distinguish individual positions.

5 Formatting (**bold**, CAPS, bullets) sets off the details.

6 Expertise and experience comes first with dates inclusive for all positions.

7 Education is listed last, due to this person's extensive workplace experience.

Using Resume Templates

The most commonly used word processing software packages (Microsoft Word for the PC; Pages for the Mac) include templates that can be used to create resumes simply and quickly. As is true whenever you let the computer program do the thinking, in some instances one of these templates will be just right and will fit the audience, purpose, and situation. Other times, however, the template may not be the best approach, and you will want to start from scratch, using a blank document.

For instance, in Microsoft Word, you can choose from several templates that come loaded with the software. Page 92 shows an example of the template called "contemporary resume":

To use these templates, simply type over the existing text, substituting your name, objective, experience, and so on.

Another option is to use what is called the resume wizard. This tool, part of Microsoft Word, walks you through a series of questions that you respond to, creating a more custom resume than the standard templates. Most of the online job search sites also use templates and wizard-like forms that you fill out, helping you create a resume that is suitable for their site. See "Scannable and digital resumes" on p. 93.

[Click here and type address]   [Put phone, fax, and e-mail here]

Deborah Greer

OBJECTIVE   [Click here and type your objective]

EXPERIENCE   1990–1994   Arbor Shoes   South Ridge, WA
National Sales Manager
  • Increased sales from $50 million to $100 million.
  • Doubled sales per representative from $5 million to $10 million.
  • Suggested new products that increased earnings by 23%.

1985–1990   Ferguson and Bardwell   South Ridge, WA
District Sales Manager
  • Increased regional sales from $25 million to $350 million.
  • Managed 250 sales representatives in 10 Western states.
  • Implemented training course for recruits—speeding profitability.

1980–1984   Duffy Vineyards   South Ridge, WA
Senior Sales Representative
  • Expanded sales team from 50 to 100 representatives.
  • Tripled division revenues for each sales associate.
  • Expanded sales to include mass-market accounts.

1975–1980   Lit Ware, Inc.   South Ridge, WA
Sales Representative
  • Expanded territorial sales by 400%.
  • Received company's highest sales award four years in a row.
  • Developed Excellence in Sales training course.

EDUCATION   1971–1975   South Ridge State University   South Ridge, WA
  • B.A., Business Administration and Computer Science.
  • Graduated *summa cum laude*.

INTERESTS   South Ridge Board of Directors, running, gardening, carpentry, computers.

TIPS   Select text you would like to replace, and type your information.

■ Sample resume template in Microsoft Word

Audience Considerations

Potential employers review thousands of resumes and skim them quickly, selecting candidates for more in-depth materials or interviews. Keep your resume to one page in length if you can; advanced degree students and candidates who have worked for several years require more than one page or a comprehensive c.v., which also includes publications, talks, and other details. A c.v. is typical in any field where you have a list of publications to include (professors, or scientists who might not be academics). Include items that are most relevant to and supportive of your career goal. Don't embellish your qualifications. Be selective and accurate. It is illegal for potential employers to ask about your race, gender, age, religion, or marital status before or during an interview. Only include the personal information that you want to make public. Include an objective only if you can use it to show self-awareness and commitment to a field.

Design Considerations

In general, you want your writing to be concise and focused on action verbs. Phrases, rather than complete sentences, are preferable for details. Making items within lists brief and grammatically parallel is also important. Use common headings such as those in the examples or other important categories: Affiliations, Publications, Papers, Licenses, Certifications, Examinations. If your resume is more than one page, make sure your name appears at the top of every page.

If you create a print resume, use clean, simple lines and reasonable amounts of white space on the page. Use light-colored, good quality 8½ × 11 paper and print only on one side of paper if you have a two-page resume. Choose font types and sizes that are readable, typically no smaller than 10 point or larger than 14 point. Your name, however, is usually much larger, perhaps up to 32 point if desired. Use formatting, such as boldfacing, underlining, and italicizing consistently and sparingly. As a rule of thumb, do not use two special formats in one phrase. If you post a resume or portfolio online, be certain all pages of your web site and email account names are suitable for employers to view. Many people choose to post an Adobe Acrobat version of their print resume (i.e., a PDF file) on their web site. Find out whether the organization requests a printed resume, an online resume, or a scannable resume. Online job services offer templates that will step you through the most important style and content features of the resume. Others prepare a scannable resume, as shown in the following example.

Scannable and Digital Resumes

Some companies require you to submit scannable or digital resumes so they can create searchable databases. Scannable resumes are usually submitted on paper or in a Word file and are scanned by a database program to select key words and phrases that help identify your skill set. These type resumes were more common in past years. Today, most digital resumes are submitted by filling out a template or form on the company or recruiting agency's web site. Monster.com and large technology companies often give you a choice of uploading a Word document, cutting and pasting, or creating a resume using their template.

The sample on page 95 is a scannable resume. There may still be instances where you need to submit a resume that can be scanned in. The key with scannable resumes is to use little to no formatting and to ensure that there are no graphics, blurred type, or other marks that will cause the scanner to make a mistake.

■ Monster.com provides several options for creating and posting resumes

Norisha M. Price

nprice@notauniv.edu
1223 Maple Lane, Springtown, MN 55666 (656) 222-1234

OBJECTIVE

A position in the software industry involving human factors, concept generation, and user interface design.

EDUCATION

University of New York, Albany, NY
BA, Biology (Neurophysiology), 2005
Upstate Polytechnic Institute, Rochester, NY
MS, Technical Communication, 2007

PROFESSIONAL EXPERIENCE

Startup Software, Springtown, MN
Staff Usability Engineer, 2007-present
- Conducted usability and system analysis activities for cross-platform enterprise software.
- Led development of a program for disabled accessibility compliance.
- Designed and supervised development of a state-of-the-art usability lab.

TECHNICAL EXPERIENCE

Rondo Corporation, Iron, NY
Intern, Online Tutorial Design, 2006
- Collaborated with a team of tutorial designers to design a multimedia online help system for a software package.

LEADERSHIP EXPERIENCE

Upstate Polytechnic Institute, Rochester, NY
Usability Project Lead, 2007
- Directed usability component of a joint project between a medical instrument manufacturer and a research team.
- Designed usability study and directed graduate students in conducting usability tests.

COMPUTER SKILLS

Programming Languages: C++, Java, Perl, Quartz Composer, XPath, XQuery
Operating Systems: UNIX, MS Windows, Macintosh
Applications: Morae, Silverback, Ovo Logger, Screen Recorder, Cam Do

AFFILIATIONS

Usability Professionals' Association
Society for Technical Communication
Usability and User Experience Community, STC

HONORS

Phi Beta Kappa
Society for Technical Communication, International Technical Writing Competition Winner, 2005
Senior Service Award, University of New York

ACTIVITIES

Swimming, mountainbiking, distance running
Habitat for Humanity, Gulf Recovery Effort

■ Sample scannable resume

FOR CREATING A RESUME

- Determine the final format for your resume (print, digital, or scannable). Sometimes, you will need to send a digital copy but follow up with a print copy by mail.

- Generate a list of all your work experience and educational credentials.

- Use a chronological format with education at the top if you are a recent graduate. Use a skills-based format if you have extensive relevant experience or are changing careers.

- Select honors, activities, and skills to include, particularly those relevant to the job or that distinguish you.

- Use action verbs to describe your work-related experience and skills

- Design a clean, organized page with plenty of white space around items.

- Follow the guidelines for resumes if any are provided by the company or organization.

- If you wish to preserve the formatting of the page, save the file as a PDF before sending.

See also

Document Design (Part 3)

Letters (Part 2)

Lab and Field Notebooks

Laboratory and field notebooks are used to record the processes involved in scientific experiments and field research. Both types of notebooks provide a means by which to record important data as well as insights and observations. While lab and field notebooks have certain features in common (such as the need for clear handwriting and dating each entry), each type of notebook has slightly different functions and features.

For a science or engineering class, students are often asked to purchase a standard format notebook. Typically, the notebook will look like this:

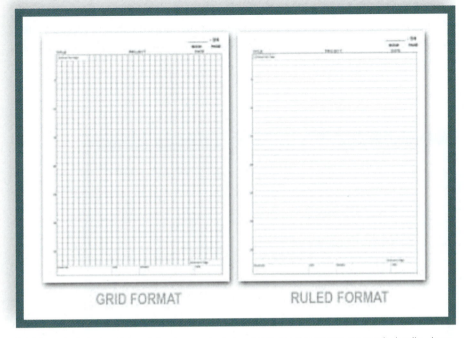

GRID FORMAT   RULED FORMAT

■ This typical student lab notebook page includes fields for signatures, research details, dates, and other information. It is printed on acid-free paper so that the pages will last if they are needed in the future.

Laboratory notebooks provide primary evidence of the research involved as scientists and engineers analyze and interpret an experiment or design. These notebooks can become a kind of diary, chronicling the steps, discoveries, observations, and inventions. They often include drawings, diagrams, and other technical illustrations. Notebooks can provide important legal backup, if any patent or other intellectual property claims arise. Researchers often return to their notebooks later to confirm the various steps and processes involved in a project. According to one physics instructor, "[t]he goal of the lab notebook is for someone who has not done the experiment before (or you, ten years later, when you're in a real research situation and want to re-remember how to build a notch filter) to be able to pick it up, walk into the lab, and duplicate your work" (Habig 2006). In other words, the success of a notebook in any experimental science means that another researcher should be able to use the notebook to reproduce or continue the research.

Elements of Lab Notebooks

Thomas Edison's lab notebooks provide a famous historical example of how essential they were for the inventor's understanding of his research. Edison's notebooks became public when he applied for patents. The pages illustrate the combination of text, drawings, captions, questions, and data that documented his research and experimental process. This example shows his 1886 technical notes and drawings for the incandescent lightbulb.

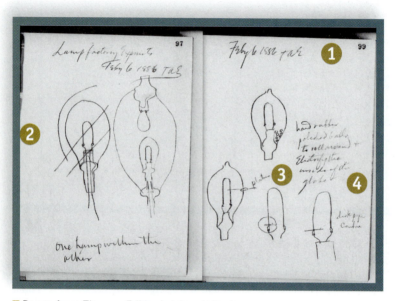

■ Pages from Thomas Edison's lab notebook

Features of the Edison Laboratory Notebook Entry

❶ Entry is dated and signed by Edison.

❸ Parts have labels and include exploded views.

❹ Handwritten notes explain design and record data.

❷ Drawings illustrate various versions of the incandescent lightbulb.

The format, entries, and other required features of an experimental notebook differ depending on the science and the context. For example, if you are working for a chemical engineering company in the research and development lab, you will need to follow the company's format and process for documenting your research. In a corporate setting, such documentation is key to protecting the patents of any new inventions. Large organizations may use electronic lab notebooks, which combine the features of digital text entry with the ability to search, archive, and retrieve entries.

A typical laboratory notebook from an undergraduate chemistry class might look like this:

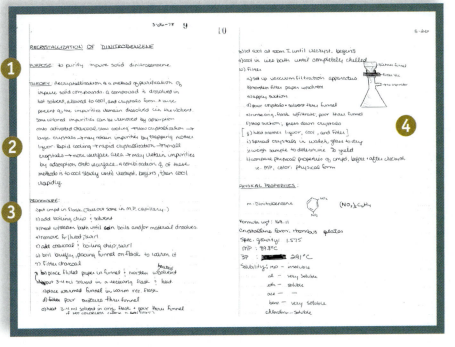

■ Student chemistry lab notebook

Features of Student Chemistry Lab Notebook Pages

1 Experiment title, purpose, and theory are clearly stated using headings.

2 Handwriting is easy to read.

3 Procedure is outlined.

4 Drawing helps illustrate the procedure.

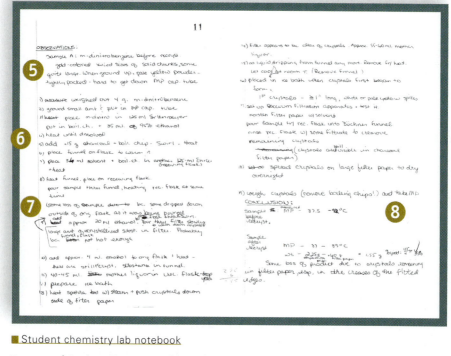

■ Student chemistry lab notebook

Features of Student Chemistry Lab Notebook Pages

5 Observations are recorded in easy-to-read outline format.

6 Measurements are recorded.

7 Details ("some loss of sample…") are noted.

8 Conclusion is easy to locate on the page and is written mathematically.

Field notebooks, like lab notebooks, must contain careful documentation and notes. Field notebooks, however, often reflect observations made on location, rather than carefully designed laboratory experiments. Plant biology, ecology, and geology are some of the disciplines that use field notebooks.

Elements of Field Notebooks

Geology is one area where field notebooks are used extensively to describe rock formations, topography, rock and mineral deposits, landscape features, and other items important to this research. Geology and other field notebooks often use a numbering system to map a particular finding to a geographic location. Increasingly, scientists also use GPS to indicate the precise location of a finding. Plant biologists might use field notebooks to record the daily or weekly observations about an experimental trial. Wildlife biologists might make sketches or drawings to accompany a page of written observations.

A geology field notebook might look like this:

■ Student geology field notebook

Features of Student Geology Field Notebook Pages

1 Day is clearly marked (actual date is recorded at the beginning of the notebook).

2 Observation and question arising from the observation provided in the assignment.

3 Location of the area is noted both by map coordinates and by reference to features in the landscape.

4 Sketches serve to document the observation and to note its defining geologic features.

stop 3
coarse grained sandstone
highly crossbedded
black grains (chert)
salt and pepper look

Kkss (KKlc) lower clastic of Kootenai

dipping east
strike along ridge

N5°E, 47°E   beware of crossbeds
when taking dip.
look at entire package.

N
red shale down to east
makes up valley

stop 3
stop 2   KK??

E

Morrison/Kootenai
contact
contact parallel to strike
|| to ridge crest

moving E and upsection
we expect to find limestone —
Kkm
limestone should make a resistant
ridge (dry climate)
the middle lime makes a hill ridge
looks like a highway.

N15°E, no clear dip
white gray w/ tan siltstones

Kkss
Jm
Kkm
all dipping east

ribs of ss
w/ shale valleys

traversing NE toward limestone
sandstone = ridges
shales = valley formers - red valleys low

■ Student geology field notebook

Features of Student Geology Field Notebook Pages *(continued)*

5 Clear handwriting is used.

6 Specific descriptions of the geologic features are given.

7 Language is concise and precise ("no clear dip").

Audience Considerations

Scientific notebooks serve many important purposes for immediate and future audiences. Your notebook, while primarily a personal record-keeping document and writing space, also serves as documentation of your thinking and discoveries during the experimental and observational process. Collaborative research groups share notebooks or review one another's entries regularly, and they sometimes respond in writing to each other. Outside audiences will see your notebook entries during formal review processes or if they become published.

Lab notebooks often become the property of the lab, not the author. If the lab notebook is part of a class, ask your instructor for feedback on the clarity and detail of your entries. For students working with a professor on a research project, the lab notebook may be part of the data associated with a larger project that has several principal investigators (PI) and grant funding. In cases like this, the lab or PI would own the notebook and its contents.

Design Considerations

Most researchers and students use paper lab notebooks. For classes in chemistry, biology, physics, or related labs, the instructor will specify the type of notebook and provide guidance on the requirements for how to document your research and use a lab notebook. For field research, practical matters such as carrying cases and waterproof covers become important.

Notebooks are both a personal record and a scientific document. The quality of your notes depends on speed and accuracy but also on completeness. Write clearly and legibly, using permanent ink or hard pencil. Sketches and free-form notes can encourage thinking and brainstorming. Detailed drawings help you recall your observations after the fact. Be sure to label these drawings.

Guidelines

FOR CREATING LAB AND FIELD NOTEBOOKS

- Use a portable medium and consistent format that is appropriate for the class, company, or location. Paper lab notebooks are still the most common medium.

- At the beginning of the notebook, state the problem, field setting, or experiment.

- For each entry, provide a chronology of the experiment or observation to document the process.

- Use a consistent title, number, or other descriptor for each entry.

- If required by your class or your employer, include your signature or initials on each entry.

- Use a clear, concise style, but do not leave out important observational details.

- Write down all details about the process, experiment, observations, physical conditions, trends, and other information related to the collection of data and setting.

- When appropriate, include drawings or sketches with details observed, multiple views, labels, calculations, tables, and other visuals.

See also

Document Design (Part 3)

Instructions (Part 2)

Maps (Part 3)

Technical Illustrations (Part 3)

Letters

In business and technical communication, letters are the written correspondence most often used for delivering news, requesting action, or documenting an agreement or understanding. Writing clear, polite letters is a vital skill in the technical workplace.

Unlike internal memos or emails, a business letter usually designates a formal communication to people both inside and outside the organization. Whatever your purpose, writing clear, polite letters is a vital skill in the technical workplace.

In an era where people increasingly choose email, instant messaging, or text messaging rather than letters, it is important to consider the reasons for writing a formal letter. Although email and texting are used widely in the workplace, most of the time these communications are quick, short, and tend toward an informal tone. Letters, on the other hand, are meant to be read more slowly and use a formal tone. Letters, especially on company letterhead, create credibility and convey the full backing of the organization. Since most letters are signed by hand, letters can also become a formal, legal document. So, for instance, you may receive a job offer on the phone or over email, but you will then receive a formal offer in the form of a letter, on company letterhead. This letter may be sent to you via surface mail or as a PDF attachment.

You will normally write a letter whenever you submit a report or request something, as in a claim or inquiry. You will also write letters to acknowledge your approval or understanding of legally binding documents. Letters need a strategic organization, style, and tone, whether you are delivering bad news or generating an important document for the record.

Content of the Letter

The following chart lists some basic items that you need to include in the content of most letters. The rest of this section offers examples of specific types of letters and provides more detail on how these items are used in various situations when a letter is appropriate.

Element	Purpose
Contact Name	Always get a contact name for addressing your letter. Get a name from the organization, the job ad, the web site, or human resources. As a last resort, use the intended recipient's job title or "Director of Human Resources."
Purpose	Your first few sentences should invoke a specific context and state the purpose for the sending the letter. Make that purpose and its importance immediately clear to readers.
Idea or Outcome	Include your central idea, intended outcome, or requested action item in the introductory paragraph of the letter. Let readers know right away what you are asking of them.
Background	While some background may help give context, it works best to use a subheading for discussions longer than a sentence or two. A background section can begin after the first paragraph.
Information	The body of your letter provides data, results, examples, or other information in short paragraphs. Discuss one topic per paragraph, using clear topic sentences or subheadings for each one.
Persuasion	Provide sound reasoning and logic to support your main idea or position. Look for ideas or values that you and your readers share, and use these to help establish the case you are making.
Action Requested	Finish your letter by emphasizing the specific action you want readers to take. If appropriate, mention future accomplishments that will result from this action.
Goodwill	Your letter should conclude by thanking readers and offering additional assistance or correspondence.
Tone	Keep the tone of your letter polite and respectful. Such an approach will convey professionalism on your end and, if you are writing as an employee of an organization, will also convey professionalism on the part of the company. Stay with the facts of the situation and avoid excessively emotional language.

Specific Types of Letters

While all letters contain similar elements, specialized letters sometimes fulfill delicate purposes. A refusal, sometimes called a "Bad News" letter, often hides the messenger and the message behind careful structure and passive voice. Positive, "Good News" letters, and also neutral letters like acknowledgments, use an active-voice, reader-centered style, and a structure based on convincing readers to take that action. A complaint, while negative, also uses active voice to persuade the company to appease or compensate the writer. Here are some common types of letters.

Acknowledgment Letters

Acknowledgment letters respond to an inquiry or briefly confirm receipt from the recipient. Also called response letters, they play a simple but important role in business and technical communication as they offer the status of a situation and can also recommend next steps.

The Best Kitchen Supply Company
1199 Million Dollar Drive
Great River, NJ 37777

October 25, 2007

R. Baron
2221 Alexandria Street
Ballston Meadows, NY 12345

Dear Ms. Baron:

(1) Thank you for sending us your letter along with the broken portion of flexible tubing from your recently purchased Kitchen Magic high power food processor. We are sorry for your inconvenience. We understand your concern and want to correct the situation immediately.

(2) We will be sending you a new tubing portion next week. This tubing is made from an improved material and normally accompanies our more expensive models. However, we want to ensure your complete satisfaction and so will be sending you this tubing plus a gift certificate for $50 that can be redeemed at any local store.

(3) Once again, please accept our apologies. Thank you for purchasing a Kitchen Magic food processor. We hope you enjoy it for many years to come. Please contact us if you have any further questions or concerns.

Yours truly,

Riley Madigan

Riley Madigan
Director, Customer Relations
(4) customerrelations@bestkitchensupply.com

■ Example of an acknowledgment letter

Features of the Acknowledgment Letter

(1) Opening paragraph offers thanks and acknowledges the original letter.

(2) Second paragraph explains what will be done to redress the problem and by what date.

(3) Closing paragraph is sincere and offers to answer any additional questions.

(4) Email address provides additional contact information.

Complaint Letters

Complaint letters explain a grievance or concern and ask for specific actions to redress that concern. Also called a claim, this letter must express displeasure but also use persuasion and accurate information to be effective. Keeping a record of details and the dates becomes important if you need to follow up or even take legal action.

DBG Construction　　　　　**Charleston, SC**　　　　　**Design Build Green**

April 10, 2007

John Johnston, President and CEO
Sudden Alarms
Charleston, SC 22222

Dear Mr. Johnston:

I have used your equipment for years since your company installed my own house alarm. But last week my confidence in your company was shaken because your alarm system failed to activate when our construction site was burglarized.

The break-in occurred at the Townehome Walk, where I work as a project manager. When I returned from a business trip on March 28, I found broken windows, vandalized equipment, and tools missing. I called your monitoring center and the service representative found in the record that a unit's living room sensor activated at 4:15 p.m. on March 25. However, I was not notified and the police were not dispatched to Townehome Walk. The burglars ran away when the alarm's indoor siren went off.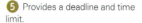

When I asked why nothing was done on March 25, your representative cited a notation in your files to ignore the sensors "until further notice." DBG Construction had no knowledge of that notice. This alarm system was not fully armed for months without our knowledge.

Sudden Alarms failed to provide its contracted service and dispatch the police. We ask that you reimburse DBG Construction for our damaged and lost equipment. We estimate the loss to be $100,000 and will submit an itemized claim next week. Sudden Alarms is known as one of Charleston's most reliable companies. Until this incident, DGB Construction has always trusted in your company to provide our security.

DBG construction would like to have this matter resolved quickly, by April 30. Please contact me if I can be of any assistance.

Cordially,

Sarah Kingston

Sarah Kingston
Project Manager

■ Example of a complaint letter

Features of the Complaint Letter

1 First paragraph establishes common ground and confidence in the reader's company.

2 Events and details in the next two paragraphs provide facts to support complaint.

3 Fourth paragraph places the responsibility on the company.

4 Ending requests a specific action or adjustment from the company.

5 Provides a deadline and time limit.

Cover Letters

Cover letters usually accompany a resume as part of a job application. For information and examples of this type of letter, see **Job Search Documents**, in Part 2.

Inquiry Letters

Inquiry letters pose a question or a request for information, including inquiries about unadvertised jobs. The letter must clearly indicate what you want, why you are asking, and what you are asking of the recipient. The goal is to receive follow-up information or an invitation for additional correspondence.

August 1, 2008

1022 North Side Lane
Albany, NY 12345

Dr. Stephanie Iversbretson
Editor, *Electronic Vehicles* magazine
18 Curveside Way
Flagstaff, AZ 86003

Dear Dr. Iverbretson:

(1) I am writing to inquire about whether it would be possible to obtain more technical information about a recent cover article from your magazine as well as the name of the photographer who shot this outstanding cover.

The cover of your most recent issue of *Electronic Vehicles* (June 2008) displayed a **(2)** photograph of a gas-electric hybrid that had been converted to electric with a vegetable oil by pass system. I am an electrical engineer creating a start-up company here in the upstate New York area, and I am interested in learning more about this technology for a report I am preparing. Unfortunately, the article in the magazine didn't provide any of the technical information I had hoped to learn about, and the author's name is not listed.

The photograph is also an outstanding example of how hybrid vehicles can look every bit as attractive as traditional gasoline powered sports cars. A photograph of this type would be great on my company's new web page.

(4) Would you be willing to provide me with contact information for the technical experts behind this story as well as the name of the photographer? I would be most grateful for this information and would be happy to send you a copy of my forth coming report.

The easiest way to contact me is via email; my email address is listed below. Thank you for your assistance. **(5)**

Sincerely,

J.J. Matthiesen, P.E.

J.J. Matthiesen, P.E. **(6)**
jjengineering@arixnext.org

■ Inquiry letter from a technical professional seeking additional information

Features of the Inquiry Letter

(1) First paragraph is puts the request up front and is short and to the point.

(2) Second paragraph provides background and introduces the writer to the reader.

(3) This paragraph compliments the editor on the cover design, creating goodwill between the writer and reader.

(4) This sentence restates the request in specific.

(5) Contact information and a polite, sincere thank you make the best closing.

Reference Letters

Reference letters support applications for a job, an award, or a grant. When asked to write a reference letter, also called a recommendation, consider what you can honestly offer before you agree. Then, write the letter from that perspective—as the coworker, supervisor, teacher, or community leader. Your goal is to assess the applicant with specific examples and as compared to other applicants. Avoid writing generic, mediocre, or formulaic letters; they will not be taken seriously and can do more harm than good.

In many cases, an employer or person providing a reference is only allowed to provide basic information (that the person worked for that company and for what dates). It is important to check on this policy before you write a reference letter using official company or organization letterhead.

University of Western Junction
Department of Technical Communication
200 South Evergreen Street
Boise, ID 98765

June 25, 2007

Mr. Arbash Somatro
Molecular Pharma-optic Corporation
77 Downing Road
Finely, CA 86655

Dear Mr. Somatro:

Ms. Jana Sendentra has asked me to provide this reference letter relative to her application for a position as junior research scientist in your organization. I am most happy to recommend Ms. Sendentra to you enthusiastically and without reservation. Ms. Sendentra was my advisee for four years in her B.S. program here at the university; she also served as my research assistant for one semester. She is without a doubt one of the most capable and talented students I have worked with in my many years of research and teaching.

In particular, Ms. Sendentra's ability to multi-task, handling many projects simultaneously and without any difficulties, will make her a valuable employee in a high technology, fast-paced environment. She works well individually and is also a strong team member. Ms. Sendentra is also an excellent researcher. In one case, my project had fallen behind, and I asked Ms. Sendentra to do a detailed review of technical literature so we could get our grant application in on time. She did an outstanding job, equal to or better than many of the graduate students or professional researchers I have worked with.

In addition to these skills, you will note that Ms. Sendentra speaks five languages, all fluently. This feature makes her a strong addition to a company that has a global market. She is also quite adept at understanding and communicating highly technical information to many different audiences.

It is hard to imagine a better candidate for your position than Jana Sendentra. I hope you will give her application your most serious consideration. Please contact me if I can provide any additional information.

Sincerely,

Laura F. Serwanski, Ph.D.

Laura F. Serwanski, Ph.D.
Professor and Department Chair
Serwanlf1224@westernuniv.edu

■ Example of a reference letter written by a professor for a student

Features of the Reference Letter

❶ First paragraph explains how the professor knows the applicant and for how long.

❷ First paragraph gets right to the point, stating that the reference is positive and without reservation.

❸ Body of the letter contains specific examples and compares candidates to others in a similar context.

❹ Closing paragraph is enthusiastic and restates the opening.

❺ Closing paragraph provides additional contact information.

Refusal Letters

Refusal letters deliver bad news when turning down a job offer or other opportunity. Refusal letters and letters of resignation, which are similar, must project clarity and confidence without offending the recipient. A short, clear message without much explanation is best: "After considering my options carefully, I have decided to decline your offer." Today, most short refusal letters are sent via email. Use a hardcopy letter to follow up or for a more formal approach.

1234 Oak Hill Boulevard
Granite Center, MI 33445

September 15, 2007

Spectrometer Research Inc.
555 High Tech Lane
Olenville, CA 99933
Attention: Mr. John Dunning

Dear Mr. Dunning:

1 **2** Thank you for encouraging me to apply for the position as bench scientist and for your help in arranging a telephone interview with the hiring manager. As I noted on the phone today, I have received and accepted an offer with another company and so am withdrawing from the search at Spectrometer Research Inc.

Thank you for your time and interest. **3**

4 Yours truly,

Pat Meany

Pat Meany

■ Example of a refusal letter

Features of the Refusal Letter

1 Text is brief and to the point.

2 Tone is polite and appreciative.

3 A brief reason is given but without too much detail.

4 The writer thanks the recipient for the opportunity.

Transmittal Letters

Transmittal letters briefly summarize and accompany a report or other substantial, self-contained item. Transmittal letters help get the document accurately to its intended audience, provide context, including the date submitted, and also indicate any action to be taken. This example illustrates a transmittal letter for an EPA Annual Report on letterhead using modified block format.

UNITED STATES ENVIRONMENTAL PROTECTION AGENCY
WASHINGTON, D.C. 20460

2

OFFICE OF THE
ADMINISTRATOR

November 15, 2004 **1**

The President
The White House
Washington, DC 20500

Dear Mr. President:

3 I am pleased to present the Environmental Protection Agency's (EPA) *Fiscal Year 2004 Annual Report*, which highlights EPA's programmatic and financial performance over the past fiscal year. This document fulfills requirements set by the Government Performance and Results Act and other management legislation. **4**

5 EPA made significant progress toward each of the five long-term goals for protecting human health and the environment that we established in our 2003-2008 *Strategic Plan*. EPA is increasing the pace of environmental improvement while keeping the nation economically competitive by focusing on results, expanding collaborative partnerships, improving technology, and strengthening market incentives.

This report evidences EPA's commitment to be accountable for results measured against the annual performance goals presented in EPA's FY 2004 *Annual Plan*. Due to the November 15 reporting date, final end-of-year performance data for several key programs are not available for publication in this report, but will be provided in future reports.

6 You have my personal assurance that the performance and financial data included in this report are complete and reliable, comporting with guidance provided by the Office of Management and Budget. Detailed information on data quality is included in Appendix B of the report. We are proud of the accomplishments that we and our state, tribal, local, and federal government partners have achieved, and we intend to build on these results to fulfill **7** our responsibility for protecting human health and the environment.

Sincerely,

Michael O. Leavitt

■ An EPA transmittal letter

Features of the Transmittal Letter

1 Date appears at the top of the letter and below the printed letterhead.

2 Letterhead and signature provide contact information for the writer.

3 First sentence identifies the enclosed report or items.

4 The first paragraph provides the context of why the report was required.

5 Second and third paragraphs summarize the report findings briefly.

6 Final paragraph builds credibility of the writer by offering his "personal assurance."

7 Final sentence builds good will with readers.

Audience Considerations

Because letters are often addressed to an unknown individual, they must be polite and somewhat formal, yet personable enough to avoid stuffiness. Good letters state requests clearly and directly, even if they include negative information like a complaint. Assume that people other than the recipient will look at your letter. Almost all business letters are read by multiple audiences such as managers and not just the primary audience. Emailed letters should be crafted just as carefully. Remember that email can be easily forwarded to many people at once and a hasty note can easily create wrong impressions. Letters, if done correctly, can also provide important legal documentation for your interactions. Most letters also use persuasion to move the recipient to some action. Letters to international audiences may present challenges of cultural difference, because the aggressive argumentative style of much western correspondence may offend colleagues from other cultures.

Design Considerations

Professional tone, a strong paragraph structure, detailed examples, and a clear, readable format define the content of a good letter. Don't go overboard and use false formality, stilted word choice, or wordiness in your business letters. Use concise, active language and directly address the needs or requests of the reader whenever possible. Formal situations like job applications require a professional letter format, preferably printed on letterhead paper from your organization or school. Always edit, proof, and spell-check a printed or emailed letter, and ask a colleague to read and proof it before sending it. It's best to print a letter to proof it because you will catch mistakes easily missed on the screen.

Letters should be properly formatted, using clear, readable type. For emailed letters, be sure to include an electronic signature. Save the file as a PDF, so that no one can modify your original letter and so that the formatting stays intact.

Many word processing programs provide wizards and templates that offer basic formatting and layout for letters. For example, in Microsoft Word you can choose from templates such as "professional letter," or "contemporary letter." Be careful using templates. Some designs may look nice when writing to a friend but may be the wrong format for a letter to a business contact. Many web sites offer templates and other ideas for writing letters. These sites offer good ideas, but the best letter is an original letter that you create specifically for the audience and purpose of your situation.

FOR CREATING EFFECTIVE LETTERS

- Get a contact name for addressing your letter.

- Determine your purpose for writing and choose the appropriate type of letter.

- Craft a succinct, persuasive letter that follows standard business letter format.

- Include current date at the top and leave room to sign the letter.

- Print and proof your letter; spell-check the text.

- Print your final version on letterhead, if possible. On blank paper, include your address in the header.

- Sign the letter.

- Fold the letter carefully and send in a typed, matching envelope; or send your formatted letter as an attachment via email.

- If appropriate, call to confirm that the letter has been received.

See also

Audience (Part 1)

Document Design (Part 3)

Email and Attachments (Part 2)

International Communication (Part 1)

Job Search Documents (Part 2)

Memos (Part 2)

Persuasion (Part 1)

Medical Communication

Medical communication is an important type of technical communication, because so much of this highly technical, specialized information is available over the Internet. According to the Pew Internet&American Life Project in 2006, 80 percent of Internet users in the United States (over 113 million adults) have used the Internet to search for medical and health information. Many medical web sites are designed for specific audiences, while others are designed in layers, with links for patients, insurance companies, or medical professionals.

Medical communication should be written and designed with a sharp focus on the intended audience and how these people will use the information. Effective medical communication uses language that is appropriate for the audience and answers questions or provides links for additional information. However, medical communication online is not a substitute for a visit to the physician's office.

The following two examples illustrate common forms of medical communication: one for patients and non-expert audiences, and the other for expert medical audiences (doctors, nurses, technicians). The first is a document, available on the Web in PDF format, intended primarily for patients. The second is a web site intended for medical professionals, to help them use a software tool for a pacemaker.

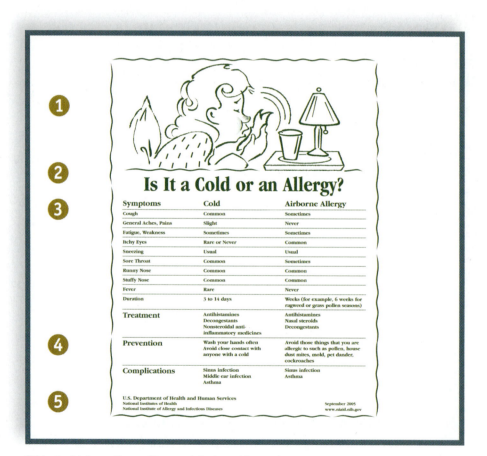

■ Medical information written and designed for patients

Features of Medical Advertisements and Information

1 Cartoon makes it easy to understand what topic is being addressed.

2 Heading is brief, in the form of a question, and to the point.

3 Information is set up in an easy-to-use table format.

4 Advice is non-technical ("wash your hands often") and accessible to people who ae not medical experts.

5 Design uses one color and does not distract readers with a lot of unnecessary visual clutter.

Communication for Medical Professionals

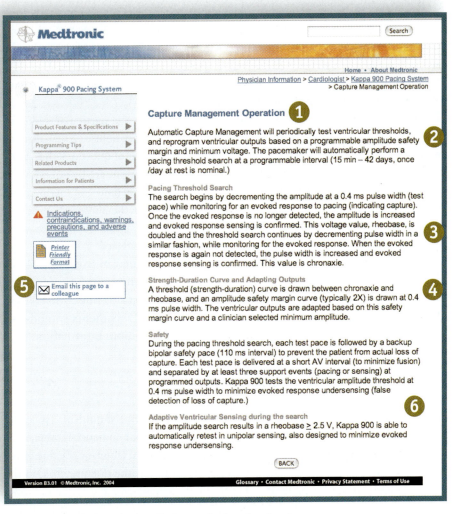

Medtronic

Search

Home • About Medtronic

Physician Information > Cardiologist > Kappa 900 Pacing System
> Capture Management Operation

Kappa® 900 Pacing System

Product Features & Specifications

Programming Tips

Related Products

Information for Patients

Contact Us

⚠ Indications, contraindications, warnings, precautions, and adverse events

Printer Friendly Format

✉ Email this page to a colleague

Capture Management Operation ❶

Automatic Capture Management will periodically test ventricular thresholds, and reprogram ventricular outputs based on a programmable amplitude safety margin and minimum voltage. The pacemaker will automatically perform a pacing threshold search at a programmable interval (15 min – 42 days, once /day at rest is nominal.) ❷

Pacing Threshold Search
The search begins by decrementing the amplitude at a 0.4 ms pulse width (test pace) while monitoring for an evoked response to pacing (indicating capture). Once the evoked response is no longer detected, the amplitude is increased and evoked response sensing is confirmed. This voltage value, rheobase, is doubled and the threshold search continues by decrementing pulse width in a similar fashion, while monitoring for the evoked response. When the evoked response is again not detected, the pulse width is increased and evoked response sensing is confirmed. This value is chronaxie. ❸

Strength-Duration Curve and Adapting Outputs
A threshold (strength-duration) curve is drawn between chronaxie and rheobase, and an amplitude safety margin curve (typically 2X) is drawn at 0.4 ms pulse width. The ventricular outputs are adapted based on this safety margin curve and a clinician selected minimum amplitude. ❹

Safety
During the pacing threshold search, each test pace is followed by a backup bipolar safety pace (110 ms interval) to prevent the patient from actual loss of capture. Each test pace is delivered at a short AV interval (to minimize fusion) and separated by at least three support events (pacing or sensing) at programmed outputs. Kappa 900 tests the ventricular amplitude threshold at 0.4 ms pulse width to minimize evoked response undersensing (false detection of loss of capture.) ❻

Adaptive Ventricular Sensing during the search
If the amplitude search results in a rheobase ≥ 2.5 V, Kappa 900 is able to automatically retest in unipolar sensing, also designed to minimize evoked response undersensing.

BACK

Version B3.01 © Medtronic, Inc. 2004   Glossary • Contact Medtronic • Privacy Statement • Terms of Use

■ A web page designed primarily for medical professionals

Features of Medical Communication

❶ Heading describes the process or technical concept.

❷ Specialized language is technical for this audience.

❸ Sentences are complex and often use the passive voice, although this usage is changing with the emphasis on active voice in all scientific writing.

❹ Descriptions are kept to a minimum.

❺ Link lets physicians email this information to a colleague, thus appealing to the community nature of medical professionals.

❻ The page design is attractive and readable, with adequate white space and links to additional information.

Audience Considerations

A clear sense of audience and purpose is important when creating medical infor-mation, which can be created for a variety of audiences: patients, medical sales re-presentatives, physicians, nurses, medical therapists, veterinary doctors, lawyers, federal and state health care regulators, and so on. Each audience will have a different set of needs. A cardiologist working with an implantable device, such as a pacemaker, wants to know how it works on a very technical level and the proper procedures for implanting the device. A patient will want to know about the impact of the surgery and long-term use.

You should structure medical communication to suit the particular audience and purpose. For example, if the purpose is to introduce a patient to a new drug or an unfamiliar medical procedure, use headings in the form of questions that read-ers might commonly ask. You can find out what these questions are by interviewing patients or those who work with patients in a usability study (and you might create a "top ten list" of frequently asked questions). No matter the audience, structure medical communication so that it's easy to access: keep paragraphs short, use white space, and allow readers to navigate to the information they need.

Medical information should be honest and not attempt to cover up risks or bury them in the small print. Since some medical communication is federally regulated, even minor mistakes can lead to legal problems or fines. Mistakes in documentation can also lead to risks for the patient. Documents should also provide contact information so that readers can obtain more information.

Design Considerations

When you write and design medical communication, you should be especially careful about the level of technical language you use. If your audience is general, as in the Allegra example, use short sentences and nontechnical language. Define terms as you go along (you can do this with a hypertext link if the document is on the Web). When writing for patients, you should use a professional yet friendly tone. When writing for other medical professionals, stay on point and use language that is appropriate to their specialty. Plain style is especially important since medical infor-mation has a global audience, and such information will often be translated into hundreds of different languages and designed for numerous different cultures.

Most medical information is available in multiple formats: hard copy, DVD, and web sites. You should design your communication with the idea that it could easily be repurposed for different media. Keep prose short and to the point. Use headings. Select colors that work well in both print and on screen. Considering using a combination of both HTML (web pages) as well as PDF (Adobe Acrobat), since, with PDF documents, you can be certain the page will print exactly as you've designed it.

Medical communication should be visually easy on the eye, employing clean document design and appropriate use of visuals. As with any complex technical information, medical content should be easy to read not only because the language is appropriate but also because there is a good balance of text, white space, and images.

Because medical communication on the Internet can be accessed by a wide variety of people, a layered approach, providing links for different audiences (for patients, for doctors, for pharmacists, for parents, and so on) is an effective approach for medical web site design.

Guidelines

FOR CREATING MEDICAL COMMUNICATION

- Identify the primary audience for the information, whether patients, medical professionals, or both.

- Keep information brief and to the point whenever possible.

- Write in short paragraphs and use appropriate headings.

- When writing for patients:
  - Keep language simple and clear by avoiding Latin and technical terms
  - When you must use a technical term, define it
  - Use headings in the form of questions that patients might ask
  - Use visuals that patients can relate to
  - Provide links to additional information.

- Use PDF format if you want the printed version to look like the printed copy, or provide a "print-friendly" version.

- When writing for medical professionals:
  - Avoid unnecessary prose
  - Use technical terms appropriate for that audience
  - Use visuals that illustrate technical concepts
  - Illustrate procedures and identify products, when necessary
  - Use PDF format if you want the printed version to look like the printed copy.

See also

Memos

A memorandum, or memo, is the most common correspondence within an organization. Memos share qualities with letters, including content, purpose, and format elements that keep them focused and easy to read. While memos can contain short, informal office communications, they also circulate policy statements and reports intended for an internal audience. Memos are most often sent from supervisors to their employees or staff and use conventional elements that identify them as interoffice communication. Most memos are created using email or as PDF documents attached to the email in a separate file. Email has many benefits over paper memos, including easy storage, forwarding, and retrieval.

While formats vary by organization, this example illustrates a typical memo format. This policy memo provides information about Title IX gender equity policy at a university and is delivered via email.

From: postmaster@mailgroup.uu.edu
Subject: University Title IX Coordinator
Date: August 24, 2007 10:38:53 PM EDT
To: employees@mailgroup.uu.edu

Email submitted by Tina Kelton, Title IX Coordinator
Please send replies to email@uu.edu

State University strives to provide an environment where faculty, staff, and students can work, educate, learn, and grow. In this endeavor, we recognize the stigmatizing affect sexual harassment can have on our campus environment. State University is bound by various state and federal regulations and laws prohibiting harassment. One such law is Title IX of the Education Amendments of 1972.

What is Title IX of the Education Amendments of 1972?

Title IX of the Education Amendments of 1972 was the first comprehensive federal law to prohibit sex discrimination against students and employees of educational institutions. Title IX states, in part:

"No person . . . shall, on the basis of sex, be excluded from participation in, be denied benefits of, or be subjected to discrimination under any education program or activity receiving federal financial assistance."

The policy of State University is to implement affirmative action and equal opportunity for all employees, students and applicants for employment or admission without regard to race, color, religion, national origin, sex, age, veteran status, or disability.

How do athletics comply with Title IX?

Educational institutions that receive federal funding are required under Title IX to provide equal opportunities for members of both sexes.

Under Title IX who is protected from sexual harassment?

Title IX prohibits sex discrimination. Sexual harassment is a form of prohibited sex discrimination. Students (male and female) and employees (faculty and staff) are protected from sexual harassment and may recover monetary damages.

Who is responsible for enforcing Title IX?

The Title IX Coordinator is responsible for enforcing the law. Faculty, staff, and students can file complaints of sex discrimination with the Title IX Coordinator. Retaliation against complainants is prohibited. For more information, see our web site: http://T9.uu.edu.

Please do not hesitate to contact the Title IX Coordinator with any questions, concerns, or claims of harassment that you feel might be in breach of Title IX.

■ Example of a policy memo sent as email

Features of the Policy Memo

1 The sender incorporates her name and reply address in the text because official messages sent out under the "Postmaster" address the entire organization.

2 A short introductory paragraph gives the context for prohibiting sexual harassment at the university and states the policy. Background on the Title IX sets the context much like a report or a letter does.

3 The memo uses a question-and-answer structure as organizational headings, making additional information user friendly and easy to skim.

4 A final paragraph offers contact information and additional services.

Memo Reports

Many kinds of internal reports are submitted in memo format—even complex, analytical ones. A recommendation memo can use subheadings to organize the reported information clearly in relation to its purpose. For example, a problem-solving memo can be organized into causes and recommended actions. The recommendation memo can then be composed in three sections each with its own persuasive purpose:

Introduce the recommendation

List the benefits

Conclude with expected results

MEMORANDUM

To: John Ray, President
        Raymetrics, Inc.
From: Jay Cee, Radiation Team
Date: May 5, 2007
Re: Request for senior research scientist

Introduction

The Radiation Team has been unable to keep up with the recent increases in workload. Given its current contracts, Raymetrics is severely understaffed. Every day that delays a finished project costs the company at least $5,000. We need to hire a senior research scientist to keep up with this workload and increase our production.

Benefits ❸

Although it will cost Raymetrics $90,000 plus benefits to add a senior researcher to our team, the positive effects on our productivity and national profile will far outweigh this cost.
1. With three senior-level researchers, the team can increase the number of projects it works on at any one time by at least 30% because of additional leadership.
2. The person hired will bring new radiation systems and techniques to our team.
3. Our tests will be more efficient and yield better results with another person on board.
4. A senior researcher will attract additional contracts in radar research.

Conclusion ❹

Given these immediate and long-term benefits, I recommend that Raymetrics recruit and hire a senior scientist as soon as possible. If we can develop the Radiation group, we can raise the national profile of our company, which is already known for excellence. If I can be of any assistance, please let me know.

■ Example of a short recommendation report in memo format

Features of Memo Report

❶ The **Header** includes four separate lines using alignment with a colon after each element:

To: With name and title of the recipient(s).

From: With name and title of writer(s) or sender(s). Often initialed by the writer(s).

Date: Current date in numeric format.

Re: Subject line (literally means regarding) succinctly represents the subject matter and purpose of the memo with a short *noun* phrase.

❷ The **Introduction:** States a clear purpose and provides context for the hiring of a senior scientist.

❸ The **Benefits** (body of the memo): Functions just like a recommendation report by providing details about costs and the benefits that out-weight that cost.

4 The **Conclusion:** The final paragraph may serve many purposes—here, the recom-mendation and specific steps for action along with conven-tional polite conclusions and offers of assistance. The end-ing tells the recipients what they can or must do next.

Audience Considerations

Memos are written to peers, group members, a supervisor, or even an entire organi-zation. The primary audience determines both the level of formality and the extent of detail you need to provide. Memos are sometimes authored by more than one per-son or even "ghostwritten" by one person for the boss or supervisor. Since memos communicate within an organization, you can assume an insider audience familiar with basic information about your business or industry. Your topic and context for writing the memo determine how much background information you must explain.

Even when writers address memos to one particular person, such as a boss, or the group of people who requested the information, memos usually circulate far beyond the addressee. Like letters, memos also often serve as official policy statements or re-ported information that provides later documentation for informational or even legal purposes. You should carefully craft your memos with these other audiences in mind.

Design Considerations

Memos open by setting the context for the subject, and typically close with resources for further information or a requested action. Like other correspondence, memos need a method of organization, such as the questions and answers in the emailed memo, or the detailed changes that organize the print memo. Depending on the length of the memo, you can use headings or bullets as document design techniques to help readers skim the content without missing the message.

The writing style of your memo depends on the audience you address directly. For a peer, informal, everyday language in second person is appropriate, as is termi-nology specific to your work. For a supervisor, a more formal style may be necessary, as well as definitions or descriptions of specific work activities. The more general and removed the audiences, the more care you must take to explain and describe your information in formal terms. Remember that memos also can provide impor-tant documentation of processes and activities and activities within an organization.

You will produce memos using word processing software or an email program. Memos are not typically printed on letterhead, but printed memos sometimes include the word MEMORANDUM in all caps at the top of the page. Memos do not include a complimentary close or signature; instead, the writer initials after the From: line. Email programs automatically contain header information, but in a dif-ferent order or format depending on the email software program, and they still required a concise subject line.

If you're producing a long or complex memo, make sure all formatting elements are clearly defined and parallel to one another before printing or attaching it to a message. Print Preview provides a good method for checking format. When you repurpose or forward memos, be aware that automatic dates inserted by your software or email program will change the date of your original document. Remember to save your memo with a datebased filename for later reuse or documentation. When creating email memos, check your memo by sending it to yourself before you distribute it to ensure that all elements are clearly identifiable and that your correct contact information appears clearly near the beginning.

Guidelines

FOR CREATING MEMOS

- Determine the purpose for the memo and its primary recipient.

- List or outline briefly the necessary information and background.

- Write a succinct subject line using a noun phrase that is neither vague nor too detailed.

- Compose an opening paragraph that offers clear context.

- Write body paragraphs using an appropriate method of organization.

- Include format, if needed, both between paragraphs (e.g., headings) and within paragraphs (e.g., bullets).

- Compose a closing that requests specific follow-up or offers contact information.

- Spell-check your text and check the date.

- Save a copy of your memo with a datebased filename for later reuse or documentation.

See also

Document Design (Part 3)

Email (Part 2)

Letters (Part 2)

Reports (Part 2)

Newsletters

Newsletters are regular communications from an organization aimed at a specific audience using a mailing list. You probably receive traditional print newsletters in the mail as well as through email. What determines that you always read the newsletter from your local Humane Society when you don't read the monthly newsletter inserted in your water bill? The content must be timely, useful, and important to the specific users who subscribe to your newsletter. Some newsletters are unsolicited and offer little more than sales information. Unlike brochures, press releases, and other sales or marketing materials, newsletters for technical communication can offer specific information to a well-defined group of product users, company employees, or organization members. Technical communicators may be responsible for researching and writing the product updates, user profiles, expert tips, and other useful information that keeps readers interested and engaged. Along with good content, the layout and medium of the newsletter will also affect how likely users are to read it. Newsletters are still mailed on paper, but increasingly they appear on web sites or blogs, in email messages, or attached to an email as a fully formatted PDF document.

In this example of an electronic newsletter delivered by email, the software company TechSmith gives users updates about their main product, SnagIt (a popular screen capture program), as well as other software tools and helpful tips from users and experts.

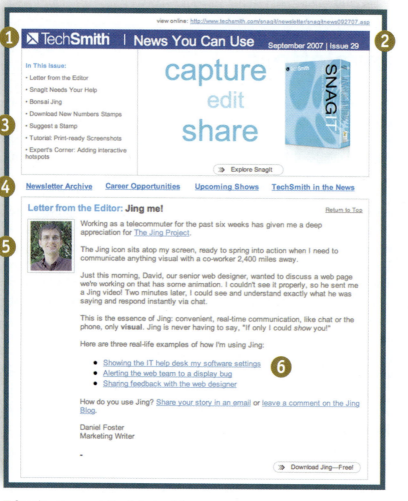

■ SnagIt user newsletter (electronic)

Features of the Electronic Newsletter

1 Banner across the top of the page (also called a *flag*) includes the title "News You Can Use" and the company logo and name.

2 The date and issue number appear at the far right of the flag.

3 Contents of this issue are listed as links to the current issue articles.

4 Standard links lead to archives and company information.

5 Photographs of the editor and of featured users add familiarity and personality to the document.

6 Links provide more information or user interaction—"watch the screencast."

The next example illustrates a more traditional print-based newsletter about medical research from the National Center for Contemporary and Alternative Medicine (NAC), which is part of the National Institutes of Health (NIH). Notice that this newsletter is available on the Internet in PDF format but could also be printed.

CAM AT THE NIH

FOCUS ON COMPLEMENTARY AND ALTERNATIVE MEDICINE

VOLUME XIII, NUMBER 2   SUMMER 2006

International Research Conference Highlights Progress, New Directions

Close to 250 posters on research added an important dimension

© Richard Siemens

The North Saskatchewan River Valley in Edmonton, Alberta, Canada, was the panoramic setting for the North American Research Conference on Complementary and Integrative Medicine, held May 24-27, 2006. Close to 650 researchers, health care practitioners, representatives of government agencies and non-government organizations, students, and other attendees came to Edmonton from 22 countries and 210 institutions to share information and perspectives on complementary and alternative medicine (CAM) and integrative medicine (IM).

The Consortium of Academic Health Centers for Integrative Medicine (CAHCIM), a group of 32 medical centers in North America affiliated with academic institutions, sponsored the conference. CAHCIM's goal is to "make a qualitative difference in people's health by advocating an integrative model of health care, incorporating mind, body and spirit."

Representing the Rich Diversity of CAM

The Edmonton event had over 300 offerings—keynote and plenary addresses,

(continued on pg. 2)

Getting To Know "Friendly Bacteria"

Lactobacillus bacteria

© Monika Wisniewska

INSIDE
9 Research Roundup
11 New NACCAM Members
11 From the Clearinghouse
12 News for Researchers

NCCAM
NATIONAL CENTER FOR COMPLEMENTARY AND ALTERNATIVE MEDICINE

NATIONAL INSTITUTES OF HEALTH

U.S. DEPARTMENT OF HEALTH AND HUMAN SERVICES

If you go to the supermarket, or look at a health magazine or commercial Web site, chances are you will find products with "probiotics"—certain types of bacteria that are also called "friendly bacteria" or "good bacteria." Probiotics are available as conventional foods and dietary supplements (for example, capsules, tablets, and powders), and in some other forms as well. While some probiotic foods date back to ancient times (fermented foods and cultured milk products), recently interest in probiotics in general has been growing. Americans' spending

on probiotic supplements, for example, nearly tripled from 1994 to 2003.

What Are Probiotics?

Experts have debated how to define probiotics more specifically. One widely used definition, developed by the United Nations Food and Agricultural Organization and the World Health Organization, calls probiotics "live microorganisms, which,

(continued on pg. 6)

■ NAC research newsletter (PDF and print)

Features of the Print Newsletter

1 Flag includes the newsletter name "CAM at the NIH" and an explanatory subtitle.

2 The volume, date, and issue number appear on a separate line below the title in the style of a research-based publication.

3 Feature articles about an international conference and new findings about "friendly bacteria" generate interest for an **audience** of researchers.

4 Articles and photographs appear in boxed areas, but sidebars and photos also extend outside the box margins, adding interest to the document design.

5 Contents of this issue and regular sections are listed in a box on the left titled "Inside."

6 The National Center for Contemporary and Alternative Medicine logo and the governmental organi-

zations that sponsor the center all appear in the lower left corner.

7 High quality photographs of research findings like the bacteria lend credibility.

8 Photos that feature researchers at the conference and other distinguished scientists add personality and interest throughout the newsletter.

International Research Conference Highlights Progress, New Directions
(continued from pg. 1)

oral abstracts, workshops, discussions, symposia, and posters—in five major areas of science: basic science, clinical studies, methodology, health services, and education. NCCAM was one of the conference's funders and participated on the planning committee.

Margaret A. Chesney, Ph.D., Deputy Director of NCCAM and Director of its Division of Extramural Research and Training, delivered one of the keynote addresses. She opened by discussing why there is intense public interest in CAM and in a new, more integrated medicine. For example:

Margaret A. Chesney, Ph.D.

- The population is aging.
- Information on health has become much more available (e.g., through the Internet).
- The consumer now has a more important role in health care.
- People have complaints about conventional care (e.g., when cures are elusive, side effects are problems, providers have very limited time, or care is fragmented among specialists).
- Many people find CAM and IM appealing (e.g., they feel these offer more "natural" treatment alternatives, emphasize patient-provider relationships, and allow individuals to take more responsibility for their health).

This interest in CAM and IM commands a "bold" research effort, Dr. Chesney said, but caution as well:

- There are many therapeutic claims that are attractive but unsupported by research. To illustrate, she presented an array of advertisements dating back to the 19th century and noted, "The plural of claims is not evidence."
- Media reports may oversimplify study findings, resulting in headlines that fail to communicate the value of the research.
- Methodological challenges and pitfalls exist in the research endeavor.

"We have a long way to go," she said, "but there is much to discover, and we have an agenda rich in research challenges."

Dr. Chesney set forth her vision for continued progress. "Be bold in what you try," she urged, "cautious in what you claim, and thoughtful about what you do. Express your purpose in a way that inspires commitment, innovation, and courage. We need you to contribute your part to the whole, as we work together to add to the fabric of knowledge about CAM and create a new, comprehensive health care." Drawing upon the Institute of Medicine's 2005 report on CAM, she described this care as being based on the best science available, recognizing the importance of compassion and caring, and encouraging people to actively participate in choices that enhance resilience, prevent illness, and improve quality of life.

In addition to Dr. Chesney, other keynote speakers were:

- Brian M. Berman, M.D., professor of family medicine at the University of Maryland School of Medicine and director of its Center for Integrative Medicine. Dr. Berman described his journey from carrying out pilot research to a major clinical trial of acupuncture for osteoarthritis (most of this work was funded by NCCAM or its predecessor, the NIH Office of Alternative Medicine).
- Richard Davidson, Ph.D., an NCCAM grantee who is professor of psychology and psychiatry at the University of

CAM at the NIH:

Focus on Complementary and Alternative Medicine is published by the National Center for Complementary and Alternative Medicine, National Institutes of Health.

Subscriptions: For a free subscription (by postal mail or e-mail), contact:

NCCAM Clearinghouse
Toll-free in the U.S.:
   1–888–644–6226
TTY (for deaf and
   hard-of-hearing callers):
   1–866–464–3615
Web site: nccam.nih.gov
E-mail: info@nccam.nih.gov

Editorial Address:
CAM at the NIH
P.O. Box 7923
Gaithersburg, MD
20898-7923, or
info@nccam.nih.gov

■ NAC research newsletter, page 2

Features of the Print Newsletter *(continued)*

9 Findings from the featured conference are bulleted and speakers are highlighted in blue text.

10 The *masthead*, which contains publishing and editorial information, is a standard element on the inner page of magazines and most newsletters.

CAM at the NIH:
Focus on Complementary and Alternative Medicine
U.S. DEPARTMENT OF HEALTH AND HUMAN SERVICES
NCCAM, NIH
31 Center Drive MSC 2182
Building 31, Room 2B-11
Bethesda, MD 20892-2182

Official Business
Penalty for Private Use $300

NATIONAL CENTER FOR COMPLEMENTARY
AND ALTERNATIVE MEDICINE

News for Researchers

NIH has begun requiring that grant applications be submitted via the Web portal Grants.gov (www.grants.gov) using Form SF 424 (Research and Related, or R&R, application). At least 2 weeks before submitting an application, institutions must register with Grants.gov, and principal investigators with eRA Commons. These changes are being phased in by grant mechanism (type of grant). For more information, see era.nih.gov/electronicreceipt.

Funding Opportunities

NCCAM has a new Web page for active funding announcements, including those below, at nccam.nih.gov/cgi-bin/grants/funding.pl.

PA-06-396 and PA-06-397: New Technologies for Liver Disease STTR and SBIR
Sponsors: NCCAM and seven other components of NIH. These opportunities—part of an NIH-wide

initiative on liver disease—will enlist members of the small business research community in developing new approaches (including from CAM) to diagnosing, treating, and preventing liver disease.

PAR-06-312: Biology of Manual Therapies
Sponsors: NIH (including NCCAM) and the Canadian Institutes of Health Research. These awards will support projects designed to gain insight on the mechanics of the body's response to manual therapies, such as chiropractic manipulation and massage therapy.

FIRCA-BB and FIRCA-BSS: International Research Collaborations—Basic Biomedical and Behavioral, Social Sciences
Sponsors: NCCAM and selected other components of NIH. This initiative seeks to establish collaborative research programs between NIH-funded investigators and others who are knowledgeable about indigenous medicine practices. ❖

NCCAM Exhibits at Upcoming National Meetings

- Society for Neuroscience Annual Meeting, October 15-18, Atlanta

- AARP National Event and Expo, October 26-28, Anaheim

- American Association for the Study of Liver Diseases Annual Meeting, October 28-30, Boston

- Society for Public Health Educators Annual Meeting, November 2-4, Boston

- Annual Biomedical Research Conference for Minority Students, November 9-11, Anaheim

D325

■ NAC research newsletter, page 12

⑪ The "News for Researchers" regular section appears on the back cover, along with funding opportunities and upcoming exhibits in the sidebar.

⑫ The top portion of the back panel can be used for mailing labels on the printed version.

Using Newsletter Templates

Rather than begin a design from scratch, many people prefer to use a template that comes with their software program or from the Internet. Newsletter templates let you quickly create newsletters for printing or saving as a PDF file. Basic document layout programs like Apple Pages and word processing programs like Microsoft Word offer many templates for newsletters. The benefit of using a template is that colors, borders, and other page layout features are preset, and all basic information is included. You simply fill in the content and make minor adjustments to the page design when needed.

This selection screen in Microsoft Word lets you choose from a range of styles, formats, and color schemes, depending on the purpose of your newsletter.

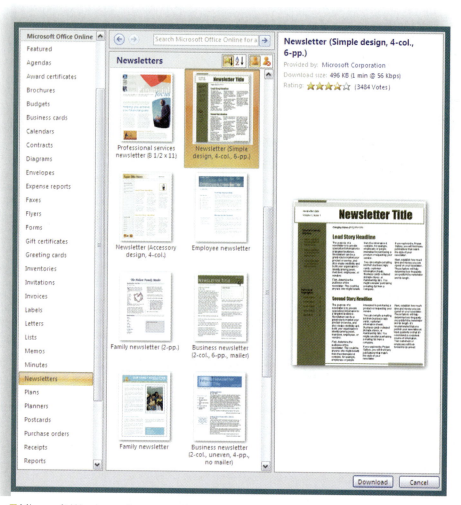

■ Microsoft Word newsletter templates

Once you select an appropriate template, you can use a wizard that prompts you to fill in the content. Here is a newsletter wizard in Microsoft Word:

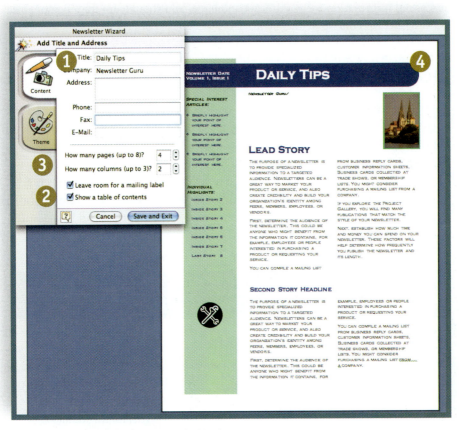

■ Newsletter template in Microsoft Word

Features of the Newsletter Template

❶ The wizard on the left prompts you for information such as the newsletter name and automatically inserts the content into a flag for you.

❷ It also can create an automatic table of contents and mailing label area, when selected.

❸ The "theme" tab allows you to adjust the color scheme and graphical features of this page design.

❹ The resulting document (on the right) automatically displays the title; you then click on text boxes to enter your own newsletter stories and pictures.

Audience Considerations

Newsletters can be effective publications for almost any well-defined audience: product users, company employees, supporters of a non-profit, or members of a professional organization. Because newsletters are subscription based, you have somewhat of a captive audience who already know the purpose of your newsletter.

But busy readers are less likely to read generic or broad information that reads like sales literature. Readers will most likely pay attention if you offer a calendar of events, usable tips, in-depth features, reader profiles, interviews, and other timely and useful information for your target audience. You should aim for a personable, enthusiastic, yet concise writing style.

Design Considerations

Newsletters follow the same principles of document design as other page-based documents for print or screen, such as the use of headings, visuals, and page layout elements. Screen-based newsletters such as TechSmith's, however, must also follow principles of good web page design. While software templates make it easy to produce a colorful and well-formatted newsletter, they also limit your choices of the color combinations, clip art, and other design elements. Be careful that the template you use does not look silly or unprofessional, but instead has features that are appropriate for the audience and your information.

Guidelines

FOR CREATING NEWSLETTERS

- Target a specific group of users or members for your newsletter and create a mailing list.
- Decide whether your newsletter will be delivered via email, on paper, on screen, or a combination.
- Determine the purpose of your document and select information that is useful, timely, and of interest to your selected audience.
- Consider featuring user tips, reader profiles, or interviews that add a human touch.
- Emphasize specifics that readers can use, such as grant opportunities or technical tasks.
- Include photographs of the editor and the organization members or users to generate interest.
- Determine an effective document design and create your layout for the printed page or screen, or use a software template.
- Create or edit articles and other content using a personable and concise writing style.
- Test your newsletter on the target audience and gather feedback for improving the content and design features.

See also

Audience (Part 1)

Purpose (Part 1)

Document Design (Part 3)

Email (Part 3)

Templates and Style Sheets (Part 1)

Web Sites (Part 2)

Web Page Design (Part 3)

Online Help

When users are trying to perform a task and have an immediate question, this question is best answered without making the person stray too far from what he or she was trying to do. For example, you are trying to create a brochure using Microsoft Word and you have a question about how to insert a photograph into one of the columns of text. If you had to get up, find the manual, and look up the answer, you would waste time and lose your sense of place in terms of what you were doing. If you had to close out the document and open a new screen of instructions or help information, the same problems would occur.

These situations are where online help can be effective: the information users need in order to answer a question or perform a task is actually part of the program and is available in the context of the task at hand. Most applications use context-sensitive help: users don't have to go searching on the "help" menu for what they need. Instead, the software recognizes that the user is trying to figure out something and offers tips or ideas.

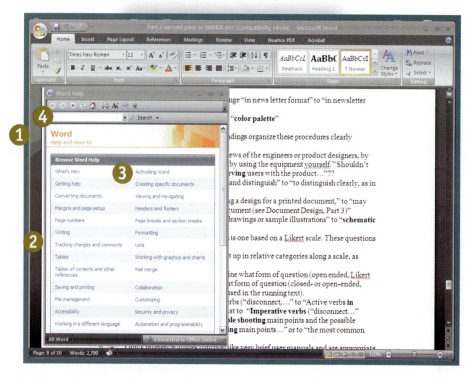

■ Online help in Microsoft Word

Features of Online Help

1 A small window opens on the side of the document without interrupting the work flow.

2 Users can choose from the items that are listed or click on "search" to go to the help menu.

3 Each item begins with an action verb, making it easy for users to find what they are looking for.

4 Users can also type in a question if theirs is not included in the items listed in the window.

Online help is often part of the actual product (as shown in this figure) and gets installed with the software. But increasingly, online help is also available on a company's customer support web site. Large teams of customer support personnel and technical writers work on maintaining online help web sites that complement what is built into the product. Web sites have the advantage of being more up to date than what comes on the product CD.

Audience Considerations

Users of online help are looking for specific answers to specific questions. They hope to be able to solve a problem or learn how to use or activate a particular software feature quickly and easily. Online help is most useful when it responds to actual user questions, not a question software designers or engineers *guess* users might have. Different types of applications require different types of help topics.

According to one expert (Weber 1999), it is best to design and write online help by clustering information into these categories:

Overview topics, describing the purpose and use of applications and their features

General help topics, providing information on an active window, or dialog box

Procedural topics, providing instructions on how to perform specific tasks

Message topics, explaining the meaning of specific error messages and recommending solutions.

Design Considerations

Online help needs to be very concise and written in language users will understand quickly and easily. As in the example above, questions ("What would you like to do?") can help orient users to the menu of help options. Online help should not require users to scroll or scan a long list of menu options. Instead, subdivide help categories into manageable lists or related questions or issues.

Online help is usually created using a help authoring software tool. Two standard applications are AuthorIT, a collaborative tool used for content management, and RoboHelp, a program created for technical communicators who need to develop contextual online help. Help files may also be created in Word, HTML, or PDF, depending on the context and needs of users. Online help projects usually go through a round of testing and usability review as the products or software they are designed to work with are quality tested. Unfortunately, technical communicators are often asked to develop online help only at the end of a software development cycle, and they are often asked to meet intense deadlines with little support. Online help is most successful when it is developed early in the development process, alongside the development of the product or application.

Guidelines

FOR CREATING ONLINE HELP

- Plan for online help from the beginning of a software or product development cycle.

- Consider how much information needs to be part of the product and how much can be kept on the customer support web site.

- Use language that is clear and to the point.

- If the purpose of the information is to help people perform a task, write this information in the form of brief, step-by-step instructions.

- Test online help with real users of the product.

- Use a style sheet to ensure consistent terminology across topics written by different authors.

See also

Instructions (Part 2)

Frequently Asked Questions (Part 2)

Style Sheets and Templates (Part 1)

Usability and User Testing (Part 4)

Presentations

Presentations are a form of oral communication conducted in person (or sometimes via video conferencing technologies) to an audience. Almost every workplace, organization, and career requires that you give presentations, formal or informal, using visuals as speaking support. Whether given in a face-to-face setting or delivered over a video conferencing system, presentations always incorporate matters of delivery; they require confidence, speaking skills, preparation, and, when necessary, strategies for dealing with apprehension. While there are numerous ways to plan and create an effective presentation, presentation software programs like Microsoft PowerPoint are the most common means for preparing visually based talks.

■ PowerPoint presentation

Many presentation programs let you organize your presentation as a series of slides that you can show to the audience as an organizational tool and as a device for emphasis. The software templates incorporate visual and textual elements, multimedia elements, and hypertext links in a consistent presentation and effective outlined format. In addition, most programs include organizational outlines, navigational tools, and speaker's notes for the presenter. You can use template-driven prompts for information on the screen, your outline, and notes simultaneously when editing your slide.

■ PowerPoint presentation; editing view

Features of Visual Presentations

Each slide in presentation mode contains the following:

1 Common background, colors, and visual elements from a template or custom-designed style.

2 Headings for each slide.

3 Bullets or other alignment techniques for brief lists and outlines.

4 Inserted graphics and other multimedia elements, when appropriate.

5 A clearly defined screen area for all the information.

In editing mode, the software displays the following:

6 Corresponding outline elements.

7 Speaker's notes area.

Audience Considerations

When preparing a presentation, learn all you can about your audience. Think about technical background, familiarity with your subject, attitudes, and information needs. Try to connect and involve the audience in your presentation. Make sure your graphics are interesting, relevant, and easy to read from the back of the room.

PowerPoint and similar presentation software automates much of the process, but don't expect the software alone to carry your talk and automatically turn it into an engaging presentation. There have been many recent studies about the use of PowerPoint in technical settings: if you let the software do all of the work, and you don't pay attention to the content, logic, or technical details, you could create a serious problem by conveying the wrong meaning or ideas. For instance, one expert recently described how the use of PowerPoint may have contributed to a lack of independent analytical thinking, which in turn may have contributed to the space shuttle *Columbia* disaster (Tufte 2005).

Remember that, when you are organizing any talk, your purpose is to inform, to request, or to persuade, and not to impress. To engage your audience, use the following techniques:

Structure your presentation. An outline structure works best to summarize and present information as you speak. Use the familiar guideline: Tell us what you are going to tell us, tell us, then tell us what you have told us.

Use speaker's notes or an outline. Practice talking from your notes. In most situations, you don't want to read a script.

Plan your opening and closing statements. Openings and closings are the most important and memorable parts of your presentation. Begin with an example or story to connect with your listeners. End with a question, call to action, or compelling image that will leave an impression on your audience. Memorizing your opening statement can help alleviate nervousness and give you a strong start.

Engage your audience. Look at them, talk to them, ask a question, take a straw poll, or give them something to examine.

Offer specific conclusions and recommendations. Conclusions and recommendations can appear throughout the presentation, though they usually work best at the end to give your audience action steps.

Dress appropriately. When the presentation is in a face-to-face setting, make sure you know the formality of the situation and will be comfortable and confident in your clothes and appearance.

Speak carefully. Make sure you project your voice to the back of the room and pronounce words clearly and completely.

Make eye contact with the audience. Look at people and smile when you begin and throughout the talk.

Design Considerations

All graphics should support your talk. Usually, a speaker will spend 2–3 minutes on a particular graphic —and presentation software can automate this for you. If you linger, your audience may become bored.

Limit each slide to eight lines. An outline form with bullets like Power-Point's templates usually works best.

Use a slide template or visuals with a consistent background and color scheme throughout the talk.

Explain data carefully. When you display data sets, tables, or numeric graphs, highlight key data and discuss them with the audience before moving on.

Avoid photocopied material. Unless you have high-quality reproductions, create your own graphics, charts, and text.

Use models or props. When appropriate, props can be good attention-getters if they are interesting and relevant.

Guidelines

FOR CREATING A PRESENTATION

- Focus your topic for the occasion and limit yourself to no more than three to five key points or concepts.

- Use an outline to keep yourself on track, but do not read from a script.

- Present your thesis, conclusion, or main idea explicitly.

- Provide appropriate visuals and supporting material based on the audience and occasion.

- Use anecdotes and examples to illustrate key points and add interest for your audience.

- Use language that is appropriate to the audience, occasion, and purpose.

- Err on the side of being more formal rather than too casual or nonchalant.

- Practice your presentation to make sure your pronunciation and vocal pitch are clear and understandable at all times and to ensure you are familiar with the software and room you are using.

- Use gestures and body language to emphasize your key points and lend energy to your presentation.

- Use your analytical, logical, and communication skills. Don't let PowerPoint create your ideas for you.

Press Releases

Press releases are time-sensitive items or specific statements that an organization wants published by various news media. Written for the general public, press releases shape an organization's messages to the public, but also must appeal to the journalists and editors who will publish or announce them. Press releases must appear newsworthy, succinct, and engaging for the targeted audience and market.

① FOR IMMEDIATE RELEASE

② Contact: Sarah Findley
Telephone: (404) 555-1212
E-mail: sarah_findley62@yahoo.com

③ **WAG-A-LOT SET TO OPEN BIGGEST AND MOST PRESTIGIOUS DOGGIE DAY CARE IN METRO ATLANTA**

PAMPERING AND FUN NOT JUST FOR POSH POOCHES

④ ATLANTA (May 11, 2005)— WAG-A-LOT, Atlanta's premier dog daycare business, has today announced the planned opening of its 2nd location. The 22,000 square feet store, opening to the public on May 16th, is the most modern and prestigious of the local dog care businesses and boasts the most 'dog-friendly' living and recreational area available in Metro Atlanta. Its gala **⑤** opening, for dogs and their people, takes place from noon until five this Saturday

The new WAG-A-LOT is situated on Northside Drive and features the latest in canine care. The location is complete with a dog day-spa, self serve dog wash, a mirrored grooming room for the latest in dog hair styles, and spacious 6' x 6' custom made large dog boarding dens, with a separate Cubbie Village of elegant brownstones for the smaller dogs. Webcams allow **⑥** owners to watch their dogs from work or home computers, and a large plasma screen in the lobby shows up to the minute images of the dogs at play.

When not being pampered, dogs roam the 5,000 square foot play area that features a freestanding waterfall and a series of mountains, platforms and benches to loll on or climb over at their own discretion. A team of highly trained, upbeat employees, referred to as 'Wag Specialists', carefully monitors all playtime. "There are an estimated 250,000 dogs in Atlanta, many of whom really don't want to stay at home," said Craig Koch, co-owner and the founder of WAG-A-LOT, Decatur in 1999. "The growth in our business shows that Atlantans want quality care for their furry friends while they are away on vacation or out at work." For more information, check out www.wagalot.com. **⑦**

■ Press release

Features of the Press Release

In this example, the business owners of a new dog day care location generate a message of excitement, fun, and luxurious pet pampering to lure locals to their grand opening in a densely populated urban area. Written for publication in a large city, the text speaks to potential dog-loving customers who either want or need dog care services.

1 The words FOR IMMEDIATE RELEASE, in capital letters, identify the piece's purpose.

2 Organization contact information, set on an indent at the top of the page, opens the document.

3 A headline addresses the readership and emphasizes local interest.

4 Location of publication and the date in parentheses open the body text.

5 First paragraph includes all key information about the event (who, what, where, when, and why).

6 Short paragraphs organize the body text.

7 Text concludes with a website address.

Audience Considerations

Press releases must speak directly to the needs and interests of the public audience where they are published. While they aren't advertisements for companies, press releases are a means for generating more attention for a company or organization and for outlining its services or values. Keep the text succinct, local, and timely. Persuade with details, but don't manipulate the audience. Tell the truth using facts and avoid statements that might mislead readers.

Press releases should be written in a journalistic style, employing the reverse pyramid style, which moves from most to least important items. The first sentence or paragraph provides all the details of this announcement (who, what, where, when, why). The writing style uses primarily active voice and emphasizes nouns and verbs: it uses simple rather than complex sentences that can be read aloud for radio. Short and succinct, 400 words at the most, the press release uses brief paragraphs for easy scanning and quick reading.

Design Considerations

For text that will be reprinted, read, or published, it's best to use simple, common fonts and to create emphasis simply by using bold or a slightly larger type size for headings. Because most press releases will be reformatted for a publication and are also published on web sites or even distributed via email, it's important to keep format and layout simple and cross-platform.

FOR CREATING PRESS RELEASES

- Target your primary audience by addressing their goals and locals interests.

- Identify all important dates, details, and factual items.

- Include complete contact information and sources for further inquiries.

- Write so that the most important information comes first.

- Use a simple, succinct writing style; avoid complex sentences, long paragraphs, and extraneous descriptions.

- Read your text aloud and test it with a member of the organization or a journalist before publication.

See also

Audience (Part 1)

Email (Part 2)

Persuasion (Part 1)

Web Sites (Part 2)

Product Descriptions

Product descriptions are commonly used in business and industry to market a product and present end users with its purpose and key features. The description might include technical illustrations and also be accompanied by technical specifications for actually installing and using the product. Your description will highlight the most important features and innovations for end users based on their needs.

This Emriver process simulator, taken from a company web site, provides a good example of a product description based on user needs and requirements.

Emriver river process simulators

A powerful new tool for river research and education

Using a moveable thermoplastic bed and recirculated water, the Emriver model simulates river processes with remarkable accuracy.

Now in wide use by river scientists and managers for education and research, the Emriver demonstrates basic principles of river behavior and subtle channel morphology and sediment transport processes.

The Emriver model uses only 27 gallons (102 liters) of water, and is easily moved for field use.

Visit the components pages for photos and details on the Emriver's construction and parts.

The Emriver model is based on nearly two decades of experience in portable river models used by Midwestern conservation organizations.

The Emriver was designed and is built by Little River Research & Design in Carbondale, Illinois.

Don't miss our recent video clips of the Emriver in action!

■ Product description for a river modeling system

Features of the Product Description

1 Title of product includes subheading that focuses on what users value—river research and education.

2 Benefits include details to market the product's "remarkable accuracy" and being "in wide use by researchers and managers."

3 Color photographs catch attention and show the product in use but are less detailed than a technical illustration.

■ Product description for a river modeling system.

Features of the Product Description Page

1 Notice that this page of the web site does not just give specifications, but instead offers links and an interactive image map for users to explore all components of the product.

2 Key details of all components and user manuals are offered with the links on the right.

Note that even a photograph or detailed technical drawing without a complete product description or specifications might mislead readers into thinking that this equipment is as simple as it looks. Take care to label an actual model of the product versus a generic version in your description, and to label each element accurately. Provide separate pages or sections for illustrations and specifications as needed.

Audience Considerations

Consider your end users or clients as the primary audience, but also keep in mind the contractors, subcontractors, architects, or engineers who will help select and actually install the product. In order to write effective product descriptions, you may need to research and compile the features, options, and key technical information—possibly by interviews of engineers or product designers, and by conducting usability research or even using the equipment yourself. Your goal is to capture the attention of these users with key words and concepts that address their needs—in the case of this example, quality, accuracy, and portability of a simulator.

Global audiences may use different kinds of electrical currents and various terms for mechanical processes, so check with a translation and localization resource. It's best to use the exact technical name of a product component and be consistent in usage, so that this name is translated consistently.

Design Considerations

Product descriptions are a marketing tool and an informational document—much like what you would find in a brochure or a white paper. The language should be lively and use subtle persuasion—as in "designed and built for durability under demanding conditions"— to make the product appealing to a potential user. For a web site document such as this one, principles of good web page design, such as headings and lists, and visual communication techniques will help you create an effective product description page.

Guidelines

FOR CREATING PRODUCT DESCRIPTIONS

- Identify your primary audiences for the product—the customers, installers, and engineers.

- List important features and innovations for end users based on their needs.

- Collect the most important features and technical details through research.

- Organize your information to catch a reader's attention.

- Use headings, subheadings, and bulleted lists to highlight what buyers want.

- Include a photograph or technical illustration if possible. Label any generic drawings as such.

- Create a separate page or section for each complete, detailed product description.

- Test your description on target users.

Proposals

Proposals are persuasive documents that invite and encourage readers to do something. For instance, the purpose of a sales proposal is to obtain a contract or make a sale; the purpose of a research proposal is to convince a scientific or technical agency to provide funding for a project. Similar to a report in scope, a proposal has one key difference: it is written to *propose a specific action or solution* to a specific set of conditions or a problem. Proposals are usually competitive in nature, with several proposals vying for funding or other support. Proposals can come from within a company or organization, or they can come from outside of the company or group evaluating the proposals.

A proposal can take on a variety of forms. Proposals can be divided into *unsolicited proposals* and *solicited proposals*. An example of an unsolicited proposal would be one that a group of students decided to write to propose a new food service for their college. A solicited proposal, which is the more common form in technical communication and business, is often written in response to an RFP or *request for proposal*.

The following are three common proposal types.

Sales proposal: Anyone involved in selling a product or service will be familiar with a sales proposal. A good sales proposal convinces the audience to purchase the product or service.

Student proposal: Students are often asked to write proposals for a class project; these assignments usually require the student or group of students to find a real problem on campus, in their community, or in some other setting, and propose a solution.

Research proposal: Research proposals are written for many reasons. Most research proposals are written by university and other researchers as a way to request funding for a specific project. These proposals are often directed at government agencies (such as the National Science Foundation) or private foundations that fund research. Other types of research proposals, some of which request funding and some of which do not, include those that propose an experiment, a dissertation, a senior thesis, or any other research project that proposes to test an hypothesis, solve a research problem, or make a new discovery. Many research proposals begin with a short proposal, called a pre-proposal. A *pre-proposal* offers a brief overview of what the researcher wishes to propose. Pre-proposals are used by large funding agencies as a way of screening out ideas that do not fit the mission or goals of a particular funding area. A winning pre-proposal usually is followed by an invitation to submit a complete proposal to the agency.

Parts of Proposals

Proposals vary widely, but most proposals include front matter, introduction, body, and back matter.

Front matter: Includes a required information form, a cover or title page, a letter of transmittal, a table of contents, list of illustrations and figures, and an abstract or executive summary.

Introduction: Normally focuses on the purpose of the proposal, the problem, and how the proposal will remedy the problem.

Body: Details what the proposal is, how much the project would cost, why you or your group is qualified to carry out what is proposed, and what kind of actions should be taken.

Back matter: Includes resumes of all involved in the project, appendixes, and references.

Requests for proposals usually outline the specific parts that are required and set limits on length for the proposal parts. It is critical that you follow the guidelines and format specified by the agency or other requester.

Sales Proposal

A good sales proposal needs to convince the audience that the proposed solution is the best choice, that the costs are appropriate, and that the company or person proposing to do the work is credible and will complete the work on time. If a sales proposal is a response to a request for proposals (RFP), you would need to follow the RFP's instructions carefully. If a proposal has been solicited (for instance, if you are a technical writer and you have been asked to prepare a proposal to do some work for a company), you may have less convincing to do. But if a proposal is unsolicited, you will have to work harder to establish the reason why the reader would want to spend the money on the job you propose.

The following is a solicited proposal for a technical writing project. The writer of this proposal, a freelance contractor, was asked to create a proposal for creating a new user manual for an engineering firm. She has worked for this company before, so the proposal did not require a lot of background information. If the writer were a new contractor for this company, she might need to include her resume, writing samples, and a longer section on the reasons why her firm is the right one to do the job. Very long sales proposals often start with an executive summary. Complex sales proposals require a budget section that is usually put into the appendix.

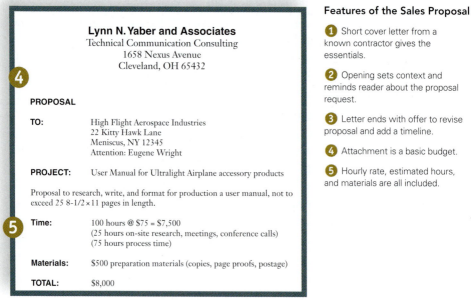

Lynn N. Yaber and Associates
Technical Communication Consulting
1658 Nexus Avenue
Cleveland, OH 65432

November 16, 2007

High Flight Aerospace Industries
22 Kitty Hawk Lane
Meniscus, NY 12345
Attention: Eugene Wright

Dear Gene:

Thank you for inviting me to submit a proposal to research, write, and format the user manual for your ultralight airplane accessory products. I have attached a short proposal that breaks down the time and estimated material expenses. As always, I am happy to meet with your technical team to fine-tune this proposal and to get a better sense of your timeline and requirements.

I appreciate the continued opportunity to work with High Flight.

Yours truly,

Lynn Yaber

■ Cover letter for sales proposal

Lynn N. Yaber and Associates
Technical Communication Consulting
1658 Nexus Avenue
Cleveland, OH 65432

PROPOSAL

TO: High Flight Aerospace Industries
22 Kitty Hawk Lane
Meniscus, NY 12345
Attention: Eugene Wright

PROJECT: User Manual for Ultralight Airplane accessory products

Proposal to research, write, and format for production a user manual, not to exceed 25 8-1/2 × 11 pages in length.

Time: 100 hours @ $75 = $7,500
(25 hours on-site research, meetings, conference calls)
(75 hours process time)

Materials: $500 preparation materials (copies, page proofs, postage)

TOTAL: $8,000

■ Sales proposal

Features of the Sales Proposal

1 Short cover letter from a known contractor gives the essentials.

2 Opening sets context and reminds reader about the proposal request.

3 Letter ends with offer to revise proposal and add a timeline.

4 Attachment is a basic budget.

5 Hourly rate, estimated hours, and materials are all included.

Student Proposal

The following annotated example of a student proposal demonstrates a community-based organization's appeal for funding in light of a new municipality structure in suburban Atlanta, Georgia. These changes in local government have left a shortfall for his youth sports organization and this student assesses the best methods of raising the yearly improvement funds. This unsolicited proposal is written as a memo to multiple boards of directors for youth sports programs.

Rick Cole
English 3000
April 14, 2006

To: Boards of Directors for Sandy Springs Youth Sports Football, Cheerleading, Softball, and Baseball
Programs
From: Rick Cole, 2004–2005 Vice President of Development
Re: Overcoming Shortfall Created by Sandy Springs Governmental Changeover

Background

Each year, Sandy Springs Youth Sport's (SSYS) programs are financial successes. Every program pays for its own uniforms, umpires, and equipment through player's fees, fundraising programs, and the immeasurable hard work of our volunteers. A part of our success is tied to playing in a municipal park. Fulton County has underwritten our program by taking responsibility for the upkeep and maintenance of our facility for four decades. They have mowed the fields, kept up the roads and parking lots, and maintained our bathroom facilities. Beyond this contribution, each year, SSYS has been able to approach the county through our friend and representative, Tom Lowe, to get special allocations. These monies have been used to remodel our snack shack, add walkways, reseed our baseball fields, add bleachers, renovate batting cages, and more. We have benefited from this relationship materially in terms of large capital infusions that we did not need to raise. But our community has changed.

We have entered a new era in Sandy Springs. Last year's historic vote created a city where there was none. The city of Sandy Springs has now assumed ownership and responsibility for Morgan Falls Park where the SSYS program lives. Fulton County is no longer involved. The new government has continued with the maintenance regime instituted by Fulton County and their crews are doing a great job. However, this year they were unable to budget for or provide the annual special project allocation that we have become accustomed to receiving. According to a representative that Keith Bradley spoke with, Sandy Springs has not yet received one dollar of tax money and accordingly they have had to operate on a shoestring budget. We are reaching out to city representatives and have been assured they will consider our park's needs in next year's budget. Hopefully, we will work with city leaders to ensure continued and greater support for our park.

The Problem

Even if Sandy Springs puts a high priority on our needs next year, for this year, we have a shortfall. This shortfall will be felt in delayed plans to build a pavilion in between field one and the lower softball field. It will also be found in downgraded improvements to our practice cages rather than the complete overhaul that was needed. We will be unable to plan future improvement projects. That is why we should consider a proposal aimed at providing alternative, independent and permanent sources of income to supplement and, if necessary, replace the money previously allocated by local government.

Features of the Student Proposal

1 Memo headings and format demonstrate a clear audience, purpose, and the writer's credibility as part of the leadership in this organization.

2 Subheads break the memo into clear sections based on a problem and on proposed solutions.

3 Background sets up the success of the organization and the current situation of a funding shortfall due to changes in local government.

4 Problem section gives the bottom line for this document— "a proposal aimed at providing alternative, independent, and permanent sources of income."

The Solution: Now and Tomorrow

This proposal sets out to accomplish the goal of providing capital for improvements to our sports programs. It will focus on three ways of raising money for SSYS: pursuing grants and matching gifts, expanding the booster club's contributions beyond present-day giving, and unifying our fundraising in the form of a golf tournament. We have experience with all of these methods, but I think we can do them smarter, better, and more efficiently.

One of the main goals of this proposal is to create an institutional memory along with the plans we implement. Most of our individual services to these boards and this park runs parallel to our children's current participation in its activities. This is only natural, as families want to serve where their children play. But eventually children's lives move us on to school sports and other interests. The baton is then passed from one generation of leadership to the next—sometimes with better results than others. If we can implement well thought out and successful fundraising programs that are easily duplicated year in and year out, we will establish financial success for our park for the long run. These three solutions may accomplish these goals. If not a complete blueprint, they are definitely a good starting point for success.

Grants from Foundations and Charitable Trusts

Many of you know that two years ago we secured a $20K grant from the Falcon's Foundation for our football program. I have had many inquires into the feasibility of getting more of this type of funding. The prospect of large block allocations that require seemingly little organization is attractive. Such grants would seem an obvious replacement for the block funding missing from our budget. However, there is more than meets the eye with this type of money.

The Falcons required, as a condition of their grant, attendance at a three-day seminar held by the Foundation Center. This is an organization focused on facilitating the needs of grant makers and recipients through information. The seminar taught not-for-profits how to access the huge amounts of money available through grant applications. I represented our program and learned three key things for our park.

First, there is a lot of money looking for a home. Countless institutions and individuals set up foundations with the purpose of giving away billions each year. There are listings and requirements available through the *Foundation Center*. It takes focused research and requests, in the form of grant applications, to receive this money. Finding the right donor is perhaps the greatest challenge a not-for-profit organization faces.

Second, the research and writing of a grant proposal requires skill. If we decide to pursue grant money, we would need to hire a professional grant writer to find the applicable monies and make the applications. This person would know specifically how to best word our needs and how to approach these foundations. A grant writer would have to be paid either per proposal or a percentage of the actual grant money obtained.

Third, and most problematic for our park, is the type of organization that benefits from this type of giving. According to the book *Foundation Fundamentals: A Guide to Grant Seekers*, "The overwhelming majority of foundation grants are awarded to non-profit organizations that qualify for 'public charity' status under Section 501(c) (3) of the Internal Revenue Code." While we are a tax-exempt organization, we are not a "public charity."

Foundations are interested in giving primarily to "charitable, religious, educational, scientific, or literary purposes." We received the Falcon's Foundation grant because of a relationship to promoting football. However, to apply annually for that money, we were told we would also have to facilitate some ongoing educational programs beneficial to our players. We investigated an after-school study program in conjunction with Sandy Springs Methodist Church, but we couldn't get it off the ground. The problem we discovered was both foundational and systemic. The people working and playing in SSYS are not oriented toward charitable outreach. When it comes to our park, we are able to band together to work toward building a strong, healthy, athletic program. However, we don't have the type of paid staff and connections necessary to build a well-run social outreach program, and few of us want to pursue that direction on a long-term basis.

In my assessment, the Falcon's Foundation is not a viable source for permanent funding. It just doesn't fit who we are. However, knowing what won't work is good for focusing on what will.

The Booster Club: Our Greatest Asset

For over forty years, children have been playing ball for SSYS in Morgan Falls Park. This is a great history and something worthy of pride. Yet, we have never completely leveraged what could be our greatest resource.

Each year, we sign up new members to our booster club. We offer premiums at different levels of giving: bleacher cushions for $50, a hat for $75 and so on. The way we run our booster club brings in thousands of dollars and is a great effort that produces results. It can be improved on though.

One of the limits to the way we run our booster club is that we are collecting money from ourselves each year. In other words, the parents who are paying the fees, paying for uniforms, buying the candy, and volunteering are being asked to fund the annual booster club, too. We are going back to the same well time after time. Instead, we should be looking beyond the gates of our park for revenue. We have a forty-year history with thousands, if not tens of thousands, of children who have played ball here. I propose we tap into that reservoir for our booster club.

The booster club can be expanded to establish a broader and more permanent membership. Instead of focusing on large-dollar ($100–$500) gifts from the 200–400 parents involved in any given year, we should build a membership of thousands of $25 members who give each year. They will be motivated to give, even though they have moved on, because of nostalgia and the benefit of giving back to the community of Sandy Springs.

To make this happen, we should mine our collective databases and establish a list of past, present, and future parents and team members and solicit them for annual contributions in the range of $25–$35. We can transition our booster club to this model over the next couple of years, continuing our in-park efforts until the permanent booster club becomes the primary generator of capital. This model will create a more stable source of revenue because it is spread over a larger group. It will also be more reliable because it doesn't depend on the efforts of any one leader or group of volunteers, which vacillate in effectiveness year to year.

The Unified Fundraiser

Those of us who have children in private schools or are involved with any other large-budget, not-for-profit organization are aware of the annual fundraising pushes that are necessary to fill revenue gaps. Often these events take the form of a silent auction, Sally Foster, sales, or a golf tournament. The benefit of one main fundraiser annually is the focus. It is easier to hold people's attention, to get them excited, and to open their wallets if it is only done once a year. It keeps the donors from suffering what professionals call "donor fatigue," brought on by being asked to contribute quarterly or more in smaller ways.

In our case, I believe the golf tournament is the best annual event. We are, after all, an athletic organization and I know many of the parents and children are avid golfers. Golf tournaments can be held in many different ways and are another way to bring outside money into the park, in the form of donations, sponsorships, and fees. I will mention two different possibilities. One is the traditional golf tournament, mostly played by adults and their friends. It has the potential to raise a lot of money, if organized well, but has some negatives. The second is a more laid back affair that could raise money, though not as much, but would have the added benefit of being fun. This is the parent/child format.

The traditional golf tournament is an organizationally demanding event. It requires renting a course, establishing catering, and selling a minimum of between 100–120 slots. To be successful, it must also be sponsored by local business, typically each sponsoring a hole. A well-run event costs $10K to put on. But the benefit can be huge. Some private schools fund all of their capital funds with well-established golf tournaments raising $300K and more for their organization.

Our goals would be more modest. A net profit of $25K to $50K from an organization like ours would be great. The downsides are the huge effort to make this happen and the tough competition with other golf tournaments for our time and money. As a golfer, I am approached all the time to participate in school, church, or charity tournaments. We would be yet another tournament, though I think we start with a really strong base of interest.

It really takes the focused efforts of one individual and a committee to organize one of these events. Companies can be hired to handle the day of event. But all of the players and sponsors would be up to us. I believe it will take a volunteer who commits to lead this effort for three years to make this a permanent fixture at our park. I have received information from *Champion Development Group* and from *Global Golf Events*, two professional fundraising organizations. If we decide to pursue this, we can have their representatives present their programs for more detail on the commitment needed.

The second type of tournament would be a parent/child event. Children in the park, along with their mom or dad, as the name suggests, would play. It would be a lot less formal and would not have the potential to raise as much money. Sponsorship would still be necessary and all fees would come from people in the park. However, the costs of the event would be far less. We could play at a much less expensive course and our catering, prizes, and other costs would be much less expensive. The cost could be held to around $3K. The net would be lower, probably $7K to $12K. Another benefit is that it would be a fun afternoon, one that families could enjoy.

Either plan could effectively build our park in two ways. It would give us an annual focus to our fundraising, increasing the amount we raise, and we would benefit as a community from the fun of a golf tournament.

Conclusion

We are in a tight spot financially this year, due to the Sandy Springs changeover. This situation can become a positive if we take this opportunity to revamp our fundraising efforts by positioning ourselves and our park for greater and more reliable sources of improvement capital. I propose the two best ways to accomplish include a much larger past, present, and future booster club and a focused annual golf tournament. Grant money is not a viable alternative, as it would move us away from our core activities and focus. Making these changes can ensure we have a steady income for years to come.

7

■ Student Proposal

Features of the Student Proposal *(continued)*

5 Three possible solutions for fundraising make up most of the proposal, with each solution evaluated separately and backed up with details and data, either information the writer knows or from quoted sources.

6 Persuasive and logical language defines the argument, as in the following: "*If* we can implement well thought out and successful fundraising programs that are easily duplicated year in and year out, *we will establish* financial success for our park for the long run" (emphasis added).

7 Conclusion gives a concise summary of the writer's recommendations: that an expanded booster club and annual golf tournament would provide the best source of ongoing funding.

Research Pre-Proposal

The following is an annotated example of a research pre-proposal proposal written to NASA. Unlike a longer, fully developed proposal, this pre-proposal must make all points quickly and succinctly in the text. The relationship between the proposed research project and NASA's request for proposals must be very clear and obvious. The idea of a pre-proposal is that the researcher must capture in a nutshell the main ideas, key research questions, proposed method, and impact of the research in a way that will advance this project to the next stage where a longer proposal is then requested.

(1) Affordable and Sustainable 21st Century Mission Operations: Human-Centered, Preference-Based Scheduling Tools

(2) Creating effective human-centered systems is not just designing good interfaces. One must delve deeply into the software systems themselves, creating additional capabilities and inventing technologies to make them possible.

Proposed Technology Project

Goals. Reduce cost and increase effectiveness of mission operations through development of an efficient, human-centered planning and scheduling assistant that incorporates users' scheduling preferences into solutions.

(3) *Relevance to H&RT*. Conducting affordable and sustainable missions that extend human presence across the solar system require the capability to create efficient, high quality, error-free mission operations schedules. Unfortunately, modern missions have become so complex that traditional manual scheduling methods are no longer sufficient; the "standing armies" of mission operations personnel required are not sustainable. Computer scheduling tools have potential to make modern mission planning affordable and sustainable. However, currently available computer tools are neither scalable nor do they meet humans' interaction needs: people cannot express all their solution constraints, reasons behind the tools scheduling decisions may not be obvious, and constraints in schedules produced are difficult to visualize and understand. The class of constraints that are not well addressed are scheduling preferences (i.e., soft constraints used to mold a feasible schedule into an excellent one). We will enhance NASA's capability to conduct affordable and sustainable operations for complex missions though joint innovations planning and human-planner interaction. Failure to address these issues now may greatly compromise NASA's return on investment in all future missions; operations planning staff size will be unwieldy, resources (robot, exploratory vehicles, test equipment, etc.) will be underutilized, less science will be conducted over the duration of the mission, and people and equipment may be endangered.

Objectives.

1. Increase quality and efficiency of mission operations schedules.
2. Increase quality and quantity of science that can be performed during a mission, maximizing NASA's return on investment.
3. Make mission operations more affordable and sustainable.
4. Maintain safety of remote equipment.

(4)

Technology development approach

Partnership approach. This project combines expertise from University of Minnesota, High Technology Laboratories, NASA Ames, and JPL. Team members are leaders in planning, decision support, human-centered design, and human-computer interaction. Researchers Roberts and Violet are experienced developers of planning and scheduling tools that have been fielded for the Mars Exploration Rover mission. Hayes' planning system, Fox-GA, has been fielded in the Army operations planning tool, Scheherazade. Davidson developed human-centered scheduling techniques for airline operations centers. High Technology Labs has an established track record in technology transition for many NASA, DARPA, and other projects. University of Minnesota will lead the project, with program management provided by Dr. Lucio who has nine years experience as Associate Program Director of NASA/Minnesota Space Grant Program.

Specific technical challenges. Creating a scheduler that allows people to express their schedule preferences, visualize constraints, and understand computer decisions is not simply a matter of adding on an interface that allows them to do so; the automated scheduler also needs to be able to incorporate them into the schedule in a computationally efficient manner. Schedulers, such as SCHED-X, explicitly do not address preferences, because its scheduling algorithm cannot efficiently traverse the enormous search space that preferences generate. Addressing the challenge of creating an efficient scheduling system that meets users' actual needs requires tight coupling of human-computer interaction and scheduling techniques.

Approach. The research will be explored in the context of operations planning for lunar and planetary rovers. We will draw on lessons learned from human-centered planning tools used in the Mars Exploration Rover mission. We will apply genetic algorithms (GAs) technology to the scheduling task because it can search large search spaces efficiently while meeting complex objective functions composed of hard and soft constraints (preferences). We will assess our tools using experienced tactical operations planners, in the context of actual operations planning, when possible.

Milestones and Technical Readiness Levels

⑤
    Year 1 (TRL 2, 3) Identify preferences schedulers need to express. Develop, evaluate, and demonstrate initial prototype scheduler.

    Year 2 (TRL 2, 4) Revise scheduler based on evaluations. Develop displays for effective visualization of schedules and constraints. Integrate with scheduler.

    Year 3 (TRL 4, 5). Evaluate usability, effectiveness of integrated system. Revise based on evaluation.

    Year 4 (TRL 5) Validate integrated system in mission context, demonstrate.

Impact

Specific benefits. Mission operations scheduling tool allowing users to rapidly generate schedules that are not only feasible but also robust, high quality, and scalable to complex 21st century missions.

Overall long-term use. Technologies and principles developed in this research are applicable to a broad range of planning applications including as robotic assembly and human-robot collaboration.

■ Research pre-proposal

Features of the Research Pre-proposal

① Title is clear and easy to understand.

② A two-sentence abstract gives the proposal argument and goals briefly.

③ Headings and subheadings are used to direct the reader to the right section.

④ Numbered list makes the objectives very clear.

⑤ Milestones provides key dates in the research process.

Audience Considerations

The examples provided in this section illustrate the different features of various proposal types based on the targeted audience. In general, people who read proposals want the details but they also want the bottom line. That is why most proposals begin with an executive summary. You need to let busy readers know the major plan, costs, and benefits in a brief and concise format.

Proposals should be as clear and accurate as possible. This document uses persuasion at its core, so you need to be direct about why your services are the best choice. Arguments must use a sound, logical, and clear presentation based on the facts. Because proposals often involve large sums of money as well as contractual obligations, they should have a strong ethical and legal basis. Do not make false promises or act as an expert in areas where you are not. Often, proposals have a global audience, so you will need to make sure your proposal is readable to readers whose first language may not be English.

Design Considerations

Especially in the case where the organization might receive a dozen or more proposals, you want yours to be written and designed in a style that is interesting to read, to the point, and accurate. Use visuals, bulleted lists, and clear charts and information about budgets and costs. Write the executive summary last, after you have drafted the rest of the proposal. Keep the executive summary short and to the point, but don't forget to convey credibility and enthusiasm for the project or plan. The rest of the proposal should be grouped into sections that are short and to the point, but thorough. Extensive data and details can be placed in an appendix. When you are responding to a request for proposal, be sure both your content and your design and format follow these instructions precisely.

Increasingly proposals (especially for academic research) are accepted by email, use Internet sites and templates, or depend on other computer programs. In order to write successful proposals that are accepted, it is important to be knowledgeable of these technological formats for proposals. Check with the research office of your company or institution, if you have one, or contact the program officer from the granting agency for more information.

Guidelines

FOR CREATING PROPOSALS

- Clearly identify the problem and objective for your proposal.

- Use the request for proposals (if solicited) or the sections required by your organization to plan material.

- Write an executive summary or abstract that condenses the problem, approach, proposed solution, and other high-level information.

- Give technical background and other context information in a separate section.

- Use language that is clear and concise but also logical and convincing.

- Include a budget in a separate section or refer to an appendix, as appropriate.

- Use visuals as appropriate.

- Put lengthy data sets and other information into an appendix.

- Make sure your proposal conforms to the guidelines in the request for proposal.

See also

Abstracts (Part 4)

Audience (Part 1)

Letters (Part 2)

Persuasion (Part 1)

Reports (Part 2)

Style Sheets and Templates (Part 1)

Reports

Many different kinds of documents are referred to as "reports." Reports can be as short as a one-page memo or as long as 200 pages. Unlike a proposal, which is expressly intended to persuade or convince readers, a report is usually more fact based, with the goal of delivering information to its intended audience. Yet there is really no such thing as an entirely neutral report. Most reports offer some conclusions or interpretation of the facts. The amount of interpretation depends entirely on the audience, purpose, and organizational context for the report. For instance, a report written to help citizens of a local community understand the possible impact of a water treatment plan on their town might be short, include many visuals, and use language that is clear and nontechnical. A report written for a group of engineers and scientists to determine the feasibility of sending a spacecraft to Pluto would probably use more technical language, include some quantitative analysis, and follow the format used at NASA.

Parts of a Report

Reports can contain many sections or parts. In general, a long report may include these parts:

Front matter
Letter of transmittal
Executive summary
Acknowledgments
Table of contents (TOC)
Introduction

Body of the report
Background
Problem statement and research questions
Methods used to produce the report
Main sections of information
Results
Cost / benefit analysis
Conclusion

Back matter
References
Appendices

For a class in technical communication, you will probably be asked to write a report with all of these parts. For a report in the workplace, you will usually follow a prescribed format based on the company, clients, domain (biology, engineering, medicine—all have their own ways of reporting information), and delivery medium. Keep in mind that reports are usually delivered simultaneously on paper and other media, such as an electronic document in PDF file format, on a CD or DVD, or as a web page available on the Internet or an organizational intranet.

Six common report types are as follows:

Consulting report: Consultants must write a variety of reports, often combining features from different report types. Consulting reports may need to provide the client with information about the feasibility, outcome, and test results of a particular project.

Feasibility report: These reports help determine whether a particular approach is feasible or practical and can be implemented. Feasibility reports assess the effectiveness, practicality, cost, benefits, outcomes, and other factors related to a problem.

Progress report: Progress reports provide an update about the progress or status of a project, grant, experiment, process, workplace task, or other items. Sometimes, these documents are written as brief weekly or biweekly status reports.

Research report: These reports offer a synthesis of the latest research, concepts, themes, and trends in a particular area, in as factual a manner as possible. Research reports may also become the basis for a feasibility report or a proposal.

Test report: Test reports are used to convey the results of specific tests, such as those done on a feature in a software application, an experimental medical procedure, or a new safety feature.

Trip report: Employees who travel for business purposes usually use these reports to document their activities and request reimbursement.

Consulting Report

Depending on the particular situation, consulting reports might combine features of feasibility reports, research reports, and test reports. In the following example, Steve Gough and his consulting firm Little River Research & Design were hired to help Commerce township in Michigan understand the geomorphology of the watershed surrounding Seeley Creek. The township needed this report in order to help solve a problem: can they release more water into the watershed, and how will that affect the stability of the soil, trees, and other factors?

Executive Summary

The Township of Commerce proposes to increase output at the Commerce WWTP in Seeley Creek's watershed. The increased discharge, which may eventually reach 8.5 mgd, will increase baseflow from nearly zero to about 13 cfs.

Stakeholders have raised concerns about channel stability and tree mortality under the higher flow regime. To address these concerns I carried out a geomorphological survey of Seeley Creek and its watershed that included observation of most of Creek's channel, surveys of channel cross sections, bank materials tests, and hydraulic analysis. We also tested bank and bed materials for particle size and strength characteristics. This report also addresses specific concerns about erosion of cohesive soils and bankfull discharge theory.

Seeley Creek's channel was largely excavated by farmers decades prior to urbanization. Its channel has changed little since then because hydraulic forces are generally low and bank and bed materials, especially when vegetated, are highly resistant to erosion. Numerous culverts and other structures have prevented channel downcutting. The channel now runs in plant root-reinforced sediments that it is unable to rework, even under flood conditions.

Increases in current velocity and hydraulic shear from the higher baseflow are small, and not high enough to initiate channel instability of any kind. Seeley Creek appears to be little affected by even high return period (e.g., 10-year) floods, which strongly indicates that it is not a self-formed channel and cannot adjust to such a small change in its flow regime. I found only two instances of significant lateral movement of the channel. In both cases, the lateral movement in the past 45 years has been less than 18 feet.

We did locate areas of high tree mortality just south of 13 Mile Road. The poor condition of trees and shrubs in this reach, however, is caused by standing water on the floodplain and, more importantly, strong competition from grasses. The increase in baseflow will have no effect on tree mortality in Seeley Creek.

Seeley Creek's bed is generally composed of gravel and sand-dominated materials. These sediments are underlain, however, by cohesive strata that are resistant to erosion. Coupled with control of vertical erosion by built structures and generally low hydraulic energy, the chances for downcutting in this system are very slight.

Seeley Creek's morphology and vegetation will be unaffected by the increase in baseflow, and the chances of significant instability or woody plant mortality are very small.

■ Consulting report

Features of a Consulting Report

❶ A one-page executive summary states the reason for the report, outlines the approach, and summarizes the findings.

❷ Use of running footers and page numbers makes for a professional style report.

Introduction

The Township of Commerce, in Oakland County, proposes to increase wastewater treatment plant output at the Commerce wastewater treatment plant (WWTP) in of Seeley Creek's watershed. The increased discharge, which may eventually reach 8.5 mgd, will increase baseflow in Seeley Creek from the current average of nearly zero to 13.2 cfs. Figure 1 shows a schematic map of the Seeley Creek watershed.

3 The Michigan Department of Environmental Quality (MDEQ) and other stakeholders have raised questions regarding the environmental effects of this increased flow. Other reports have addressed these impacts, but the MDEQ remains concerned about effects on channel stability and tree mortality. On behalf of Commerce Township and under contract with Giffels-Webster Engineers (GWE), I have done a geomorphological reconnaissance of Seeley Creek to address these questions. My findings and conclusions are given in this report.

Scope of work and data sources

In this report I address two questions: First, will the addition of 8.5 mgd of flow to Seeley Drain's baseflow significantly affect channel stability? Second, will the added flow cause tree mortality?

4 To address these questions I reviewed existing data sources, past analyses of Seeley Creek, and made measurements and observations in the field.

I surveyed Seeley Creek in 1992 when this issue arose, and consulted with Tom Coon at Michigan State University regarding fish habitat. I wrote a report (Gough 1993) on channel stability and possible impacts to the redside dace (*Clinostomus elongates*) then. I will address fish habitat questions only briefly here, because Dr. Coon's work (Coon 1992a and 1992b) on that topic is very thorough.

Recently, I have consulted Dr. Coon's reports and letters from the Michigan DEQ. I met with Joe Rathbun and Charlie Hill of MDEQ, and talked with Steve Verhoff, who is conduting a nutrient assessment study on Seeley Creek. I used historical and recent aerial photography and maps to determine past watershed and channel conditions. I reviewed the scientific literature on erosion of cohesive soils at MDEQ's request. I also used the Oakland County USDA Soil Survey along with local publications on geology and groundwater.

During October and November of 2003, I observed most of Seeley Creek from its headwaters to its intersection with Interstate 696. I also looked at sites between this point and Seeley Creek's confluence with the Upper Rouge River downstream of Drake Road. I

■ Consulting report

Features of the Consulting Report (*continued*)

3 The introduction is brief and gives context for this report as the completion of research already conducted.

4 Scope of work and data sources explains what the report looks at and what data the consultant is using to do the analysis. This explanation is especially important for consultants, since not all readers will be familiar with the consultant or his credentials.

	Average velocity, ft/s			Shear, lbs/ft2		
	baseflow	13 cfs	10-year flood	baseflow	13 cfs	10-year flood
Halstead below I-696	1.7	2.5	5.9	0.463	0.775	2.568
12 Mile	0.3	1.0	4.3	0.105	0.327	1.147
13 Mile	0.2	1.5	5.8	0.225	0.522	1.518
Haggerty	1.5	2.7	6.4	0.014	0.116	1.030

Table 2. Hydraulic characteristics at three sites on Seeley Creek. Values are from 1992 GWE HEC modeling.

Geomorphology and Channel Stability Theory and Overview of Seeley Creek

⑥ I have been asked by the Michigan DEQ to assess the probable effects on stability of an increase in baseflow and to support my conclusions with the best available science. Any discussion of stream stability should begin with a definition of the term. There are many ways to approach and define stability with respect to river channels. A widely accepted view among geomorphologists is that river reaches are stable when the net sediment load entering the reach is balanced by that leaving. In other words, the reach is stable if no net erosion or deposition occurs. Under this definition, a river channel may be stable but still move laterally, vertically, and change its cross-sectional shape. But as the channel moves laterally, for example, erosion of outside banks is balanced by deposition on point bars. Scour at one point is balanced by deposition at another. Professionals outside the field of geomorphology tend to define stability in terms of spatial sameness: they assume that a stable river does not change its size, shape, or position.

Channel stability is influenced by many variables, including erodability of bank and bed materials. Water and sediment moving through a reach interact with these materials to cause erosion and deposition. Important bank and bed material characteristics include particle size distribution, channel shape, and plant root density. Discharge and channel slope determine the erosive power of water in reach. High sediment loads may cause channels to aggrade (fill with sediment). Reductions in sediment load often lead to incision or downcutting in channels, a common problem in urbanized areas, which are often overloaded with sediment during development and then sediment starved once build-out is complete.

■ Consulting report

Features of the Consulting report (continued)

⑤ This page from the body of the report used a table of quantitative data to illustrate a point and to break up the text.

⑥ The use of the first person ("I") is perfectly appropriate here, given that this is a report authored by a sole consultant. Readers move more quickly through text in active voice.

Leopold, Wolman and Miller (1964, page 266) define equilibrium (or stability) in river systems, arguing that a stable river is one in which, over a period of years, fluvial characteristics such as sediment load, channel size and sinuosity are adjusted so that just the velocities needed for transportation of the sediment load from the drainage basin exist. They note that a diagnostic feature of such a system is that "any change in any of the controlling factors will cause a displacement of the equilibrium in a direction that will tend to absorb the effect of the change. Such equilibrium channels are necessarily alluvial channels, defined as channels that flow through sediments they have previously transported (Church 1996). But a river system can adjust to changes imposed on it only if it is able to.

Our analysis of Seeley Creek's stability centers on its response to changes in hydrology. Rivers that are truly alluvial will always show some response to significant changes in the sediment load or water regime imposed on them. From my reconnaissance of this watershed, however, it is clear that Seeley Creek is not such a system, at least within the range of flows in question. To begin with, Seeley Creek is mostly a constructed channel. This in no way precludes it from being valuable habitat or an alluvial stream. It is not, however a self-formed alluvial stream. It does not flow through sediments that it has transported, but in almost all cases runs through plant root stabilized sediments that it is unable to rework.

In a recent paper, Trush, McBain and Leopold (2000) list the essential attributes of an alluvial river:

1. Spatially complex channel morphology.
2. Flows and water quality are predictably variable.
3. Frequently mobilized channelbed surface.
4. Periodic channelbed scour and fill.
5. Balanced fine and coarse sediment budgets.
6. Periodic channel migration.
7. A functional floodplain.
8. Infrequent channel resetting floods.
9. Self-sustaining diverse riparian plant communities.
10. Naturally fluctuating groundwater table.

Seeley Creek's drainage network shows a few of these. It does not, however, meet key geomorphic criteria for an alluvial river, specifically attributes 3, 4, 5, 6, and 8.

One might argue that Seeley Creek's channel does have a frequently mobilized bed surface (3) and shows channel bed scour and fill (4). While these processes do occur to some degree, their magnitude is very small. Key redside dace habitat in Seeley--the reach between 12 and 13 Mile Roads--is cut off from any natural bedload supply by the sediment-trapping wetlands upstream. Any significant bedload mobilization would cause downcutting of the channel. This reach, shown in Figure 15, is mildly incised, but incision

■ Consulting report

Features of the Consulting Report (continued)

7 This use of a numbered list helps break up the text and directs the reader's attention to this key information from outside research findings.

Appendix – Notes and Supplemental Information

Units

Readers will notice mixed SI and English units in this report. Though most geomorphologists now use SI units, civil engineers in the United States still largely use English units. I have tried to accommodate standard practice in both fields as possible. Units of shear, both from hydraulic forces and with respect to materials strength can become confusing when mixed units are used. Hydraulic shear as reported in here is calculated using the formula $\tau = \gamma ds$ where τ is shear in kg/m^2, γ is the unit weight of water (1000kg/m³), d is depth in meters, and s is the energy slope in m/m. Energy slope is assumed to be equal to the water surface slope. For those accustomed to shear values in psf, 4.88 kg/m² = 1psf (lb/ft²).

Cohesive Soils Erosion

A few authors give values for maximum allowable velocities for clay-dominated soils. These values appear to be based on three original sources Bear in mind that the allowable values are *very conservative and do not consider the stabilizing effects of vegetation.* Values and sources are shown in Table A1 below. Note that only one value for the augmented flow in Seeley Drain, (3.99 fps, GWE HEC modeling, 1992, at 14 Mile Road), exceeds the most conservative published velocity value.

"permissible velocities," fps	notes	source
4.5	conservative, for "lean clays"	Mirtskhulava, 1936, cited in Simons and Senturk, 1992
3.0 to 4.5	stiff clay	Chang, 1988 and Julien, 1995, cited in Fischenich, 2001
5.0	"stiff clay," water transporting silt	Fortier and Scobey, 1926, cited in Chow 1959, p. 165 and in Simons and Senturk, 1992, p. 433.

Table A1. Values for "permissible velocities" in unvegetated constructed canals. Note that these values do not consider the stabilizing effects of vegetation.

■ Consulting report

Features of the Consulting Report (*continued*)

8 An appendix allows the author to detail some of his technical choices without bogging down the body text of the report.

References

Aichele, S.S. 2000. Ground-Water Quality Atlas of Oakland County, Michigan US Geological Survey Water-Resources Investigation 00-4120 Lansing, Michigan.

Arulanandan, K. Gillogley, E., and Tully, R. 1980. Development of a quantitative method to predict critical shear stress and rate of erosion of natural undisturbed cohesive soils. Report GL-80-5, US Army Engineers, Waterways Experiment Station, Viscksburg, Mississippi.

Chang, H.H. (1988). *Fluvial Processes in River Engineering*, John Wiley and Sons, New York and other cities.

Chow, V.T. 1959. Open channel hydraulics. McGraw Hill.

Coon, T.G. 1992a. Habitat suitability criteria for redside dace, *Clinostomus elongatus*. Unpublished report for Giffels-Webster Engineers, Inc., Rochester Hills, MI. 14 pp.

Coon, T.G. 1992b. Projected impact of added discharge in Seeley Drain for redside dace, *Clinostomus elongatus*, final report. Unpublished report for Giffels-Webster Engineers, Inc., Rochester Hills, MI. 13 pp.

Fischenich, C. 2001. Stability thresholds for stream restoration materials. US Army Corps of Engineers, Ecosystem Management and Restoration Program (EMRP) Technical Note EMRRP-SR-29 May 2001. Available at http://www.wes.army.mil/el/emrrp/tnotes.html

Fortier, S., and Scobey, F.C. (1926). "Permissible canal velocities," *Transactions of the ASCE*, 89:940-984.

Giffels-Webster Engineers, Inc. 1992. Commerce Township wastewater treatment plant analysis of impact on Seeley Drain when discharging 8.5 mgd plant effluent. #13106.01/US3/Report, Feb. 3, 1992, 8 pp. (Includes HEC-2 analysis).

Gough, S.C. 1993. Impact of added discharge in Seeley Drain, Oakland County Michigan, on habitat for redside dace (Clinostomus elongates): Effects of fluvial process and channel morphology. Unpublished report, April 11, 1993.

Gough, S.C. 1997a. Stream classification and assessment. The Nature Conservancy, Peoria, Illinois Field Office. 72 pages.

Geomorphology of Seeley Creek

■ Consulting report

Features of the Consulting Report *(continued)*

9 Any report that cites sources must include a full list of credible references. Sources range from published books to the author's unpublished reports and are listed alphabetically in a common scientific citation style.

Gough, S.C. 1997b. Geomorphic stream habitat assessment, classification, and management recommendations for the Mackinaw River Watershed, Illinios. The Nature Conservancy, Central Illinois Field Office, Peoria, Illinois.

Michigan DNR 1990. A biological survey of Seeley Drain, Oakland County, Michigan, May 10, 1989. Michigan DNR, Surface Water Quality Division Staff Report. 11 pp.

Miktskhulava, T.E. Studies on permissible velocities for soil and facings. No date available. (this is the complete citation as given in Simons and Senturk, 1992).

Schumm, S.A. 1960. The shape of alluvial channels in relation to sediment type. U.S. Geological Survey Prof. Paper 352-B.

Simons, D.B. and Senturk, F. 1992. Sediment transport technology: Water and sediment dynamics. Water Resources Publications, Littleton, Colorado.

Smith, D.G. Effect of vegetation on lateral migration of anastomosed channels of a glacier meltwater river. Geological Society of America Bulletin v. 87, pp. 857-860, June 1976.

Thorne, C.R., R.D. Hey, and M.D. Newson. 1997. Applied fluvial geomorphology for river engineering and management. Wiley and Sons.

Trush, W.J., S.M. McBain, and L.B. Leopold. 2000. Attributes of an alluvial river and their relation to water policy and management. PNAS 97(22) 11858-11863.

US Geological Survey 1980. Northville, MI 7.5' map, first published 1969, photorevised in 1973 and 1980.

US Geological Survey 1983. Walled Lake, MI 7.5' map, first published 1969, photorevised 1983.

■ Consulting report

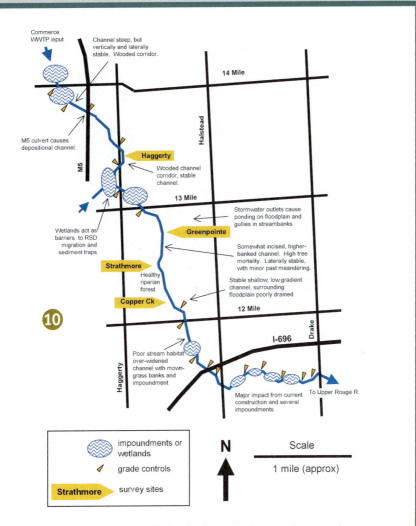

Commerce WWTP input

Channel steep, but vertically and laterally stable. Wooded corridor.

14 Mile

Halstead

M5 culvert causes depositional channel.

M5

Haggerty

Wooded channel corridor, stable channel.

13 Mile

Stormwater outlets cause ponding on floodplain and gullies in streambanks

Wetlands act as barriers to RSD migration and sediment traps

Greenpointe

Somewhat incised, higher-banked channel. High tree mortality. Laterally stable, with minor past meandering.

Strathmore

Healthy riparian forest

Stable shallow, low gradient channel, surrounding floodplain poorly drained

Copper Ck

12 Mile

⑩

I-696

Drake

Poor stream habitat over-widened channel with mown-grass banks and impoundment

Haggerty

Major impact from current construction and several impoundments

To Upper Rouge R.

Legend:
- impoundments or wetlands
- grade controls
- **Strathmore** survey sites

N

Scale
1 mile (approx)

Figure 1. Schematic map of Seeley Creek, showing important features and survey sites. Scale and position are approximate. Grade controls are culverts and other built structures that prevent channel downcutting.

Geomorphology of Seeley Creek

Page 20 of 35

■ Consulting report

Features of the Consulting Report *(continued)*

⑩ Figures can convey complex information by clearly integrating visual and textual data. The map and the aerial photographs (next page) are explained with brief captions.

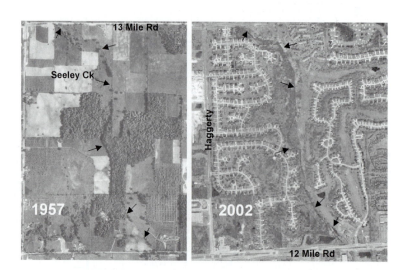

Figure 2. Aerial photographs of Seeley Creek between 12 Mile and 13 Mile roads in 1957 (left) and 2002. Arrows point to the channel and are in the same geographic point in both photographs. The photographs cover about one mile from top to bottom. Most of Seeley Creek retains a forested corridor in the 2002 photograph, but trees are less obvious in this dormant season image. Note the lack of lateral movement of the channel over this 45-year period.

■ Consulting report

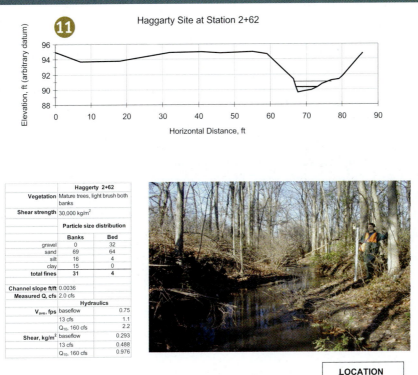

Haggarty Site at Station 2+62

Haggerty 2+62		
Vegetation	Mature trees, light brush both banks	
Shear strength	30,000 kg/m^2	
	Particle size distribution	
	Banks	**Bed**
gravel	0	32
sand	69	64
silt	16	4
clay	15	0
total fines	31	4
Channel slope ft/ft	0.0036	
Measured Q, cfs	2.0 cfs	
	Hydraulics	
V_{ave}, fps	baseflow	0.75
	13 cfs	1.1
	Q_{10}, 160 cfs	2.2
Shear, kg/m^2	baseflow	0.293
	13 cfs	0.488
	Q_{10}, 160 cfs	0.976

Figure 4. Haggerty site, transect 2+62, surveyed in November, 2003. The cross section plot shows the water surface at the time of the survey, at the measured discharge shown in the table at left. The upper line shows the water surface at 13 cfs. With the exception of a few very small depositional bars, all bank materials showed very dense root colonization. Flow is into page for the cross section, and out of the page for the photograph.

Geomorphology of Seeley Creek

Page 23 of 35

■ Consulting report

Features of the Consulting Report *(continued)*

⑪ This page effectively combines a chart, a table, and a photograph, explaining them all with a brief caption. This information appears in the Appendix.

REPORTS **167**

Feasibility Report

Feasibility reports are designed to help people and organizations determine whether a particular approach is feasible or practical and can or should be implemented. The main purpose of most feasibility reports is to assess the effectiveness, practicality, cost, benefits, outcomes, and other factors related to one or several possible solutions to a problem. A feasibility report should be based on evidence and data and usually concludes with recommendations for action or for further research.

The example here is from NASA and the task was large: to determine the feasibility of extending the scope of current programs that look for asteroids and comets that may cause problems for Earth (such as asteroids that might enter Earth's atmosphere, for example). This report is long (166 pages), which is appropriate given the scope of the project.

EXECUTIVE SUMMARY

A Study to Determine the Feasibility of Extending the Search for Near-Earth Objects to Smaller Limiting Diameters

① In recent years, there has been an increasing appreciation for the hazards posed by near-Earth objects (NEOs), those asteroids and periodic comets (both active and inactive) whose motions can bring them into the Earth's neighborhood. In August of 2002, NASA chartered a Science Definition Team to study the feasibility of extending the search for near-Earth objects to smaller limiting diameters. The formation of the team was motivated by the good progress being made toward achieving the so-called Spaceguard goal of discovering 90% of all near-Earth objects (NEOs) with diameters greater than 1 km by the end of 2008. This raised the question of what, if anything, should be done with respect to the much more numerous smaller, but still potentially dangerous, objects. The team was tasked with providing recommendations to NASA as well as the answers to the following 7 specific questions:

②
1. What are the smallest objects for which the search should be optimized?
2. Should comets be included in any way in the survey?
3. What is technically possible?
4. How would the expanded search be done?
5. What would it cost?
6. How long would the search take?
7. Is there a transition size above which one catalogs all the objects, and below which the design is simply to provide warning?

Team Membership

The Science Definition Team membership was composed of experts in the fields of asteroid and comet search, including the Principal Investigators of two major asteroid search efforts, experts in orbital dynamics, NEO population estimation, ground-based and space-based astronomical optical systems and the manager of the NASA NEO Program Office. In addition, the Department of Defense (DoD) community provided members to explore potential synergy with military technology or applications.

Analysis Process

③ The Team approached the task using a cost/benefit methodology whereby the following analysis processes were completed:

Population estimation – An estimate of the population of near-Earth objects (NEOs), including their sizes, albedos and orbit distributions, was generated using the best methods in the current literature. We estimate a population of about 1100 near-Earth objects larger than 1 km, leading to an impact frequency of about one in half a million years. To the lower limit of an object's atmospheric penetration (between 50 and 100 m diameter), we estimate about half a million NEOs, with an impact frequency of about one in a thousand years.

i

■ Feasibility report

Features of a Feasibility Report

① Like most reports, this one begins with an executive summary.

② The first paragraph uses a numbered list to call attention to the context: seven questions that the team needed to address.

③ Because the full report is 166 pages, this executive summary is longer than some (five pages).

TABLE OF CONTENTS

i

■ Feasibility report

Features of a Feasibility Report *(continued)*

4 The scope and length of this report requires it to have a thorough, detailed table of contents. The full TOC is four pages long.

5 Subsections are clearly named and numbered in scientific format.

6 This approach will help readers find the sections that most interest them.

1   INTRODUCTION

1.1   Background

In a 1992 report to NASA (Morrison, 1992), a coordinated Spaceguard Survey was recommended to discover, verify and provide follow-up observations for Earth-crossing asteroids. This survey was expected to discover 90% of these objects larger than one kilometer within 25 years. Three years later, another NASA report (Shoemaker, 1995) recommended search surveys that would discover 60-70% of short-period, near-Earth objects larger than one kilometer within ten years and obtain 90% completeness within five more years. In 1998, NASA formally embraced the goal of finding and cataloging, by 2008, 90% of all near-Earth objects (NEOs) with diameters of 1 km or larger that could represent a collision risk to Earth (Appendix 1). The 1 km diameter metric was chosen after considerable study indicated that an impact of an object smaller than 1 km could cause significant local or regional damage but is unlikely to cause a worldwide catastrophe (Morrison, 1992). The impact of an object much larger than 1 km diameter could well result in worldwide damage up to, and potentially including, extinction of the human race. The NASA commitment has resulted in the funding of a number of NEO search efforts that are making considerable progress toward the 90% by 2008 goal. At the current epoch, more than 50% of the expected population included in the goal has been discovered and the subject objects continue to be discovered at impressive rates. While the current goal covers the larger objects, which could cause global devastation, it is silent on the much more numerous smaller objects (between 50 meters and 1 km diameter) that could cause local or regional damage in an impact. Given the steeply increasing population of near-Earth objects with decreasing diameter, it is much more likely that civilization will experience the impact of an object smaller than 1 km than experience an impact from a larger one. Indeed, the significance of small impactors is beginning to be appreciated by the broad public and by scientists alike. Current NEO surveys are dedicated to finding the largest objects. They also serendipitously find some that are sub-kilometer, but are not optimized to do so and, consequently, are inefficient in finding these objects.

The vast majority of near-Earth objects (NEOs), and the roughly 20% subset of potentially hazardous objects (PHOs) that can closely approach the Earth's orbit, are near-Earth asteroids. However, a small fraction of the NEOs and PHOs are active and inactive short-period comets. Throughout this report, we will most often refer to NEOs and PHOs, generally meaning the set of near-Earth asteroids and inactive short-period comets and excluding long-period comets. However, near-Earth asteroids (NEAs) and potentially hazardous asteroids (PHAs) will also be used when appropriate. Since it is likely that the numbers of asteroids completely dominates the cometary members of the NEO and PHO groups, the reader can normally assume that the populations of NEOs and NEAs are nearly identical, as are the populations of PHOs and PHAs.

1.2   Science Definition Team Formation and Charter

Given the fact that the existing search programs are making good progress toward meeting the current goal, and the emerging discussion of smaller objects, it is natural to ask what, if any, action should be taken to catalog or warn against potential impacts of objects smaller than 1 km

1

---

in diameter. In August of 2002, NASA initiated the formation of a Science Definition Team with a charter to develop an understanding of the threat posed by near-Earth objects smaller than one kilometer and to assess methods of providing warnings of potential impacts. The Team was instructed to provide recommendations to NASA and to outline an executable approach to addressing any recommendations made. Specifically, the team was instructed to address the following questions:

1. What are the smallest objects for which the search should be optimized?
2. Should comets be included in any way in the survey?
3. What is technically possible?
4. How would the expanded search be done?
5. What would it cost?
6. How long would the search take?
7. Is there a transition size above which one catalogs all the objects, and below which the design is simply to provide warning?

The complete formal charter for the Science Definition Team is contained in Appendix 2 of this document.

1.3 Team Membership

The Science Definition Team, henceforth referred to as the "Team", was chaired by Grant H. Stokes from MIT Lincoln Laboratory. Vice Chair of the team was Donald K. Yeomans from the NASA Jet Propulsion Laboratory. The Team members, carefully chosen to represent the breadth and depth of expertise required to address the questions posed in the charter, are listed in Table 1-1, along with their institutions and technical specialties.

Table 1-1. The Science Definition Team Membership

Name	Institution	Technical Specialty
Dr. Grant H. Stokes	MIT Lincoln Laboratory	Asteroid Search, PI for LINEAR
Dr. Donald K. Yeomans	NASA Jet Propulsion Laboratory	Manager, NASA Near-Earth Object Program Office
Dr. William F. Bottke, Jr.	Southwest Research Institute	Asteroid and comet population models
Dr. Steven R. Chesley	NASA Jet Propulsion Laboratory	Hazard assessments and search strategies
Jenifer B. Evans	MIT Lincoln Laboratory	Search system simulations, Co-I for LINEAR
Dr. Robert E. Gold	Johns Hopkins University Applied Physics Lab	Space-based detector systems
Dr. Alan W. Harris	Space Science Institute	Hazard assessments and search strategies
Dr. David Jewitt	University of Hawaii	Visual detectors and search strategies

2

Features of a Feasibility Report *(continued)*

9 The body of this report does a good job of combining text with a variety of visual information. Data display includes distribution plots, tables, and charts.

Table 1-1 (cont.). The Science Definition Team Membership

Col. T.S. Kelso	USAF/AFSPC	DoD assets
Dr. Robert S. McMillan	Spacewatch, University of Arizona	Ground-based NEO Survey, PI for Spacewatch
Dr. Timothy B. Spahr	Smithsonian Astrophysical Observatory	Small body astrometry and orbit determination
Dr./Brig. Gen. S. Peter Worden	USAF/SMC	Space-based detectors and DoD assets
Ex Officio Members:		
Dr. Tomas H. Morgan	NASA Headquarters	Manager, NASA NEO Program
Lt. Col. Lindley N. Johnson (USAF, ret.)	NASA Headquarters	Surveillance of space and DoD space capabilities
Team Support:		
Don E. Avery	NASA Langley Research Center	Study Lead
Sherry L. Pervan	SAIC	Executive Secretary
Michael S. Copeland	SAIC	Cost Analyst
Dr. Monica M. Doyle	SAIC	Cost Analyst

1.4 Study Approach

Providing authoritative answers to the questions posed for the Team requires an understanding of the relationships between the costs of implementing a search effort for smaller NEOs and the benefits accrued. Thus, the study process was constructed along the lines of a cost/benefit analysis as shown in Figure 1-1.

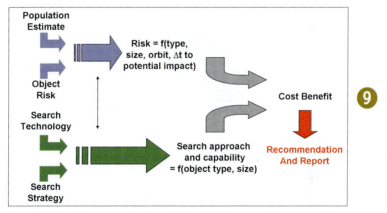

Figure 1-1. Study Process to Develop Cost/benefit Estimate and Recommendations

3

belt, and 6 ± 4% come from the Jupiter-family comet region. The model results were constrained in the JFC region by several objects that are almost certainly dormant comets. For this reason, factors that have complicated the discussions of previous JFC population estimates (e.g., issues of converting cometary magnitude to nucleus diameters, etc.) are avoided. Note, however, that the Bottke et al. (2002a) model does not account for the contribution of comets of Oort cloud origin. This issue will be discussed in Sections 2.5 and 2.6.

Figure 2-1 displays the debiased *(a,e,i)* NEO population as a residence time probability distribution plot. To display as much of the full (a,e,i) distribution as possible in two dimensions, the i bins were summed before plotting the distribution in (a,e), while the e bins were summed before plotting the distribution in (a,i). The color scale depicts the expected density of NEOs in a scenario of steady state replenishment from the main belt and transneptunian region. Red colors indicate where NEOs are statistically most likely to spend their time. Bins whose centers have perihelia q > 1.3 AU are not used and are colored white. The gold curved lines that meet at 1 AU divide the NEO region into Amor (1.0167 AU < q < 1.3 AU), Apollo (a > 1.0 AU; q < 1.0167 AU) and Aten (a < 1.0 AU; Q > 0.983 AU) components. IEOs (Q < 0.983 AU) are inside Earth's orbit. The Jupiter-family comet region is defined using two lines of constant Tisserand parameter 2 < T < 3. The curves in the upper right show where T=2 and T=3 for i=0 deg.

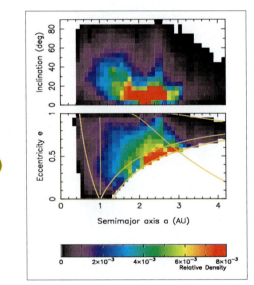

Figure 2-1. A representation of the probability distribution of residence time for the debiased near-Earth object (NEO) population.

11

Features of a Feasibility Report *(continued)*

⑨ The body of this report does a good job of combining text with a variety of visual information. Data display includes distribution plots, tables, and charts.

Table 3-2. Expected Damage from Impacts onto Land

$\langle D \rangle$ km	R_D km	F_D	F_1	F_{yr} Min.	F_{yr} Nom.	F_{yr} Max.	$(1-C)F_{yr}$ Min.	$(1-C)F_{yr}$ Nom.	$(1-C)F_{yr}$ Max.
0.031	0.00	0.00E+00	0.00E+00	0.00	0.00	0.00	0.00	0.00	0.00
0.039	0.00	0.00E+00	0.00E+00	0.00	0.00	3.64	0.00	0.00	3.64
0.050	7.30	3.28E-07	1.97E+03	0.00	2.11	11.85	0.00	2.11	11.85
0.062	17.30	1.84E-06	1.10E+04	1.23	6.88	14.03	1.23	6.88	14.03
0.079	24.70	3.75E-06	2.25E+04	3.99	8.14	12.33	3.99	8.14	12.33
0.099	30.40	5.68E-06	3.41E+04	4.73	7.16	9.54	4.72	7.15	9.53
0.125	35.10	7.57E-06	4.54E+04	4.16	5.54	7.09	4.14	5.52	7.06
0.157	39.70	9.69E-06	5.81E+04	3.22	4.11	5.31	3.20	4.10	5.29
0.198	45.10	1.25E-05	7.50E+04	2.39	3.08	4.34	2.35	3.03	4.27
0.250	53.50	1.76E-05	1.06E+05	1.79	2.52	3.97	1.75	2.46	3.89
0.315	67.22	2.78E-05	1.67E+05	1.46	2.31	3.66	1.41	2.23	3.54
0.397	84.69	4.41E-05	2.64E+05	1.34	2.13	3.37	1.25	1.98	3.14
0.500	106.70	7.00E-05	4.20E+05	1.23	1.96	3.11	1.11	1.77	2.80
0.630	134.43	1.11E-04	6.66E+05	1.14	1.80	2.87	0.94	1.49	2.37
0.794	169.38	1.76E-04	1.06E+06	1.05	1.66	2.64	0.78	1.24	1.96
1.000	213.40	2.80E-04	1.68E+06	0.97	1.53	2.43	0.58	0.92	1.45
1.260	268.87	4.44E-04	2.67E+06	0.89	1.41	2.24	0.42	0.67	1.06
1.587	338.75	7.05E-04	4.23E+06	0.82	1.30	2.07	0.28	0.45	0.71
2.000	426.80	1.12E-03	6.72E+06	0.76	1.20	1.90	0.19	0.29	0.47
2.520	537.73	1.78E-03	1.07E+07	0.70	1.10	1.75	0.11	0.18	0.28
3.175	677.50	2.82E-03	1.69E+07	0.64	1.02	1.62	0.03	0.05	0.08
4.000	853.60	4.48E-03	2.69E+07	0.59	0.94	1.49	0.00	0.00	0.00
5.040	1075.47	7.11E-03	4.26E+07	0.54	0.86	1.37	0.00	0.00	0.00
6.350	1355.01	1.13E-02	6.77E+07	0.50	0.80	1.26	0.00	0.00	0.00
8.000	1707.20	1.79E-02	1.07E+08	0.46	0.73	1.17	0.00	0.00	0.00
10.079	2150.94	2.84E-02	1.71E+08	0.43	0.68	1.07	0.00	0.00	0.00
			Total all size bins	35.01	60.98	106.13	28.48	50.65	86.12

From Table 3-2, it can be seen that the total fatality rate from land impacts of NEOs below the threshold for global catastrophe is only about 60/year, a tiny fraction of the hazard from larger, globally hazardous events. Figures 3-2 and 3-3 are plots of the total land impact hazard and the fraction remaining after five more years at the current search survey efficiency.

In performing the above calculations, we have, as noted, ignored the uneven distribution of population on the Earth. In addition to the simple fact that people live on the land and not on the sea, the population on land is also very unevenly distributed. As a result, most sub-global events are unlikely to cause any fatalities, and only the extraordinarily rare events over heavily populated areas will result in large numbers of fatalities. To investigate this quantitatively, we obtained a digital population map of the world from data compiled by the National Center for Geographic Information and Analysis (NCGIA, 2003). This map gives the population within each cell of 5 arc minutes of latitude and longitude on the Earth's surface from latitude $-57°$ to $+72°$ (there is essentially no population outside of this range). This works out to about 6.7×10^6 cells covering most of the Earth's surface.

24

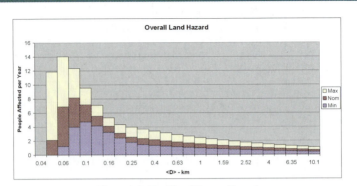

Figure 3-2. Total Land Impact Hazard

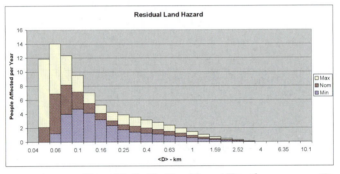

Figure 3-3. Residual Land Impact Hazard

To estimate the number of fatalities expected from events of each size, we extend the 5 arc minute grid over the full latitude range $-90°$ to $+90°$ by inserting zero population in each of the additional cells, and simulate an impact centered on each of the total of 9.3×10^9 cells. At each cell location, we map the fraction of neighboring cells included in a circle of radius equal to the Hills & Goda (1993) radius of destruction for the event size considered, and sum up the total casualties expected from an event at that location. We then sum over all possible events, weighted by the cosine of latitude of each event, since the grid with constant steps in longitude has equal numbers of events at each latitude whereas the actual surface area of the Earth at each latitude decreases as cosine of latitude. We thus tabulate the fraction of events of a given impactor size versus expected fatalities. Table 3-3 lists the results, using the same factor-of-two-in-mass bins as Table 3-1. The first two columns repeat the mean impactor diameter and impact frequency as in Table 3-1. The remaining pairs of columns list the fraction Φ_N of events of that impactor size that result in $>N$ fatalities, and the frequency, $f_N = f(n)\Phi_N$ that such an event is

25

surveys have rather large pixels (1 to 2.5 arcsec), and even the best large-scale stellar reference catalogs have rms position uncertainty values of 0.2-0.4 arcsec. Nyquist sampling the point-spread function for smaller-pixel designs will immediately result in a much better-determined image centroid. A few new and dense star catalogs are in the works and should provide complete star catalogs with rms position uncertainties of near 0.1 arcsec or even 0.05 arcsec. These key improvements will result in dramatic improvements for NEO identification, orbital linkage, orbit improvement and hazard evaluation. Astrometric observations accurate to about 0.1 arcsec represent a 5-fold increase on average over several reference star catalogs that are in use today. These improvements may well verify that most new discoveries, even those with the shortest arcs, are not hazardous objects.

5.5 Cadence Requirement for Linkage and Orbits

Sections 5.2-5.4 provide fairly solid requirements on the search strategy. Observations of individual objects must be spaced closely enough to link their orbits in the first place and also closely enough that linkages will be fairly easy from a computational standpoint. These requirements, coupled with a NEO refresh rate of a few weeks, provide sky coverage and repeat visit requirements. For the analysis conducted by the Team, the assumption was made that each object must be observed over a total arc of at least 21 nights, and two nights must be spaced by less than 7 nights in order to be placed in the catalog. A pair of nights is necessary for linkage purposes, and the third night (before or after the pair) is needed for computing the unique general solution for the orbital elements. The resulting demand on the sky coverage also places requirements on the design specifications of the telescope and spacecraft; systems incapable of covering the entire observable sky 4-5 times per month may not be competitive.

5.6 Search Strategies for Warning

In the above discussion, special attention was paid to locating and cataloging objects in order to eliminate the hazard from each individual object on a case-by-case basis. There could, however, be objects that are detected that pose an immediate threat to the Earth on timescales from days to months. Unfortunately, objects on their final approach to the Earth can come from any direction of the sky. Preferentially, some may approach from morning or evening twilight, while others will come from the day-time sky and be almost undetectable until just a few days before impact. In order to provide an adequate warning system, the entire observable sky should be surveyed at least a few times per month, with special attention paid to areas around 90 degrees from the Sun. Fortunately, this search strategy for warning differs little from our initial search strategy for cataloging all PHOs.

5.7 Conclusions

The systems and strategies discussed in this report represent a dramatic improvement over existing NEO search technology and capability. More frequent sky coverage to greater limiting magnitudes, multiple search systems, and fusing near-Sun and opposition-based surveys will

56

Features of a Feasibility Report *(continued)*

🔟 Even pages with full text use white space effectively and keep the information in readable chunks.

⓫ The writing style is technical but clear. Most sentences are short and in active voice.

⓬ When passive voice is used, it is not obstructive to the meaning ("special attention was paid to locating and cataloging").

13 Chesley, S.R. and S.N. Ward, 2003. A Quantitative Assessment of the Human and Economic Hazard from Impact-generated Tsunami, submitted to *Environmental Hazards.*

Chyba, C. F., P. J. Thomas, and K. J. Zahnle, 1993. The 1908 Tunguska Explosion - Atmospheric Disruption of a Stony Asteroid. *Nature,* v. 361, 40-44.

D'Abramo, G., A. W. Harris, A. Boattini, S. C. Werner, and G. B. Valsecchi, 2001. A Simple Probabilistic Model to Estimate the Population of near-Earth Asteroids. *Icarus,* v. 153, 214-217.

Duncan, M., T. Quinn and S. Tremaine, 1987. The Formation and Extent of the Solar System Comet Cloud. *Astron. J.* v. 94, 1330-1338.

Duncan, M., T. Quinn, and S. Tremaine, 1988. The Origin of Short-period Comets. *Astrophysical Journal Letters,* v. 328, L69 – L73.

Duncan, M.J. and H.F. Levison, 1997. A Scattered Comet Disk and the Origin of Jupiter Family Comets. *Science,* v. 276, 1670-1672

Everhart, E., 1967. Intrinsic Distributions of Cometary Perihelia and Magnitudes, *Astronomical Journal,* v. 72, 1002-1011.

Farinella, P., L. Foschini, C. Froeschle, R. Gonczi, T. J. Jopek, G. Longo, and P. Michel, 2001. Probable Asteroidal Origin of the Tunguska Cosmic Body. *Astronomy and Astrophysics,* v. 377, 1081-1097.

Gladman, B.J., F. Migliorini, A. Morbidelli, V. Zappala, P. Michel, A. Cellino, C. Froeschl'e, H.F. Levison, M. Bailey and M. Duncan, 1997. Dynamical Lifetimes of Objects Injected into Asteroid Belt Resonances. *Science,* v. 277, 197-201.

Harris, A.W., 2002. A New Estimate of the Population of Small NEAs. Bulletin of the American Astronomical Society, #34, 835.

Hills, J.G. and M.P. Goda, 1993. The Fragmentation of Small Asteroids in the Atmosphere, *Astronomical J.,* v. 105, 1114-1144.

Ivezic, Z. and 32 colleagues, 2001. Solar System Objects Observed in the Sloan Digital Sky Survey Commissioning Data. *Astronomical Journal,* v. 122, 2749-2784.

Jedicke, R., 1996. Detection of Near-Earth Asteroids Based upon Their Rates of Motion. *Astron. J.* v. 111, 970.

Jedicke, R. and T.S. Metcalfe, 1998. The Orbital and Absolute Magnitude Distributions of Main Belt Asteroids. *Icarus,* v. 131, 245-260.

122

■ Feasibility report

Features of a Feasibility Report *(continued)*

13 A complete reference section provides links to those citations that are available online.

Jedicke, R., J. Larsen, and T. Spahr, 2002. Observational Selection Effects in Asteroid Surveys. In *Asteroids III* (W.F. Bottke, A. Cellino, P. Paolicchi, and R.P. Binzel, Eds), U. of Arizona Press, Tucson, 71-87.

Jedicke, R., A. Morbidelli, J.-M. Petit, T. Spahr, and W.F. Bottke, 2003. Earth and Space-based NEO Survey Simulations: Prospects for Achieving the Spaceguard Goal. *Icarus*, v. 161, 17-33.

Jet Propulsion Laboratory, Near Earth Object Program website, http://neo.jpl.nasa.gov, retrieved August 7, 2003.

Kenkel, D., 2000, Using Estimates of the Value of a Statistical Life in Evaluating Regulatory Effects, http://www.ers.usda.gov/publications/mp1570/mp1570d.pdf, retrieved July 16, 2003.

Kresak, L., 1979. Dynamical Interrelations Among Comets and Asteroids. In *Asteroids* (T. Gehrels, Ed.), U. of Arizona Press, Tucson, 289-309.

Krinov, E. L., 1963. The Tunguska and Sikhote-Alin Meteorites. In the Moon, Meteorites, and Comets (eds. B. M. Middlehurst and G. P. Kuiper). U. of Chicago, Chicago, IL, 208-243.

Lambour, R., E. Rork, and E. Pearce. 2003. Modeling the Sky Background and Its Impact on Electro-Optic Sensors. MIT Lincoln Laboratory, in press.

Levison, H.F., 1996. Comet Taxonomy. In *Completing the Inventory of the Solar System.* (T. W. Rettig and J. M. Hahn, Eds.) ASP Conf. Series 107, 173-191.

Lamy, P., I. Toth, Y.R. Fernandez, H.A. Weaver, 2003. The Sizes, Shapes, Albedos, and Colors of Cometary Nuclei. In *Comets II*, U. of Arizona Press (in press).

Levison, H.F. and M.J. Duncan, 1994. The Long-term Dynamical Behavior of Short-period Comets. *Icarus*, v. 108, 18-36.

Levison, H.F. and M.J. Duncan, 1997. From the Kuiper Belt to Jupiter-family Comets: The Spatial Distribution of Ecliptic Comets. *Icarus*, v. 127, 13-32.

Levison, H.F., L. Dones, and M.J. Duncan, 2001. The Origin of Halley-Type Comets: Probing the Inner Oort Cloud. *Astron. J.*, v. 121, 2253-2267.

Levison, H.F., A. Morbidelli, L. Dones, R. Jedicke, P.A. Wiegert, and W.F. Bottke, 2002. The Mass Disruption of Oort Cloud Comets. *Science*, v. 296, 2212-2215. [For a detailed treatment, see http://www.boulder.swri.edu/~hal/PDF/disrupt.pdf]

Magnier, E. and D. Jewett, 2003, Personal communication.

Marsden, B.G., 1992. To Hit or Not to Hit. Proceedings, Near-Earth Objects Interception Workshop. (G.H. Canavan, J.C. Solem, J.D.G. Rather, eds.). Los Alamos National Laboratory, Los Alamos, NM. 67-71.

123

■ Feasibility report

Mr. Chairman and Members of the Subcommittee:

I am pleased to have this opportunity to appear before the Subcommittee today to discuss NASA's current efforts and future plans to inventory and characterize the population of Near Earth Objects (NEOs).

Background

This Committee has been a leader in focusing attention on the importance of cataloging and characterizing Earth-approaching asteroids and comets. In 1992, the Committee on Science directed that NASA sponsor two workshop studies, the NEO Detection Workshop, which was chaired by NASA, and the NEO Interception Workshop, which was chaired by the Department of Energy. In March 1993, the Science Committee held a hearing to review the results of these two workshops. In 1995, at the Committee's request, NASA conducted a follow-up study which was chaired by the late Dr. Gene Shoemaker. Each of these studies stressed the importance of characterizing and cataloging NEOs with diameters larger than 1 km within the next decade. We have taken steps to put us on a path to achieving this goal. I am here today to tell you about those steps, as well as to bring you up to date on the rich program of space missions to NEOs and related objects.

The NEO population is derived from a variety of scientifically interesting sources including planetessimal fragments and some Kuiper belt objects. Indeed, the Office of Space Science Strategic Plan includes as a specific goal "...to complete the inventory and characterize a sample of Near Earth Objects down to 1 km diameter". While the threat of a catastrophic collision is statistically small, NASA has a vigorous program of exploration of NEOs planned, including both asteroids and comets.

(14) There has been much recent discussion about the potential threat posed by NEOs, but NASA has long been interested in them from a scientific standpoint. NEO investigations have had to compete for support against a number of other compelling science programs; funding selection

128

■ Feasibility report

Features of a Feasibility Report *(continued)*

(14) Several appendices provide background and other information that is helpful but not needed in the body of the report.

Progress Report

The term *progress report* can be applied to several kinds of reports. In general, all of these reports would provide an update about the progress or status of a project, grant, experiment, process, workplace task, or other items. In some companies, employees turn in status or activity reports on a regular basis. Often these reports are done as quick memos, following a standard format and sent via email. Progress reports tend to be longer than status reports and tend to be part of a project or grant where regular updates are required. Depending on the circumstance, a progress report may include a timeline for the project, illustrating where on the timeline the current report is referring. Timelines and other project management visuals for use in progress reports can be generated using specialized software (such as Microsoft Project). The progress report shown here is brief and in the format required by the Environmental Protection Agency (EPA).

2004 Progress Report: Development of Environmental Indicators of Condition, Integrity, and Sustainability in the U.S. Great Lakes Basin

EPA Grant Number: R828675
Title: Development of Environmental Indicators of Condition, Integrity, and Sustainability in the U.S. Great Lakes Basin
① **Center:** Great Lakes Environmental Indicators Project
Center Director: Gerald J. Niemi
Investigators: Richard P. Axler[1], JoAnn M. Hanowski[1], George E. Host[1], Robert W. Howe[2], Lucinda B. Johnson[1], Carol A. Johnston[1], John C. Kingston[1], Euan D. Reavie[1], Ronald R. Regal[3], Carl Richards[4], Deborah L. Swackhamer[5]
Cooperators: John R. Kelly[6], Janet Keough[6], David Mount[6], Paul Bertram[7], John Schneider[7]
Institutions: [1]Center for Water and the Environment, Natural Resources Research Institute, University of Minnesota Duluth; [2]University of Wisconsin, Green Bay; [3]University of Minnesota Duluth; [4]Minnesota Sea Grant College Program; [5]University of Minnesota Twin Cities; [6]U.S. EPA Mid-Continent Ecology Division, Duluth; [7] U.S. EPA Region 5, Chicago, IL
EPA Project Officer: Barbara Levinson
Project Period: January 10, 2001 to January 9, 2005 (Extended to January 9, 2006)
Project Period Covered by this Report: January 11, 2004 to January 9, 2005
RFA: Environmental Indicators in the Estuarine Environment Research Program (2002)
Research Category: Ecological Indicators/Assessment/Restoration

② **Description:**

Objective: The major question being addressed is "What environmental indicators can be developed to efficiently, economically, and effectively measure and monitor the condition, integrity, and long-term sustainability of the coastal region?"

■ Progress report

Features of a Progress Report

① The first page of this progress report follows the format and content required by the EPA.

② The report uses headings well (description, objective, approach) to break up the text and direct the reader's attention to these key sections.

Our specific objectives include:

- identification of environmental indicators that will be useful to define the condition, integrity, and change of the ecosystems within the coastal region,
- testing these indicators with a rigorous combination of existing data and field data to link stressors of the coastal region with environmental responses, and
- recommendation of a suite of hierarchically structured indicators to guide managers toward informed management decisions.

The final product will provide information for managers to communicate with the public on the condition and integrity of the coastal region, to guide development of monitoring programs to measure change, to identify areas in need of restoration or conservation strategies, and to use as key indicators for input into modeling efforts to predict the future of the coastal region.

Approach: The primary focus during the past year has been to complete the processing of field samples, data compilation, data analysis, and the preparation of publications resulting from these efforts. The bird and amphibian group sampled over 600 sites in 2002 and 2003. The fish and macroinvertebrate group sampled 112 sites in 2002 and 2003. The diatom and water quality team has sampled 240 sites from 2001 to 2003. The wetland vegetation group has sampled 86 sites from 2001 to 2003. The contaminants group has sampled 22 sites across the Great Lakes basin. In addition, over 40 sites were visited by each of four project components: fish and macroinvertebrates, wetland vegetation, bird and amphibian, and diatom groups. Most of the sites sampled by the contaminant group were also sampled by the remaining four groups.

The compilation of data for each subcomponent has been developed through central administration of the project to insure data compatibility and ease of analysis among the study components. In addition, all study components have documented points of sampling using current geopositioning (GPS) instruments to insure spatial integrity and allow visualization of sample sites and overlap. Over 12,000 GPS points have been recorded during field sampling within the coastal region of the Great Lakes.

This co-operative agreement with U.S. EPA Office of Research and Development includes regular conference calls and individual face-to-face meetings on an as-needed basis for each subcomponent. This generally has occurred on a monthly basis and more frequently during the field sampling period (April to September). U.S. EPA Mid-Continent Ecology Division has also coordinating their sampling of the Great Lakes to overlap with our study design. Their primary focus, however, was to examine nutrient gradients in the coastal region of the Great Lakes.

Additional Information: The investigators have given a host of presentations which collectively totals more than 50 this past year. In particular, GLEI investigators made presentations at four major national/international science meetings. These included meetings of the American Society of Limnology and Oceanography, International Association of Great Lakes Research, North American Benthological Society, and the Society of Environmental Toxicology and Chemistry. PI Niemi in collaboration with Dr. Hans Paerl of ACE INC (U of North Carolina) were successful in having an Organized Oral Discussion accepted for the upcoming International Ecological Society and Ecological Society of America meeting in Montreal, Quebec in August 2005. This was a highly competitive process. In addition, about ten presentations were made to managers or the general public.

A major emphasis of current efforts is the preparation of peer-reviewed publications. To date, GLEI investigators have published or have in press 12 papers in peer-reviewed journals and 24 manuscripts are in preparation. In particular, PI Niemi coordinated the publication among all EaGLe lead investigators on a "Rationale for a new generation of ecological indicators for coastal waters (*Environmental Health Perspectives* 112: 979-986, 2004. In addition, PI Niemi was invited to complete a review paper on the "Application of ecological indicators" for *Annual Review of Ecology, Evolution, and Systematics* 35:89-111, 2004. These reviews are among the most-cited papers in ecology in the world. Seventeen graduate students have been involved in GLEI-associated research and nine of these have successfully defended.

■ Progress report

Features of a Progress Report (*continued*)

3 An "additional information" section goes beyond the requirements, providing a brief summary of other activities.

Each of the components has been adhering to and refining its quality assurance and quality control objectives. We have requested and have been granted a one-year no-cost extension to the project to January 9, 2006. Budget targets for women and minority owned businesses are also being met.

Progress Summary: The primary focus of the fourth year in what is now a five year effort has been the data compilation, analysis, and summarization of the results to test the hypotheses of stress and response relationships. Field data have continued to verify many of the stress gradients (e.g., nutrients) that we had identified *a priori* with our experimental design. The results of this experimental design was recently published in the journal *Environmental Monitoring and*

Assessment (Vol. 102:41-65, 2005). The details of these efforts are described under each subproject included in this report.

Among the major activities that we have been involved with during the past 6 months is the integration of data among the GLEI subprojects. We have been using a procedure known as hierarchical partitioning to identify the variance explained for several dependent variables (potential ecological indicators) by ecotype (e.g., habitat), lake (e.g., one of the five Great Lakes), ecoprovince (2 in our region), stressors (a combination of land use, pollution, and human population density), and basinwide. We have been exploring this partitioning for each of the subcomponents from the perspective of amphibians, birds, diatoms, fish, insects, and wetland vegetation. We also have incorporated indicators at the species level, compositional level, and function-based indicators. These analyses are critical for a variety of reasons, including the following: 1) to identify the relative merit of using different types of ecological indicators, 2) what is the scale upon which these indicators can be applied such as basinwide or lake-specific, or 3) are these indicators related to stress and at what scale. The preliminary analyses indicate that lake is very important to consider when applying an indicator and ecotype has not been as important as we would have expected.

An all-investigators EaGLe meeting was hosted by the GLEI group this past fall on September 30 to October 2, 2004 in Duluth. An all-investigators meeting of the GLEI investigators was held in September 2004 and another meeting is planned for June 2-3, 2005 in Duluth, MN.

Future Activities: The primary emphases over the next year will include the following 1) complete the compilation and synthesis of the data gathered, 2) continue with the analysis of hierarchical partitioning and other integration analysis among the subcomponents, and 3) prepare presentations and manuscripts for peer-reviewed publications. GLEI investigators will be making presentations at several national/international science meetings this year. These include the Ecological Society of America, the Estuarine Research Federation, International Association of Great Lakes Research, the Society of Environmental Toxicology and Chemistry, and North American Benthological Society.

Publications and Presentations: Total count: 12 published or accepted, 24 submitted or in preparation; presentations; over 102.

■ Progress report

Features of a Progress Report *(continued)*

4 Progress reports typically provide a summary of major activities.

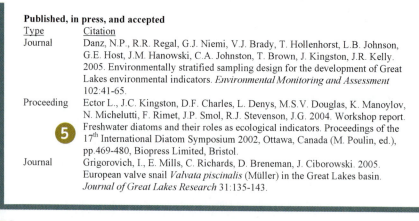

Published, in press, and accepted

Type	Citation
Journal	Danz, N.P., R.R. Regal, G.J. Niemi, V.J. Brady, T. Hollenhorst, L.B. Johnson, G.E. Host, J.M. Hanowski, C.A. Johnston, T. Brown, J. Kingston, J.R. Kelly. 2005. Environmentally stratified sampling design for the development of Great Lakes environmental indicators. *Environmental Monitoring and Assessment* 102:41-65.
Proceeding	Ector L., J.C. Kingston, D.F. Charles, L. Denys, M.S.V. Douglas, K. Manoylov, N. Michelutti, F. Rimet, J.P. Smol, R.J. Stevenson, J.G. 2004. Workshop report. Freshwater diatoms and their roles as ecological indicators. Proceedings of the 17th International Diatom Symposium 2002, Ottawa, Canada (M. Poulin, ed.), pp.469-480, Biopress Limited, Bristol.
Journal	Grigorovich, I., E. Mills, C. Richards, D. Breneman, J. Ciborowski. 2005. European valve snail *Valvata piscinalis* (Müller) in the Great Lakes basin. *Journal of Great Lakes Research* 31:135-143.

■ Progress report

Features of a Progress Report *(continued)*

5 This progress report details the status of publications that have been accepted and are in press.

Research Report

Many reports are based on research. The phrase *research report* can have a wide-ranging definition. Often, when one is asked to write a research report, the person making the request wants to see the latest studies, concepts, themes, and trends in a particular area, in as factual a manner as possible. The information in a research report could eventually become the basis for a feasibility report or a proposal. In a college class, a research report is often a required assignment, and normally the instructor provides guidelines as to the format, page length, topics, and research requirements. In the workplace, an individual or a team would write a research report for highly technical audiences and also more general audiences.

Two research reports are presented here. The "Tobacco Addiction" report is published by the National Institutes on Drug Abuse (NIDA) and uses a newsletter approach to make the information accessible to a wide audience. The report "Investigation of the Low-Temperature Fracture Properties of Three MnROAD Asphalt Mixtures" is written for a technical audience of transportation and materials engineers. It uses visuals and professional formatting but its content and layout are more appropriate for the expert audience.

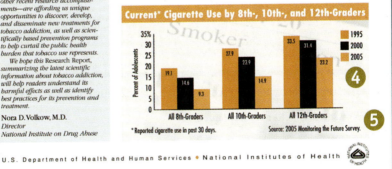

NATIONAL INSTITUTE ON DRUG ABUSE
Research Report
SERIES

TOBACCO
Addiction

Tobacco use kills nearly half a million Americans each year, with one in every six U.S. deaths the result of smoking. Smoking harms nearly every organ of the body, causing many diseases and compromising smokers' health in general. Nicotine, a component of tobacco, is the primary reason that tobacco is addictive, although cigarette smoke contains many other dangerous chemicals, including tar, carbon monoxide, acetaldehyde, nitrosamines, and more.

An improved overall understanding of addiction and of nicotine as an addictive drug has been instrumental in developing medications and behavioral treatments for tobacco addiction. For example, the nicotine patch and gum, now readily available at drugstores and supermarkets nationwide, have proven effective for smoking cessation when combined with behavioral therapy.

Advanced neuroimaging technologies further assist this mission by allowing researchers to observe changes in brain function that result from smoking tobacco. Researchers have also identified new roles for genes that predispose people to tobacco addiction and predict their response to smoking cessation treatments. These findings—and many other recent research accomplishments—are affording us unique opportunities to discover, develop, and disseminate new treatments for tobacco addiction, as well as scientifically based prevention programs to help curtail the public health burden that tobacco use represents.

We hope this Research Report, summarizing the latest scientific information about tobacco addiction, will help readers understand its harmful effects as well as identify best practices for its prevention and treatment.

Nora D. Volkow, M.D.
Director
National Institute on Drug Abuse

(vertical heading: from the director)

What is the extent and impact of tobacco use?

According to the 2004 National Survey on Drug Use and Health, an estimated 70.3 million Americans age 12 or older reported current use of tobacco—59.9 million (24.9 percent of the population) were current cigarette smokers, 13.7 million (5.7 percent) smoked cigars, 1.8 million (0.8 percent) smoked pipes, and 7.2 million (3.0 percent) used smokeless tobacco, confirming that tobacco is one of the most widely abused substances in the United States. While these numbers are still unacceptably high, they represent a decrease of almost 50 percent since peak use in 1965.

NIDA's 2005 Monitoring the Future Survey of 8th-, 10th-, and 12th-graders, used to track drug use patterns and attitudes, has also shown a striking decrease in smoking trends among the Nation's youth. The latest results indicate that about 9 percent of 8th-graders, 15 percent of 10th-graders, and 23 percent of 12th-graders had used cigarettes in the 30 days prior to the survey. Despite cigarette use being at the lowest levels of the survey since a peak in the mid-1990s, the past few years indicate a clear slowing of this decline. And while perceived risk and disapproval of

Current* Cigarette Use by 8th-, 10th-, and 12th-Graders

Percent of Adolescents (y-axis: 0, 5, 10, 15, 20, 25, 30, 35%)

Legend: ■ 1995 ■ 2000 ■ 2005

- All 8th-Graders: 19.1, 14.6, 9.3
- All 10th-Graders: 27.9, 23.9, 14.9
- All 12th-Graders: 33.5, 31.4, 23.2

* Reported cigarette use in past 30 days.

Source: 2005 Monitoring the Future Survey.

U.S. Department of Health and Human Services • National Institutes of Health

---

■ Research report for a more general, non-technical audience in newsletter format

Features of a Research Report

1 This report, intended to make research accessible to a wide audience, uses a newsletter format.

2 This heading is in the form of a question that the average reader might ask.

3 The vertical heading "from the director" provides an interesting alternative for an important section.

4 This bar chart provides a quick and easy way to illustrate quantitative data.

5 The NIH logo provides the organization's credibility.

smoking had been on the rise, recent years have shown the rate of change to be dwindling. In fact, current use, perceived risk, and disapproval leveled off among 8th-graders in 2005, suggesting that renewed efforts are needed to ensure that teens understand the harmful consequences of smoking.

Moreover, the declining prevalence of cigarette smoking among the general U.S. population is not reflected in patients with mental illnesses. For them, it remains substantially higher, with the incidence of smoking in patients suffering from post-traumatic stress disorder, bipolar disorder, major depression, and other mental illness twofold to fourfold higher than the general population, and smoking incidence among people with schizophrenia as high as 90 percent.

Tobacco use is the leading preventable cause of death in the United States. The impact of tobacco use in terms of morbidity and mortality costs to society is staggering. Economically, more than $75 billion of total U.S. healthcare costs each year is attributable directly to smoking. However, this cost is well below the total cost to society because it does not include burn care from smoking-related fires, perinatal care for low birth-weight infants of mothers who smoke, and medical care costs associated with disease caused by secondhand smoke. In addition to healthcare costs, the costs of lost productivity due to smoking effects are estimated at $82 billion per year, bringing a conservative estimate of the economic burden of smoking to more than $150 billion per year.

How does tobacco deliver its effects? 6

There are more than 4,000 chemicals found in the smoke of tobacco products. Of these, nicotine, first identified in the early 1800s, is the primary reinforcing component of tobacco that acts on the brain.

Cigarette smoking is the most popular method of using tobacco; however, there has also been a recent increase in the sale and consumption of smokeless tobacco products, such as snuff and chewing tobacco. These smokeless products also contain nicotine, as well as many toxic chemicals.

The cigarette is a very efficient and highly engineered drug-delivery system. By inhaling tobacco smoke, the average smoker takes in 1 to 2 mg of nicotine per cigarette. When tobacco is smoked, nicotine rapidly reaches peak levels in the bloodstream and enters the brain. A typical smoker will take 10 puffs on a cigarette over a period of 5 minutes that the cigarette is lit. Thus, a person who smokes about 1-1/2 packs (30 cigarettes) daily gets 300 "hits" of nicotine to the brain each day. In those who typically do not inhale the smoke—such as cigar and pipe smokers and smokeless tobacco users—nicotine is absorbed through the mucosal membranes and reaches peak blood levels and the brain more slowly.

Immediately after exposure to nicotine, there is a "kick" caused in part by the drug's stimulation of the adrenal glands and resulting discharge of epinephrine (adrenaline). The rush of adrenaline stimulates the body and causes a sudden release of glucose, as well as an increase in blood pressure, respiration, and heart rate. Nicotine also suppresses insulin output from the pancreas, which means that smokers are always slightly hyperglycemic (i.e., they have elevated blood sugar levels). The calming effect of nicotine reported by many users is usually associated with a decline in withdrawal effects rather than direct effects of nicotine.

■ Research report for a more general, non-technical audience in newsletter format

Features of a Research Report *(continued)*

6 The continued use of questions as headings is a good choice for research aimed at a general audience.

⑦ Is nicotine addictive?

Yes. Most smokers use tobacco regularly because they are addicted to nicotine. Addiction is characterized by compulsive drug seeking and use, even in the face of negative health consequences. It is well documented that most smokers identify tobacco use as harmful and express a desire to reduce or stop using it, and nearly 35 million of them want to quit each year. Unfortunately, only about 6 percent of people who try to quit are successful for more than a month.

Research has shown how nicotine acts on the brain to produce a number of effects. Of primary importance to its addictive nature are findings that nicotine activates reward pathways—the brain circuitry that regulates feelings of pleasure. A key brain chemical involved in mediating the desire to consume drugs is the neurotransmitter dopamine, and research has shown that nicotine increases levels of dopamine in the reward circuits. This reaction is similar to that seen with other drugs of abuse, and is thought to underlie the pleasurable sensations experienced by many smokers. Nicotine's pharmacokinetic properties also enhance its abuse potential. Cigarette smoking produces a rapid distribution of nicotine to the brain, with drug levels peaking within 10 seconds of inhalation. However, the acute effects of nicotine dissipate in a few minutes, as do the associated feelings of reward, which causes the smoker to continue dosing to maintain the drug's pleasurable effects and prevent withdrawal.

Nicotine withdrawal symptoms include irritability, craving, cognitive and attentional deficits, sleep disturbances, and increased appetite. These symptoms may begin within a few hours after the last cigarette, quickly driving people back to tobacco use. Symptoms peak within the first few days of smoking cessation and may subside within a few weeks. For some people, however, symptoms may persist for months.

While withdrawal is related to the pharmacological effects of nicotine, many behavioral factors can also affect the severity of withdrawal symptoms. For some people, the feel, smell, and sight of a cigarette and the ritual of obtaining, handling, lighting, and smoking the cigarette are all associated with the pleasurable effects of smoking and can make withdrawal or craving worse. While nicotine gum and patches may alleviate the pharmacological aspects of withdrawal, cravings often persist. Other forms of nicotine replacement, such as inhalers, attempt to address some of these other issues, while behavioral therapies can help smokers identify environmental triggers of withdrawal and craving so they can employ strategies to prevent or circumvent these symptoms and urges.

Are there other chemicals that may contribute to tobacco addiction?

Yes, research is showing that nicotine may not be the only psychoactive ingredient in tobacco. Using advanced neuroimaging technology, scientists can see the dramatic effect of cigarette smoking on the brain and are finding a marked decrease in the levels of monoamine oxidase (MAO), an important enzyme that is responsible for the breakdown of dopamine. This change is likely caused by some tobacco smoke ingredient other than nicotine, since we know that nicotine itself does not dramatically alter MAO levels. The decrease in two forms of MAO (A and B) results in higher dopamine levels and may be another reason that smokers continue to smoke—to sustain the high dopamine levels that lead to the desire for repeated drug use.

Recently, NIDA-funded researchers have shown in animals that acetaldehyde, another chemical constituent of tobacco smoke, dramatically increases the reinforcing properties of nicotine and may also contribute to tobacco addiction. The investigators further report that this effect is age-related, with adolescent animals displaying far more sensitivity to this reinforcing effect, suggesting that the brains of adolescents may be more vulnerable to tobacco addiction.

What are the medical consequences of tobacco use?

Cigarette smoking kills an estimated 440,000 U.S. citizens each year—more than alcohol, cocaine, heroin, homicide, suicide, car accidents, fire, and AIDS combined. Since 1964, more than 12 million Americans have died prematurely from smoking, and another 25 million U.S. smokers alive today will most likely die of a smoking-related illness.

⑧

■ Research report for a more general, non-technical audience in newsletter format

Features of a Research Report *(continued)*

⑦ The questions address issues of relevance to everyday readers, not just medical experts.

⑧ Simple contrast of one color with black ink is visually appealing.

Glossary

Addiction: A chronic, relapsing disease characterized by compulsive drug seeking and abuse and by long-lasting neurochemical and molecular changes in the brain.

Adrenal glands: Glands located above each kidney that secrete hormones, e.g., adrenaline.

Craving: A powerful, often uncontrollable desire for drugs.

Dopamine: A neurotransmitter present in regions of the brain that regulate movement, emotion, motivation, and feelings of pleasure.

Emphysema: A lung disease in which tissue deterioration results in increased air retention and reduced exchange of gases. The result is difficulty breathing and shortness of breath.

Hyperglycemic: The presence of an abnormally high concentration of glucose in the blood.

Neurotransmitter: A chemical that acts as a messenger to carry signals or information from one nerve cell to another.

Nicotine: An alkaloid derived from the tobacco plant that is responsible for smoking's psychoactive and addictive effects.

Pharmacokinetics: The pattern of absorption, distribution, and excretion of a drug over time.

Rush: A surge of euphoria that rapidly follows administration of some drugs.

Tobacco: A plant widely cultivated for its leaves, which are used primarily for smoking; the N. tabacum species is the major source of tobacco products.

Withdrawal: A variety of symptoms that occur after chronic use of an addictive drug is reduced or stopped.

NIDA
NATIONAL INSTITUTE
ON DRUG ABUSE

NIH Publication Number 06-4342
Printed July 1998. Reprinted August 2001.
Revised July 2006.
Feel free to reprint this publication.

References

Belluzzi JD, Wang R, Leslie FM. Acetaldehyde enhances acquisition of nicotine self-administration in adolescent rats. Neuropsychopharmacology 30:705–712, 2005.

Benowitz NL. Pharmacology of nicotine: addiction and therapeutics. Ann Rev Pharmacol Toxicol 36:597–613, 1996.

Breslau N. Psychiatric comorbidity of smoking and nicotine dependence. Behav Genet 25:95–101, 1995.

Buka SL, Shenassa ED, Niaura R. Elevated risk of tobacco dependence among offspring of mothers who smoked during pregnancy: a 30-year prospective study. Am J Psychiatry 160:1978–1984, 2003.

Centers for Disease Control and Prevention (CDC). Morbidity and Mortality Weekly Report (MMWR) 49(33):755–758, 2000.

CDC. CDC Surveillance Summaries, June 9, 2000. MMWR 49, SS-5, 2000.

CDC. Projected Smoking-Related Deaths Among Youth—United States. MMWR 45:971–974, 1996.

Evins AE, Mays VK, Rigotti NA, Tisdale T, Cather C, Goff DC. A pilot trial of bupropion added to cognitive behavioral therapy for smoking cessation in schizophrenia. Nicotine Tob Res 3:397–403, 2001.

Fowler JS, Volkow ND, Wang GJ, Pappas N, Logan J, MacGregor R, Alexoff D, Shea C, Schlyer D, Wolf AP, Warner D, Zezulkova I, Cilento R. Inhibition of monoamine oxidase B in the brains of smokers. Nature 22:733–736, 1996.

Giovino GA, Henningfield JE, Tomar SL, Escobedo LG, Slade J. Epidemiology of tobacco use and dependence. Epidemiol Rev 17(1):48–65, 1995.

Henningfield JE. Nicotine medications for smoking cessation. New Engl J Med 333:1196–1203, 1995.

Hughes JR. The future of smoking cessation therapy in the United States. Addiction 91:1797–1802, 1996.

Lasser K, Boyd JW, Woolhandler S, Himmelstein DU, McCormick D, Bor DH. Smoking and mental illness. A population-based prevalence study. JAMA 284:2606–2610, 2000.

Lynch BS, Bonnie RJ, eds. Growing up tobacco free. Preventing nicotine addiction in children and youths. Committee on Preventing Nicotine Addiction in Children and Youths. Division of Biobehavioral Sciences and Mental Disorders, Institute of Medicine, 1995.

Perkins KA et al. Cognitive-behavioral therapy to reduce weight concerns improves smoking cessation outcome in weight-concerned women. J Consult Clin Psychol 69(4):604–613, 2001.

Martin WR, Van Loon GR, Iwamoto ET, Davis L, eds. Tobacco smoking and nicotine. New York: Plenum Publishing, 1987.

National Institute on Drug Abuse. Monitoring the Future, National Results on Adolescent Drug Use, Overview of Key Findings 2005. NIH Pub. No. 01-4923, 2005.

Pomerleau OF, Collins AC, Shiffman S, Pomerleau CS. Why some people smoke and others do not: new perspectives. J Consult Clin Psychol 61:723–731, 1993.

The Smoking Cessation Clinical Practice Guideline Panel and Staff. The Agency for Health Care Policy and Research smoking cessation clinical practice guidelines. JAMA 275:1270–1280, 1996.

Stepanov I, Jensen J, Hatsukami D, Hecht SS. Tobacco-specific nitrosamines in new tobacco products. Nicotine Tob Res 8:309–313, 2006.

Substance Abuse and Mental Health Services Administration. Results from the 2004 National Survey on Drug Use and Health: National Findings. DHHS Pub. No. SMA 05-4062, 2005.

U.S. Department of Health and Human Services. Reducing Tobacco Use: A Report of the Surgeon General. Atlanta, Georgia: U.S. Department of Health and Human Services, Centers for Disease Control and Prevention, National Center for Chronic Disease Prevention and Health Promotion, Office on Smoking and Health, 2000.

U.S. Department of Health and Human Services. The Health Consequences of Smoking: A Report of the Surgeon General. Atlanta, Georgia: U.S. Department of Health and Human Services, Centers for Disease Control and Prevention, National Center for Chronic Disease Prevention and Health Promotion, Office on Smoking and Health, 2004.

Access information on the Internet

- What's new on the NIDA Web site
- Information on drugs of abuse
- Publications and communications (including NIDA Notes)
- Calendar of events
- Links to NIDA organizational units
- Funding information (including program announcements and deadlines)
- International activities
- Links to related Web sites (access to Web sites of many other organizations in the field)

■ Research report for a more general, non-technical audience in newsletter format

Features of a Research Report *(continued)*

9 Even in a report written for a nontechnical audience, it is important for any research report to have complete citation and reference information.

10 The use of a glossary makes the research terms accessible to a wide audience.

11 Box colors that use shades within the same color palette help organize text in a visually effective way.

12 Reverse type makes the organization logo and access links stand out on the bottom of the page.

Technical Report Documentation Page

1. Report No. MN/RC-2006-15	2.	3. Recipients Accession No.	
4. Title and Subtitle Investigation of the Low-Temperature Fracture Properties of Three MnROAD Asphalt Mixtures		5. Report Date May 2006	
		6.	
7. Author(s) Xinjun Li, Adam Zofka, Xue Li, Mihai Marasteanu, Timothy R. Clyne		8. Performing Organization Report No.	
9. Performing Organization Name and Address Department of Civil Engineering University of Minnesota 500 Pillsbury Dr. S.E. Minneapolis, MN 55455		10. Project/Task/Work Unit No.	
		11. Contract (C) or Grant (G) No. (c) 81655 (wo) 108	
12. Sponsoring Organization Name and Address Minnesota Department of Transportation Office of Research Services 395 John Ireland Boulevard Mail Stop 330 St. Paul, Minnesota 55155		13. Type of Report and Period Covered Final Report	
		14. Sponsoring Agency Code	

15. Supplementary Notes

http://www.lrrb.org/PDF/200615.pdf

16. Abstract (Limit: 200 words)

In this research effort, field cores were taken from cells 33, 34 and 35 at the MnROAD facility to determine the fracture properties of the field mixtures, to compare them with the laboratory-prepared mixtures analyzed in a previous study, and to evaluate the effect of aging at different depths in the asphalt layer. In addition, the properties of the recovered binders from the field cores as well as the properties of the original binders aged in laboratory conditions were investigated.

The test results and the analyses performed indicate that the fracture tests performed on asphalt binders and asphalt mixtures have the potential to predict the field performance of asphalt pavements with respect to thermal cracking. The binder results confirm the predictions of the current performance grading system; however, it appears that the fracture resistance of the PG-34 asphalt mixture is better than the fracture resistance of the PG-40 mixtures, which is the opposite of what the PG system predicts.

| 17. Document Analysis/Descriptors
asphalt binder, asphalt mixture, fracture energy, fracture toughness, PG system, aging, Superpave, field performance, low temperature cracking | | 18. Availability Statement
No restrictions. Document available from: National Technical Information Services, Springfield, Virginia 22161 | |
| 19. Security Class (this report)
Unclassified | 20. Security Class (this page)
Unclassified | 21. No. of Pages
74 | 22. Price |

■ Research report for a more technical audience

Features of a Research Report, Technical Audience

1 This report is required by the organization (Department of Transportation) to begin with a technical documentation page.

INVESTIGATION OF THE LOW-TEMPERATURE FRACTURE PROPERTIES OF THREE MnROAD ASPHALT MIXTURES

Final Report

Prepared by:

Xinjun Li
Adam Zofka
Xue Li
Mihai O. Marasteanu
Timothy R. Clyne

University of Minnesota
Department of Civil Engineering

May 2006

Published by:

Minnesota Department of Transportation
Research Services Section, MS 330
395 John Ireland Boulevard
St. Paul, MN 55155

■ Research report for a more technical audience

Features of a Research Report, Technical Audience *(continued)*

2 A simple but effective title page includes essential author information and required statements.

ACKNOWLEDGEMENTS

The authors would like to thank Jim McGraw and Roger Olson, at the Minnesota Department of Transportation (Mn/DOT) for their technical assistance during the project. Their guidance and assistance is greatly appreciated. The authors would also like to thank Ben Worel and the staff at the MnROAD facility for providing the pavement cores used in the research performed in this study.

3

■ Research report for a more technical audience

Features of a Research Report, Technical Audience *(continued)*

3 A long, technical research report is never developed by one writer acting alone. Many people contribute, and a brief acknowledgments statement provides a way to thank those who contributed but may not be named on the author page.

TABLE OF CONTENTS

■ Research report for a more technical audience

Features of a Research Report, Technical Audience *(continued)*

④ A detailed table of contents helps readers of long research reports navigate to sections they want to read.

LIST OF TABLES

■ Research report for a more technical audience

Features of a Research Report, Technical Audience *(continued)*

⑤ Lists of tables and figures give readers an overview of the document and help them navigate through the material.

LIST OF FIGURES

⑤

■ Research report for a more technical audience

Features of a Research Report, Technical Audience *(continued)*

⑤ Lists of tables and figures give readers an overview of the document and help them navigate through the material.

EXECUTIVE SUMMARY

In a previous research effort sponsored by MnDOT, the fracture properties of asphalt mixtures used in the construction of cells 33, 34 and 35 of the MnROAD facility were determined using a new experimental protocol based on a Semi Circular Bend (SCB) test. In recent years, a number of research efforts have shown that the field compacted asphalt mixture samples are different than the Superpave Gyratory Compactor (SGC) laboratory compacted asphalt mixture specimens. In addition, the effects of aging play a critical role in the fracture resistance of field asphalt mixtures.

In this research effort, field cores were taken from cells 33, 34 and 35 at the MnROAD facility to determine the fracture properties of the field mixtures, to compare them with the laboratory-prepared mixtures, and to evaluate the effect of aging at different depths in the asphalt layer. In addition, the binders were extracted and recovered from the 25 mm slices cut along the depth of the cores. The properties of the recovered binders as well as the properties of the original binders aged in laboratory conditions were investigated, using standard testing procedures part of the current specifications as well as additional test methods.

The analysis of the mixture experimental results indicates that the fracture energy of the field samples is significantly affected by the temperature and the type of binder used, while the fracture toughness of the field samples is significantly affected only by the type of binder used. The comparison between the field samples and the laboratory specimens shows that the fracture energy of the field samples is lower than the fracture energy of the laboratory specimens and the fracture toughness of the field samples is lower than the fracture energy of the laboratory specimens. However, the evolutions of the fracture toughness and fracture energy with type of binder and temperature are similar for the field samples and laboratory specimens. The difference in properties with the location within the pavement does not indicate a consistent pattern although in some cases the surface seemed to be less cracking-resistant (possibly indicating more aging) than the middle and the bottom layer.

The analysis of the experimental data obtained from tests performed on the binders recovered from cores and on the laboratory-aged original binders indicates that the properties of the field binders are different than the laboratory-aged binders. In particular, significant change is observed in the m-value limiting temperatures, which clearly indicates a substantial change in the relaxation properties of the recovered binder compared to the laboratory-aged binder in spite of the less significant change in stiffness. The difference in properties with the location within the pavement did not follow a consistent pattern although in some cases the surface seemed to be stiffer (possibly indicating more aging) than the middle and the bottom layer.

The limited low temperature distress data collected at MnROAD indicates the fracture tests performed on asphalt binders and asphalt mixtures have the potential to predict the field performance of asphalt pavements with respect to thermal cracking. The binder and mixture fracture toughness results indicate that cells 34 and 35 have better fracture resistance than cell 33, and the mixture results suggest that cell 34 has better resistance than cell 35, which is the opposite of what the PG system predicts.

■ Research report for a more technical audience

Features of a Research Report, Technical Audience (*continued*)

6 An executive summary describes the background, research process, analysis, and results. For many busy managers and others who are not interested in all the data, the executive summary might be all they read of the report.

CHAPTER 1

INTRODUCTION

Background

    In a previous research effort sponsored by MnDOT, the fracture properties of asphalt mixtures used in the construction of cells 33, 34 and 35 of the MnROAD facility were determined using a new experimental protocol based on a Semi Circular Bend (SCB) test. In this previous study the original loose mix was compacted to the desired parameters in the laboratory using the Superpave gyratory compactor (SGC), and 1" thick test specimens were cut from the SGC specimen.

    In recent years, a number of research efforts have shown that the field-compacted asphalt mixture samples are different than the SGC laboratory-compacted asphalt mixture specimens. In addition, the effects of aging play a critical role in the fracture resistance of field asphalt mixtures. To date, there is little agreement with respect to the magnitude of the depth, from the asphalt pavement surface, at which aging effects are still significant.

    In this research effort, field cores were taken from cells 33, 34 and 35 at the MnROAD facility to determine the fracture properties of the field mixtures and to evaluate the effect of aging with depth. To better understand the relationship between the properties of these mixtures and the field performance, the historical data on these three cells needs to be documented.

    Chapter 1 documents the history and performance of the aforementioned MnROAD cells. Most of the information contained in this task was provided by Ben Worel, with assistance from Ron Mulvaney and other MnROAD staff.

MnROAD Layout

    The Minnesota Department of Transportation (Mn/DOT) constructed the Minnesota Road Research Project (MnROAD) between 1990 and 1994. MnROAD is an extensive pavement research facility consisting of two separate roadway segments: the Mainline Test Road (Mainline) and Low Volume Road (LVR), containing a total of 51 distinct test cells. Each MnROAD test cell is approximately 500 feet long. The subgrade, aggregate base, and surface materials as well as the roadbed structure and drainage methods vary from cell to cell. The layout and the designs used for the cells part of the LVR are shown in Figure 1.1.

1

■ Research report for a more technical audience

Features of a Research Report, Technical Audience *(continued)*

7 Each chapter begins with a similar heading, providing clear signposts for readers who wish to page through the document.

CHAPTER 2

⑧

Mixture Testing

Testing Equipment and Setup

A MTS servo-hydraulic testing system was used to perform the Semi Circular Bend (SCB) tests to determine the low temperature fracture properties of the field mixtures. The SCB test is similar to the three-point bending beam test except that the SCB specimen is semicircular instead of a beam. The SCB specimen is symmetrically supported by two rollers and has a span of 120mm. Teflon paper strips were placed between the specimen and the rollers to reduce the friction on the interface. The Indirect Tension test (IDT) loading plate was used to load the SCB specimens. The load line displacement (LLD) was measured using a vertically mounted Epsilon extensometer with 38 mm gauge length. The crack mouth opening displacement (CMOD) was measured by an Epsilon clip gauge with 10 mm gauge length. The test setup is shown in Figure 2.1.

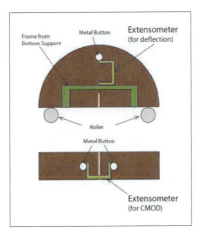

Figure 2.1. Experimental Setup

All tests were performed inside an environmental chamber. Liquid nitrogen was used to cool down the temperature. The temperature was monitored by a build-in thermocouple and an

---

Features of a Research Report, Technical Audience *(continued)*

⑧ Each chapter opens with a subheading followed by other sub-headings, breaking up the text into a logical flow of ideas.

CHAPTER 3

Asphalt Binder Testing

Introduction

One of the factors that significantly affects the performance of asphalt pavements during their service life is the aging of the asphalt binder used in the construction of the pavement. In particular, the aging process negatively affects the fracture resistance of the asphalt mixtures against low temperature cracking and thermal and traffic induced fatigue cracking. In spite of the considerable advancements made in the past years in understanding the evolution of the aging process and in simulating field aging in laboratory conditions for material specification purposes, there are a number of critical issues that are still not well understood. Among them, the issue of the pavement depth at which aging effects become less significant and the issue of the reasonableness of the laboratory aging methods to simulate field aging, in particular for polymer modified binders, are generating a strong debate in the research community.

Objectives

In the previous task, the fracture properties of asphalt mixtures used in the construction of the MnROAD facility were investigated. Field cores were taken from the cells and 1″ test samples were cut and tested using the semi circular bend (SCB) test to determine the fracture properties of the field mixtures and to evaluate the effect of aging with depth. The tested samples were then used to extract and recover the binders used in their preparation for further investigation.

This task presents the results of this investigation. The rheological properties of the binders extracted and recovered from the three MnROAD cells are compared with the properties of the original binders aged in laboratory conditions, and the effect of the sample location within the pavement on the rheological properties of the extracted binders is evaluated.

Materials

The three binders studied in this paper were obtained from MnRoad cells 33, 34 and 35. The cores were taken from MnROAD approximately 5 years after the construction of the cells. Three

37

■ Research report for a more technical audience

Features of a Research Report, Technical Audience *(continued)*

9 Chapters are of similar length and style.

sweeps from 1 to 100 rad/s were performed over a temperature range from 4°C to 76°C. At lower temperatures (4° to 34°C) the tests were performed on 8-mm plates with a 2.0 mm gap. At high temperatures (34° to 76°C) the tests were performed on 25-mm plates with a 1.0 mm gap.

The bending beam rheometer (BBR) testing was performed on a Cannon thermoelectric rheometer, according to AASHTO T 313 (10). Tests were performed at two temperatures. For the extracted binders testing was performed at higher temperatures than the original PG grade called for. For example, a PG 58-28 binder was tested at -12°C and -18°C instead of -18°C and -24°C.

Direct tension (DT) testing was carried out on a Bohlin Direct Tension with Neslab chilling system, according to AASHTO T 314-02 (11). The binders were tested at the same test temperatures as the BBR tests.

In addition, the DT procedure was modified to perform Double Edge Notch Tension (DENT) tests based on previous work performed by Gauthier and Anderson (12). The DT molds were modified to prepare DENT test specimens according to the geometry shown in Figure 3.1. The modified molds also allowed the use of a razor blade to generate 1.5mm pre-cracks on both sides of the test specimens. The strain rate was lowered to 1.8%/min due to the rapid failure occurring at the lower temperature.

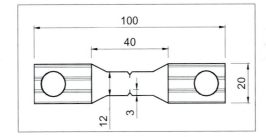

Figure 3.1. Double Edge Notched Tension (DENT) specimen dimensions (mm)

Experimental Data Analysis

The experimental data obtained for the original binders aged in laboratory conditions and for the extracted and recovered binders from the filed cores was used to investigate the effect of aging on the properties of these binders. The change in properties at high and low service temperatures was investigated by calculating limiting temperatures based on the current PG

39

■ Research report for a more technical audience

Features of a Research Report, Technical Audience *(continued)*

10 Unlike the tobacco report, this report is written for a technical audience. The language, technical detail, and use of quantitative formulas and technical diagrams are appropriate for this audience, because they illustrate the method and analysis of the primary research presented by the authors.

specifications, as well as on the new fracture test previously mentioned. In addition, complex modulus and phase angle master curves were generated to investigate the change in behavior over a wider range of frequencies and temperatures.

High Temperature

The DSR data obtained at 10rad/s was used to determine the limiting temperature at which $|G^*|/\sin\delta$ (the absolute value of the complex modulus divided by the sin of the phase angle) equals 2.2 kPa. A number of observations can be made based on the results presented in Table 3.2.

Table 3.2. Temperature Values at which $|G^*|/\sin d = 2.2$kPa, °C

Cell	Condition				
	RTFOT	**PAV**	**Extracted Top**	**Extracted Middle**	**Extracted Bottom**
33	63.4	72.7	75.1	72.6	73.1
34	63.9	73.4	72.2	70.1	70.0
35	66.1	76.5	74.0	68.9	68.5

⑪

For all three cells, the top layer has higher limiting temperatures than the other two layers, which indicates that more aging occurred in this layer. This is in particular true for the PG -40 binder for which the differences were as high as 5.5°C. The differences between the middle and the bottom layers were not significant.

With respect to laboratory versus field aging, the results indicate that for cell 33 the three extracted binders were stiffer than the PAV-aged binder. However, for cells 34 and 35 the extracted binders were softer than the PAV-aged binder. This was more pronounced for cell 35 binder. It is not clear if this trend is related to the presence of modifiers in the PG -34 and -40 binders.

DSR Master Curves

The frequency sweep data from the DSR testing was used to obtain master curves of the complex modulus and phase angle, as shown in Figures 3.2 to 3.7. For comparison purposes all master curves were obtained at a reference temperature of 34°C. A commercial computer program was used to fit the Christensen-Anderson-Marasteanu (CAM) model to the $|G^*|$ test data (13):

40

■ Research report for a more technical audience

Features of a Research Report, Technical Audience (continued)

⑪ Technical tables and equations are appropriate for this audience.

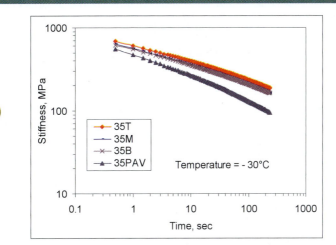

Figure 3.10. BBR stiffness comparisons, Cell 35

Fracture Toughness

The data obtained using the DENT testing procedure described previously was used to calculate the fracture toughness K_{IC} based on the following equation (16):

$$K_{IC} = \frac{P}{B\sqrt{W}} f(a/W)$$

$$f(a/W) = \frac{\sqrt{\frac{pa}{2W}}}{\sqrt{1-\frac{a}{W}}}\left[1.122 - 0.561\left(\frac{a}{W}\right) - 0.205\left(\frac{a}{W}\right)^2 + 0.471\left(\frac{a}{W}\right)^3 + 0.190\left(\frac{a}{W}\right)^4\right] \quad \textbf{(3.1)}$$

where

      P - peak load

      B - specimen thickness (equal to 6mm)

      W - half width of the specimen (equal to 6mm)

      a – notch width (equal to 3mm)

      A comparison of the fracture toughness values for the PAV-aged and the recovered binders is shown in Figure 3.11 to 3.13. Figure 3.11 indicates that for cell 33 there are no

48

CHAPTER 4

Data Analysis

Objectives

In the previous two chapters the fracture properties of the asphalt binders and asphalt mixtures from the three MnROAD cells were determined. In this chapter correlations between the binder and mixture fracture toughness and between the fracture properties and the cells field performance are investigated.

Fracture Toughness Investigation

Fracture tests were performed on the extracted asphalt binders and on the field asphalt mixtures using the DENT and the SCB test procedures, respectively. In addition, DENT tests were also performed on the PAV-aged original binders used in the construction of the three cells. A summary of the fracture toughness results is shown in Table 4.1. For comparison purposes, the values are also plotted in Figures 4.1 to 4.6.

Table 4.1. Toughness values for asphalt binders and mixtures

Source	Binder				Mixture					
	T_1 (°C)	K_{IC} (KPa.m$^{0.5}$)	T_2 (°C)	K_{IC} (KPa.m$^{0.5}$)	T_1 (°C)	K_{IC} (KPa.m$^{0.5}$)	T_2 (°C)	K_{IC} (KPa.m$^{0.5}$)	T_3 (°C)	K_{IC} (KPa.m$^{0.5}$)
33-T		48		67		740		820		780
33-M	-12	63	-18	63	-18	770	-24	910	-30	810
33-B		70		69		770		780		820
34-T		85		88		880		950		900
34-M	-18	123	-24	88	-24	900	-30	970	-36	930
34-B		91		68		900		1020		960
35-T		76		100		800		820		830
35-M	-24	105	-28	120	-24	800	-30	850	-36	970
35-B		113		140		740		860		960
33PAV	-18	67	-24	54						
34PAV	-24	133	-28	132						
35PAV	-28	168								

(12)

53

■ Research report for a more technical audience

Features of a Research Report, Technical Audience (*continued*)

(12) Data analysis is usually followed by the conclusion.

CHAPTER 5

Conclusions and Recommendations

Conclusions

A number of conclusions can be drawn from the research performed in this study. The analysis of the experimental results from the SCB fracture tests performed on the field samples cored from cells 33, 34, and 35 indicated that the fracture energy of the field samples is significantly affected by the temperature and the type of binder used, while the fracture toughness of the field samples is significantly affected only by the type of binder used. The field samples and the laboratory specimens (tested in a previous study) are statistically different with respect to both the fracture toughness and the fracture energy; the fracture energy of the field samples is lower than the fracture energy of the laboratory specimens and the fracture toughness of the field samples is lower than the fracture energy of the laboratory specimens. However, the evolutions of the fracture toughness and fracture energy with type of binder and temperature are similar for the field samples and laboratory specimens. The difference in properties with the location within the pavement did not indicate a consistent pattern although in most cases the surface seemed to be less crack-resistant (possibly indicating more aging) than the middle and the bottom layer.

The analysis of the experimental data obtained from tests performed on the binders recovered from cores taken out of MnRoad cells 33, 34 and 35 and on the laboratory-aged original binders used in the construction of the three cells indicated that the properties of the field binders are different than the laboratory-aged binders. In particular, significant differences are noticed in the phase angle master curves. It is not clear if this is a consequence of the degradation of the polymer network due to aging or due to the recovery and extraction process. At low temperatures, the most significant change, 15.2°C, is observed in the m-value limiting temperatures for the PG -40 binder, which clearly indicates a substantial change in the relaxation properties of the recovered binder compared to the laboratory-aged binder in spite of the less significant change in stiffness. The difference in properties with the location within the pavement did not indicate a consistent pattern although in most cases the surface seemed to be stiffer (possibly indicating more aging) than the middle and the bottom layer.

60

■ Research report for a more technical audience

REFERENCES

(1) Lim, I. L., Johnston, I. W. and Choi, S.K., "Stress Intensity factors for Semi-Circular Specimens under Three-Point Bending," *Engineering Fracture Mechanics*, vol. 44, no. 3, (1993), 363-382.

(2) RILEM Technical Committee 50-FMC, "Determination of the Fracture Energy of Mortar and Concrete by Means of Three-point Bend Tests on Notched Beams," *Materials and Structures*, no. 106, (Jul-Aug 1985), 285-290.

(3) Li, X., Marasteanu, M.O., "Evaluation of the Low Temperature Fracture Resistance of Asphalt Mixtures Using the Semi Circular Bend Test," *Journal of the Association of Asphalt Paving Technologists*, vol. 73, (2004), 401-426.

(4) Li, X. *Investigation of the Fracture Resistance of Asphalt Mixtures at Low Temperatures with a Semi Circular Bend (SCB) Test.* (Ph.D. Thesis, University of Minnesota, 2005).

(5) American Association of State Highway and Transportation Officials (AASHTO), *Standard Method of Test for Quantitative Extraction of Asphalt Binder from Hot Mix Asphalt*, AASHTO Designation: T 164-05i, ASTM Designation: D 2172-01.

(6) Strategic Highway Research Program, *Extraction and recovery of asphalt cement for rheological testing*, SHRP-A-370, Binder Characterization and Evaluation, Vol. 4: Test Methods, Appendix E, National Research Council, Washington, DC 1994.

(7) American Association of State Highway and Transportation Officials (AASHTO), *Standard Method of Test for Effect of Heat and Air on a Moving Film of Asphalt (Rolling Thin-Film Oven Test)*, Designation T 240-00, "Standard Specifications for Transportation Materials and Methods of Sampling and Testing, Part 2B: Tests," 22nd Edition, 2002.

(8) American Association of State Highway and Transportation Officials (AASHTO), *Standard Practice for Accelerated Aging of Asphalt Binder Using a Pressurized Aging Vessel (PAV)*, Designation R 28-02, "Standard Specifications for Transportation Materials and Methods of Sampling and Testing, Part 1B: Specifications," 22nd Edition, 2002.

(9) American Association of State Highway and Transportation Officials (AASHTO), *Standard Method of Test for Determining the Rheological Properties of Asphalt Binder*

62

■ Research report for a more technical audience

Features of a Research Report, Technical Audience *(continued)*

13 This technical research report uses a numbered citation style, common in engineering and science.

Test Report

Test reports are used to convey the results of very specific, focused tests. These tests may be conducted on a new product design, a feature in a software application, an experimental medical procedure, or a new safety feature. Some test reports are very short, especially if the questions that generated the test were quite specific. Other test reports can be long if the intent of the test was to study a large set of questions, especially if these questions were looked at over any period of time. The following sample is part of a long test report from the U.S. Consumer Product Safety Commission, reporting on the testing of several types of swimming pool alarms.

EXECUTIVE SUMMARY

On average each year 350 children under the age of five years drown in swimming pools, with most deaths occurring in residential settings. Also each year, on average, another 2,600 children under five years of age are treated in hospital emergency rooms for near-drowning incidents in swimming pools. About 79 percent of these incidents occur at a home location. These numbers have remained relatively unchanged over the past several years.

During 1999 and early 2000, the U.S. Consumer Product Safety Commission (CPSC) staff conducted a review of commercially available swimming pool alarm systems designed to detect water disturbance or displacement. There are no voluntary standards that define applicable performance requirements for disturbance or displacement type products.

The CPSC staff evaluated four water disturbance alarms and a wristband. Two of the disturbance systems used surface wave detection circuitry, while the other two detected subsurface disturbances. The fifth device was a wristband (to be worn by a child) intended to alarm when exposed to water. All of the products incorporated remote alarm receivers, some at an additional cost.

Test results showed that the subsurface pool alarms generally performed better. They were more consistent in alarming and less likely to false alarm than the surface alarms. When a test object, intended to simulate the weight of a small child, was pushed into the pool, the subsurface sensors detected it most reliably. The subsurface alarms can also be used in conjunction with solar covers, whereas the surface alarms cannot.

One surface alarm performed almost as well as the subsurface alarms. The wristband alarmed when submerged in pool water or exposed to another water source, such as tap water.

ii

■ Test report on safety of pool alarms

Features of a Test Report

1 Because this test report was prepared for the federal government, it tends to be more formal and includes all parts, such as an executive summary and table of contents. A more brief report, for use inside a company, may not include these parts.

TABLE OF CONTENTS

■ Test report on safety of pool alarms

INTRODUCTION

Each year, on average, 350 children under five years of age drown in swimming pools, with most deaths occurring in residential settings. Also each year, on average, 2,600 children under five years of age are treated in hospital emergency rooms for near-drowning incidents in swimming pools. About 42 percent of the incidents require hospitalization of the child. About 79 percent of the near-drowning incidents occur in a residential setting. These numbers have remained relatively unchanged for the past several years (see graph below).

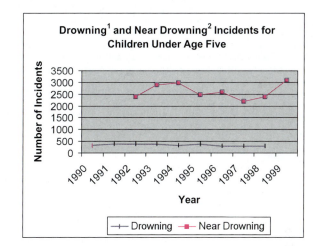

Drowning[1] and Near Drowning[2] Incidents for Children Under Age Five

[1]Rounded to the nearest ten – CPSC Death Certificate and Injury or Potential Injury Incident Files, 1990-1998
[2]CPSC National Electronic Injury Surveillance System (NEISS), 1992-1999

During 1999 and early 2000, the U.S. Consumer Product Safety Commission (CPSC) staff conducted a review of commercially available swimming pool alarm systems designed to address pool-related injuries and drowning. A similar assessment was conducted in 1987[3]. Product designs have not changed significantly from that time.

The staff's review focused on water intrusion alarm systems designed to sense disturbance or displacement of the pool water. In addition, the staff assessed a wristband that alarms remotely when exposed to water.

MARKET INFORMATION

Water intrusion devices including a remote alarm are priced from $149 to $200 for surface wave sensor alarms and between $190 and $250 for subsurface disturbance sensor alarms. The wristband alarm system, an intrusion detector recently introduced, costs $179 for one wristband and the remote alarm. It is estimated that about 24,000 pool alarms are sold annually and that sales have doubled since 1994.

[3] Memo from Ronald L. Medford to the Commission, Contract Report on Testing of Swimming Pool Alarms, April 21, 1987.

1

■ Test report on safety of pool alarms

Features of a Test Report *(continued)*

❷ An introduction and overview of the various types of alarm systems is included. Since this is a government report and widely accessible on a web site, background information is important. Not all test reports contain this much background information.

According to recent estimates by the National Spa and Pool Institute (NSPI), there are about 7 million residential pools in the continental United States. Approximately 3.8 million pools are inground (constructed of gunite, vinyl, or fiberglass), and 3.2 million are above ground. Pool sales are from 120,000 to170,000 annually for inground pools and 300,000 to almost 500,000 for above ground pools, with the majority of inground pools being of gunite construction.

❸ ALARM SYSTEM TECHNOLOGIES

There are no voluntary standards addressing performance or design requirements for pool water disturbance or displacement type alarm systems. Staff evaluated four different alarm systems and a wristband with remote alarm to determine their effectiveness in detecting water intrusions. Two of the displacement/disturbance devices detected surface waves while the other two detected subsurface disturbances. A fifth device was a wristband (to be worn by a child) which detected exposure to water. All of the products incorporated remote alarm receivers.

All of the alarm systems required 9-Volt batteries for operation. The remote alarm receivers required a 120 Volt outlet (step-down transformers were supplied with the power cords). One product offered the option of operating on a 9-Volt battery as well as on a 120-Volt source, which would provide continued operation during a power outage.

Surface Wave Sensors

Two surface wave sensors were evaluated. This type of sensor floats on the water surface. The sensor incorporates an electrical circuit that includes two contacts. One contact rests in the water; the second contact (above-surface contact) is adjusted so that it is resting above the water. When the above-surface contact is touched by water (from a surface wave), the electrical circuit is completed and an alarm sounds. The sensitivity of the device can be adjusted by positioning the above-surface contact closer or further away from the water surface. Sensitivity increases as the contact is positioned closer to the water surface (see Figures 1 and 2).

Subsurface Disturbance Sensors

Two subsurface disturbance sensors were evaluated. These sensors mount on the side of a pool, with portions of the devices being located 1/2 to 12 inches below the water surface. Each device relies on a wave-induced pressure change to activate alarm circuitry. One type of sensor (see Figure 3) uses a pressure-sensitive switch located at the top of a sensing throat. Water movement creates pressure changes within the sensing throat, which activates a switch to initiate an alarm. The other device relies on movement of a magnetic float below a magnetic sensor to create a signal that activates the alarm (see Figure 4).

Both sensors use electrical adjustments to control circuit response to stimuli. One of the subsurface sensors also uses a mechanical adjustment for the depth of the sensor to increase or decrease sensitivity. Sensitivity increases when the sensor is located closer to the water surface; the device is less responsive when it is placed farther below the surface.

2

■ Test report on safety of pool alarms

Features of a Test Report *(continued)*

❸ This page and the pages that follow describe the test in detail (facilities, methods used, testing conditions, and all procedures). Subheadings organize these procedures clearly throughout the report.

Wristband

A wristband with a remote alarm was also evaluated (see Figure 5). The wristband must be placed on a child's arm by a parent or caregiver; a locking key prevents the child from removing the band. When the sensor on the band becomes wet, the remote alarm is activated, warning the parent that the wristband has been exposed to water. There are no sensitivity adjustments for the wristband. The wristband is battery operated, and the unit is sealed. Once the battery is discharged, the wristband sensor must be replaced. The user must be cognizant of the receiver "turtle" color and be sure the replacement wristband matches that color to insure proper alarm operation.

TEST FACILITIES

Alarm tests were conducted at six different pool sites. The pools were both indoors and outdoors and differed in size, shape, and depth. The reason for different test sites was to determine whether the shape and depth of the pool had any influence on the effectiveness of the alarms. The pools varied in width from 14 feet to 30 feet and length from 35 feet to 50 feet. All but one pool were of the "spoon" design, starting at a depth of 3-4 feet and going to a depth of 6 to 10 feet. The one exception was an indoor/outdoor pool of the "hopper" design where both ends were 3½ feet deep and the depth at the center of the pool was 6 feet.

TEST METHODOLOGY

The manufacturers' instruction manuals were reviewed prior to testing. Alarm systems were assembled and placed around the test pools (diagrams provided in appendices) according to the instructions.

The surface wave sensors (labeled A[*] and B in the figures of appendices A-F) were tied to tethers (between two and six feet long) and centrally located along the perimeter of the pools. An effort was made to maintain a distance of at least six inches between the sensor and the pool wall, according to the manufacturers' instructions. The subsurface disturbance sensors (labeled C[*] and D[*]) were placed at the deep end of the pools. The sensitivity of each unit was adjusted during preliminary testing to obtain the best performance results. The remote alarm function for each sensor was also tested.

The systems were tested under two conditions:

- Detection

  - To determine whether the surface and subsurface wave sensors would alarm when a test object entered the pool, and

  - To determine whether the wristband would alarm when exposed to pool water.

---

[*] A=PoolSOS/Allweather Inc., C=Poolguard/PBM Industries, D=Sentinel LINK/Lambo Products Inc.

7

TEST RESULTS – DETECTION

First Round Testing – Pool 1 (Appendix A)

This test facility was a rectangular shaped pool measuring 22'x 42' with various alcoves molded into the gunite. The pool depth ranged from 3 1/2 – 4 feet on the shallow end up to 8 feet at the opposite end. A schematic of the pool, along with the locations of the various alarms can be found in Appendix A. As this was the first pool tested, and there are currently no performance requirements, the manufacturer-recommended three-gallon object was used for testing.

From Location "X" (Shallow End)

The first surface wave sensor, labeled A (Sensor A), alarmed in 6 out of 10 trials, with response times ranging from 16.1 seconds to 23.9 seconds. The second surface wave sensor, Sensor B, failed to alarm after two minutes in all trials. (Sensor B was set to its most sensitive detection position.)

Both subsurface wave sensors alarmed 100 percent of the time. Response times for Sensor C ranged from 10.4 seconds to 76.3 seconds. Response times for Sensor D ranged from 13.3 seconds to 116.6 seconds. Detection results and individual trial response times are shown in Tables A1 and A2 of Appendix A.

From Location "Y" (Mid-point)

Sensor A alarmed in 6 out of 10 trials, with response times ranging from 4.0 seconds to 25.1 seconds. Sensor B alarmed in 1 of 10 trials, with a response time of 5.7 seconds.
Sensor C alarmed in all 10 trials, with response times ranging from 7.7 seconds to 10.4 seconds. Sensor D alarmed in 7 out of 10 trials. Response times ranged from 7.8 seconds to 13.4 seconds. Detection results and individual trial response times are shown in Tables A3 and A4 of Appendix A.

Wristband

The wristband alarmed immediately upon submersion into the pool water. Once the wristband was completely dried, it was tested again. It alarmed immediately upon submersion.

Subsequent Testing – Pools 2-6 (Appendices B-F)

Pool 2 (Appendix B)

This test facility was a rectangular pool measuring 18'x 36' with a steel and concrete shell covered with a vinyl liner. The pool was unique in that the depth ranged from 3 1/2 feet on either end up to 6 feet in the middle, a "hopper" design rather than the typical "spoon" design, shallow on one end and gradually deeper towards the opposite end. A schematic of the pool, along with the locations of the various alarms can be found in Appendix B.

From Location "X" (Right-side of Entrance Steps)

Surface wave sensor A alarmed in 5 out of 5 trials using the three-gallon test object and 5 out of 5 trials using the two-gallon test object. The response times ranged from 19.1 to 22.4 seconds

11

for the three-gallon object and from 21.3 to 24.7 seconds for the two-gallon object. The second surface wave sensor, Sensor B, responded in 5 out of 5 trials for the three-gallon test object and 4 out of 5 trials for the two-gallon test object. The response times were from 7.59 to 9.71 seconds and 9.86 to 12.18 seconds respectively.

Both subsurface disturbance sensors alarmed 100 percent of the time. Response times for Sensor C ranged from 15.1 to 15.7 seconds with the three gallon test object and 15.0 to 19.6 seconds when the two-gallon test object was used. Response times for Sensor D ranged from 15.7 to 20.5 seconds and 20.8 to 23.5 seconds respectively.

From Location "Y" (Left-side of Entrance Steps)

Using the three-gallon object, Sensor A alarmed in 5 out of 5 trials, with response times ranging from 17.8 to 40.7 seconds. Sensor B alarmed in 3 of 5 trials, with a response time of 12.6 to 13.3 seconds. With the two-gallon object, Sensor A alarmed in 5 out of 5 trials, as did Sensor B. The response times ranged from 11.3 to 20.7 seconds for Sensor A and 11.3 to 14.4 seconds for Sensor B.

Subsurface sensors C and D alarmed in all 5 trials using each object. The response times ranged from 7.4 to 12.1 seconds and 16.9 to 17.6 seconds with the three-gallon object. For the two-gallon object, the times ranged from 13.0 to 20.0 seconds and 16.6 to 27.6 seconds respectively. A summary of the detection results and individual trial response times is shown in Appendix B.

Pool 3 (Appendix C)

This test facility was a rectangular pool measuring approximately 14'x 40' and appeared to be a typical gunite shell construction. The pool ranged from 4 feet at the entrance steps to 6 feet at the opposite end. There was a small, built-in hot tub at the shallow end adjacent to the steps. A schematic of the pool, along with the locations of the various alarms and the test results can be found in Appendix C. Due to the size and shape of the pool, and the results of the first test trials, presumed to be the worst-case scenario, there was only one drop point used for the test objects.

From Location "X" (Shallow End)

Surface wave sensors A and B alarmed in all 10 trials using both the three-gallon and the two-gallon test objects. For Sensor A, the response times ranged from 7.1 to 29.8 seconds for the three-gallon object and from 7.4 to 27.0 seconds for the two-gallon object. Sensor B response times were from 9.7 to 15.1 seconds and 8.8 to 12.8 seconds, respectively.

Both subsurface sensors alarmed in all 10 trials as well. Response times for Sensor C ranged from 5.8 to 15.7 seconds with the three-gallon test object and 8.8 to 22.1 seconds when the two-gallon test object was used. Response times for Sensor D ranged from 15.9 to 19.0 seconds and 16.6 to 20.6 seconds, respectively.

Pool 4 (Appendix D)

This test facility was a rectangular pool measuring 30'x 50' and appeared to be a typical gunite shell construction. The pool ranged in depth from 4 feet at the entrance steps to 10 feet at the opposite end. A schematic of the pool, along with the locations of the various alarms and the test results can be found in Appendix D. Due to the pool size, two surface Sensor B alarms were

12

Test report on safety of pool alarms

used. As a result of the first test trials, presumed to be a worst-case scenario, and the use of two Sensor B units (a second Sensor A was not available), there was only one drop point used for the test objects.

From Location "X" (Shallow End)

Surface wave Sensor A alarmed in all 10 trials using the three-gallon test object and in 9 out of 10 trials using the two-gallon test object. For Sensor A, the response times ranged from 25.0 to 33.4 seconds for the three-gallon object and from 14.2 to 54.6 seconds for the two-gallon object. For the two Sensor B units, there was a single response during the 10 trials using the three-gallon object. The response time was from the unit located farthest from the test object entrance point and occurred over one minute after the test object was introduced into the pool. There were two responses from each unit when using the two-gallon test object. The response times were 7.2 and 54.1 seconds from the farthest unit and 33.2 and 53.2 seconds from the nearer unit.

Both subsurface sensors alarmed in all 10 trials involving the three-gallon object. Response times varied from 10.1 to 24.5 seconds for Sensor C and from 24.8 to over two minutes for Sensor D. For the two-gallon object, Sensor C alarmed in 8 out of 10 trials with response times ranging from 17.3 to 43.0 seconds while Sensor D responded in 7 out of 10 trials with response times between 31.0 and 49.3 seconds.

Pool 5 (Appendix E)

This test facility was a rectangular pool measuring 17.5'x 26' with additional 6 foot semi-circles on either end of the rectangle. The pool appeared to be a typical gunite shell and ranged in depth from 3 feet at the stairs to 8 feet at the opposite end. A schematic of the pool, along with the locations of the various alarms and the test results can be found in Appendix E. Due to the pool size and alarm locations, there was only one drop point used for the test objects.

From Location "X" (Shallow End)

Surface wave Sensor A alarmed in all 10 trials using the three-gallon test object and in 9 out of 10 trials using the two-gallon test object. The response times ranged from 14.1 to 21.4 seconds for the three-gallon object and from 15.1 to 20.2 seconds for the two-gallon object. The Sensor B unit responded once during the 10 trials using the three-gallon test object. The response time was 23.0 seconds. There were four responses when using the two-gallon test object. The response times ranged from were 17.1 to 23.6 seconds.

Both subsurface wave sensors alarmed in all 10 trials involving both the three-gallon and the two-gallon test objects. Response times with the three-gallon test object varied from 6.3 to 10.6 seconds for Sensor C and from 9.4 to 18.1 seconds for Sensor D. For the two-gallon object, Sensor C response times were from 6.9 to 13.6 seconds while Sensor D response times were between 7.1 and 31.6 seconds.

Pool 6 (Appendix F)

This test facility was a kidney shaped pool measuring approximately 35.5' in length with the widest area being about 16.5' in width and 3 feet depth. The deeper end of the pool was about 13.5' in width and 8 feet in depth. This pool also appeared to be a typical gunite shell. A schematic of the pool, along with the locations of the various alarms and the test results can be

13

found in Appendix F. Due to the pool size, shape, and location of the alarms, there was only one drop point used for the test objects.

From Location "X" (Shallow End)

Surface wave Sensor A alarmed in all 10 trials using the three-gallon test object and in 8 out of 10 trials using the two-gallon test object. The response times ranged from 10.9 to 21.2 seconds for the three-gallon object and from 12.8 to 25.1 seconds for the two-gallon object. Surface wave Sensor B responded in 7 of the 10 three-gallon object trials and 5 out of 10 two-gallon trials. The response times were from 8.1 to 16.3 seconds and 9.7 to 11.1 seconds, respectively.

Both subsurface wave sensors alarmed in all trials involving both the three-gallon and the two-gallon test objects. However, there was one test involving the three-gallon object where Sensor C responded just as the test object was introduced into the pool. It was determined to be a false alarm and that trial was not counted for that sensor. The response times with the three-gallon test object were from 9.7 to 19.7 seconds for Sensor C and from 11.2 to 25.8 seconds for Sensor D. For the two-gallon object, Sensor C response times were from 8.8 to 22.2 seconds while Sensor D response times were between 16.6 and 20.6 seconds.

 TEST RESULTS – FALSE ALARM

During the first round of testing, surface wave Sensor A alarmed in conditions of simulated wind, simulated rain, and during a rainstorm. Surface wave Sensor B did not alarm during either of the environmental simulations nor during actual weather events. Neither surface wave sensor alarmed when objects were tossed into the pool unless the objected landed within approximately five feet of the sensor. Subsurface wave Sensor C did not false alarm in any of the environmental simulations or during actual weather related disturbances. Subsurface wave Sensor D did not alarm in simulated wind or rain conditions; it did alarm during the rainstorm. Neither subsurface wave sensor alarmed when objects were tossed into the pool. A summary of False Alarm Test Results is shown in the various appendices where false alarm tendencies were investigated.

The wristband alarmed immediately when subjected to a stream of running tap water.

CONCLUSION

Test results showed that the subsurface pool alarms generally performed better. They were more consistent in alarming and less likely to false alarm than the surface alarms. When a test object, intended to simulate the weight of a small child, was pushed into the pool, the subsurface sensors detected it most reliably. The subsurface alarms can also be used in conjunction with solar covers, whereas the surface alarms cannot.

One surface alarm performed almost as well as the subsurface alarms. The wristband alarmed when submerged in pool water or exposed to another water source, such as tap water.

A pool alarm can be a good additional safeguard in that it provides an additional layer of protection against child drownings in swimming pools. Since pool alarms rely on someone remembering to activate them each and every time the pool is in use, they should not be relied upon as a substitute for supervision or for a barrier completely surrounding the pool. A remote

14

■ Test report on safety of pool alarms

Features of a Test Report (continued)

④ Results and conclusion offer brief summaries with definitive findings.

alarm feature that will sound inside the house is important to have with a pool alarm. Some alarms include this; with other alarms it has to be purchased separately.

The wristband can also provide an additional layer of protection. However, it relies on someone putting it on the child and, since children often reach the pool unexpectedly from the house, it would be important for a child to wear the wristband all the time. This may present some difficulties since it alarms when exposed to any water, e.g., when washing hands.

15

■ Test report on safety of pool alarms

Trip Report

Employees returning from business trips use trip reports to summarize their activities. These reports may also be used to request reimbursement. Trip report formats often follow company or organization format, can be short or long, and may be attached to required expense forms. The following is a sample of a brief trip report.

TRIP REPORT

All-Purpose Technology Firm, Inc.

Submitted by: Zachary Specter

1 **Dates**: October 25–31, 2007

Describe the travel activity: Approved visit to High Point Cylinder company as part of our acquisition. Met with each team leader and plant manager. Assessed current work flow, systems, design, and product output.
Held a focus group with local customers. **4**

2 **Describe how information will be disseminated:** I have posted a summary of my report on our intranet site. I will present my findings at a meeting on November 6.

3 **Any additional findings or information:** Service team needs to follow up with a web conference to learn more from High Point service technicians.

5 *PLEASE COMPLETE THE ONLINE TRAVEL EXPENSE FORM AND SUBMIT RECEIPTS TO CENTRAL ACCOUNTING.*

■ Trip report

Features of the Trip Report

1 Uses dates of actual trip taken.

2 Includes brief summary of activities.

3 Refers to a report provided elsewhere and plans for follow-up.

4 Uses first person and active voice.

5 Usually accompanies a form for travel reimbursement.

Audience Considerations

The examples provided in this section illustrate features of many report types based on the intended audience and purpose. People who read reports may want the details, but they also want the big picture and major findings. That is why most longer and all formal reports begin with an executive summary. Busy readers and executives may read only the executive summary to find the major plan, findings, costs, and benefits in a brief and concise format. The rest of the report should appear in appropriate sections of the body and back matter. The amount of context and background you provide depends on how specialized or technical the primary audience may be and the scope of your report.

Reports of all types should be as clear and accurate as possible. These documents should use data, logic, and sound reasoning to demonstrate the validity of activities, findings, conclusions, and recommendations. Your arguments should be based on logical and clear presentation of the facts, as well as a strong ethical basis. Your report cannot make false promises or draw premature conclusions, and when you cannot claim to be an expert in an area, you need to use extensive credible outside sources. Often, reports have a global audience, so use international communication techniques to make sure your document is readable to readers whose first language may not be English.

Design Considerations

Reports need to be written and designed in a style that is interesting to read, to the point, and accurate. Use appropriate document design techniques such as headings, visuals, bulleted lists, and data display, including clear charts of information about data, research, and other key findings. Save writing the executive summary for after you have drafted the rest of the report. Extensive data and other details can be placed in an appendix. If you are writing for a government office or other agency, be sure your content, design, and format follow these instructions precisely.

Increasingly reports are delivered by email, on Internet sites, and use templates and other computer programs for formatting. If you are not sure of these technological formats for proposals, check with a manager or an officer from the agency for more information.

- Use the type of report that is best suited for your purpose and audience.

- For longer reports, make information easy to access with a table of contents or index.

- Use visuals as appropriate.

- Include separate lists of figures and tables.

- Keep language simple, clear, and direct.

- If appropriate, begin with an executive summary that condenses the problem, approach, and major findings of the report.

- Include a letter of transmittal as the cover for your report, if appropriate.

- Give technical background, lengthy data sets, any required forms, such as a budget, and other information in a set of separate appendices.

See also

Specifications

Specifications, or "specs," list all the specific details required to construct or install a product. When written for engineering applications or computer technologies, your specs give the exact technical specifications for re-creating that technology. When written for product descriptions, site installations, or construction, your specs should include every individual part needed to actually install the system. Specs may also include technical illustrations and even parts pricing as part of the company's contract. This list for the Emriver Process Simulator (see Product Descriptions, Part 2) includes all installation specs.

Emriver : Components : Specifications

Feature	Description
Box	Engineered reinforced 3/16" sheet aluminum box, 7 x 3 x 0.5 ft., 65 pounds (29 kg) TIG welded and riveted, PVC standpipe and drain fixture
Supports	Two aluminum horses, folding, each with four adjustable broad-based feet, about 13 pounds (5.8 kg) each
Reservoir	27 gal. (110 l) polypropylene, graduated, with supports for sediment filter and pump Sediment trapping system
Pump	Sealed 800 gph (3028 lph) 12-volt 3.8 amp submersible pump, with intake filter
Power Supply	B+K Precision 12-volt power supply, input standard 115-volt 60 HZ. (12 volt battery is not included) Wire harnesses for both laboratory and 12-volt battery or auto power
Modeling Media	180 pounds (82 kg) cryogenically-ground melamine plastic, specially manufactured with proprietary mix of sizes and colors
Gage	Notch gage and energy-dissipation unit, machined acrylic, capable of gaging flows from 15ml/s to 325ml/s
Valve Array	Nylon glass-filled industrial half-union ball valves, fully field serviceable, 3/4" vinyl tubing supply to notch gage, nylon hose disconnects, garden hose fittings for field filling and emptying of system
Measurement Tools	Vertical measuring rod and bracket (for leveling) Custom-built measuring tape for long profile surveys
Modeling Accessories	riprap stones acrylic shapes for bridge culverts, piers, and other flow obstructions bars for low water crossings, "slab" bridges, and other built structures roughness elements, including simulated riparian vegetation
Sediment Handling	Two large scoops Coarse sieve for removing riprap and other elements from sediment Scrapers for moving and shaping media
Support	Comprehensive use and maintenance manual Comprehensive demonstration manual, updated via email and website Full parts and supplies support

■ Specifications for a product

Features of Specifications

1 Product name and type of specifications appear as a title across the top.

2 Each feature or part is identified in the list subheading.

3 The design specs include all measurements needed by the user or contractor.

4 Technical details of the product are usually included in lists specified by each part—the aluminum box, water pump size, type of power supply, and so forth.

These details are given in separate pages on each section of the equipment and also in a schematic of the water flow through the Emriver.

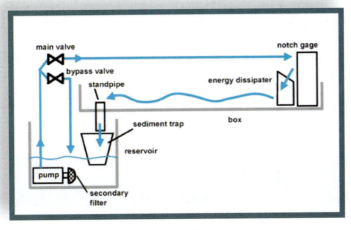

■ Drawing is included in these specifications

Technical illustrations or flow charts such as this one help users visualize the details of equipment and in this case give the water flow directions for this installation. Drawings with labels like these could mislead the reader into thinking the equipment comes in these exact sizes or other details, so take care to make absolutely clear and distinguish an exact requirement or an actual model from simply a generic sample or a schematic drawing in your specs.

Audience Considerations

In order to write effective specs, you will conduct research to collect all the necessary technical information—possibly by conducting interviews of the engineers or product designers, by observing users with the product, or by using the equipment yourself. You must know whether your specs will be read by contractors, subcontractors, engineers, research scientists, consumers, or, most likely, some combination of all these audiences. These documents do assume a captive audience that must read the details carefully and follow them in order to complete the work.

Global audiences may use different kinds of electrical currents and different terms for certain processes, so be sure to check on translation of those terms with a translation and localization resource. Always use the exact technical names—"supports," not sawhorses; "modeling media," not ground plastic—for the product components across all your documents to minimize any language confusion.

Design Considerations

A spec sheet like the one pictured above typically uses a dense list of information. Arranged with subheadings with the most important or largest features first, these specs cover necessary measurements, weights, and other product details. Since specs use technical language that focuses on parts and processes, you need the exact, specific identifying nouns to create your lists.

This example appears on a web site and may require a different design for a printed document (see Document Design, Part 3), but it illustrates the dense information typical of most spec sheets. The use of photographs as illustrations throughout the components pages gives readers the option of looking back and forth between the technical details described in the text and the visuals that illustrate those details. Headings, subheadings, and links are the primary means for dividing up the details in the web page design.

Guidelines

FOR CREATING SPECIFICATIONS

- Consider the many kinds of readers who will use your specs to actually make the product.

- Collect all the specific technical details needed through research.

- Include every part you would need to build or install this product.

- Organize your information sequentially as a list from most to the least important.

- Provide schematic drawings or sample illustrations if possible.

- Lay out the document using sections, headings, and subheadings.

- Try out the specs yourself to make sure no steps are missing.

- Test your specs on target users before publishing them.

See also

Interviews (Part 4)

Photographs (Part 3)

Product Descriptions (Part 2)

Technical Illustrations (Part 3)

Visual Communication (Part 3)

Web Page Design (Part 3)

Surveys

Surveys are tools used to collect data about a particular subject, idea, or product. Many surveys, such as those used by psychologists or social scientists, are sophisticated tools that may involve many months or years of planning and testing. Other surveys can be focused but less complicated. Some so-called surveys, such as the ones you might see on a TV news web site ("vote here on your favorite movie star!"), are less surveys than they are questionnaires. True surveys are designed and implemented so that the number and type of subjects (people who take the survey) will yield results that are useful for predicting how larger segments of this population will respond. Surveys must be designed carefully, so that the wording of questions and use of language makes sense to the audiences who will take the survey. More and more surveys are being delivered online; it is easier, less expensive, and provides access to a wider population to do a survey over the Internet.

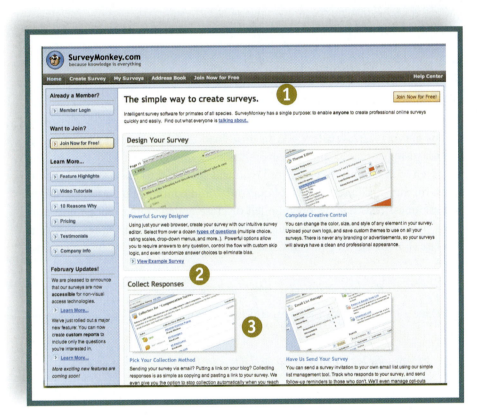

■ Online survey tool

Features of Online Survey Tool

1 An introductory section explains what the survey tool can provide.

2 Types of survey questions include:
Open ended
Likert scale (degrees)
Yes/No

3 Electronic surveys can be much quicker to distribute than paper.

Audience Considerations

When using the Internet for a survey, you need to ensure that each person can only take the survey one time. You can do this by providing a user ID and password that is only good for one use. If you create a survey as part of a class or part of a research study, you will need approval from the human subjects review board (also called institutional review board, or IRB), to ensure that participants are protected from undue risk. For most surveys, you will need to ask for informed consent from the people taking the survey. Write your consent form and the survey in brief phrases and sentences. Use language that is simple and concise. Keep the style friendly yet professional.

Design Considerations

Many online survey sites, such as surveymonkey.com, have templates that are well designed and easy to use. Use one of these rather than designing your own. Online surveys usually allow you to download the data into a format you can examine using common spreadsheet and database applications.

Even with an online survey tool, you will still need to make careful choices about the survey's length, types of questions, and content. For instance, should you ask questions that are open ended or close ended? Open-ended questions allow people to give you more feedback, but the answers are much harder to tabulate. For instance, you could ask the following question in an open-ended way:

> *"Please tell us why you decided to order from this web site."*

and leave a blank box for people to fill in. Or you could ask it in a close-ended way:

> *"Please select the main reason you decided to order from this web site:*
>
> _____ *Price*
> _____ *Web site is easy to use*
> _____ *Good reputation for customer service*
> _____ *A friend recommended this web site"*

These questions are set up in relative categories along a scale, as follows:

> *"Please indicate how strongly you agree or disagree with the following statement:*
>
> *I had an excellent experience using this web site*
>
> — *Strongly agree* — *Agree* — *Do not agree or disagree* — *Disagree* — *Strongly disagree"*

All survey questions require you to consider certain ethical issues. You need to ensure that the questions are worded in a neutral manner and, in the case of closed-ended questions, that the choice of answers does not bias the person taking the survey.

Guidelines

FOR CREATING SURVEYS

- Design the survey with care; make sure questions are easy to understand.

- Write questions that are unambiguous (have only one meaning).

- Determine what form of question (closed- or open ended, Likert scale) is appropriate.

- Know the audience for this survey and aim the questions at this group.

- Avoid terms that would be confusing or technical.

- If you want results that are scientifically sound, work with a survey research expert to design and implement the survey.

- Use online surveys to capture the widest audience.

- Determine what amount of time you expect people to spend on the survey.

- Test your survey on real audience members.

See also

Ethical Issues (Part 4)

Focus Groups (Part 4)

Forms (Part 2)

Usability and User Testing (Part 4)

Surveys and Questionnaires (Part 4)

User Manuals

User manuals are documents that help people use the features of a product or tool. There are many types of user manuals, but they all include task-based documents. These manuals and guides are available in a variety of formats: embedded with the product (such as a user manual that is part of an online help system for spreadsheets or a word processing program), online via a web site, online or on a CD in PDF format, printed and bound like a book, printed on a small card or booklet. Given the number of electronic devices and other products in the world, user manuals are an important and significant category of technical communication.

User manuals can be long or short, detailed or brief, but all are intended for a specific purpose: to provide users with an overview and details on how to use, maintain, and work with a product. Most user manuals are available in both print and PDF file formats. Many can be accessed through the organization's web site. The following excerpts are from a manual produced by the U.S. Environmental Protection Agency (EPA) for use with a software tool called BASIN. BASIN stands for "better assessment science integrating point and non-point sources)" and uses a geographical information system to allow users to analyze and model data about watersheds throughout the U.S. While users of this tool may be experts in geology and related fields, the software itself will be new to them, so this manual needs to cover the details on how it is used.

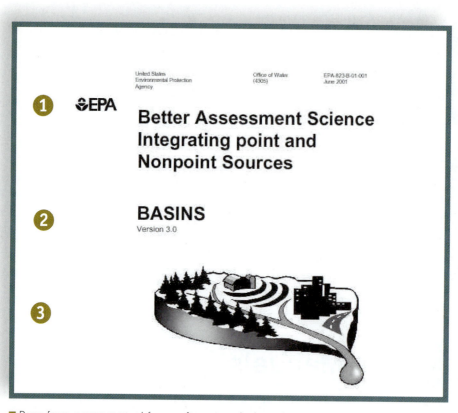

■ Page from a user manual for a software analysis tool

Features of a User Manual

❶ Organization's logo is in the upper left hand corner, making it easily accessible to readers.

❷ Product name (BASINS) is surrounded by white space and is easy to locate. The version number is set in smaller type but is still obvious.

❸ Graphic of a river/watershed makes the topic more interesting.

④ Contents

■ Page from a user manual for a software analysis tool

Features of a User Manual (continued)

④ This detailed manual contains a lot of information, so the headings and sub-headings use a clear numbering system.

⑤ First level headings are set aligned with the left margin, while second level headings are indented.

⑤ Listings are worded with enough information to guide readers but are not too lengthy ("data extraction," "grid projector").

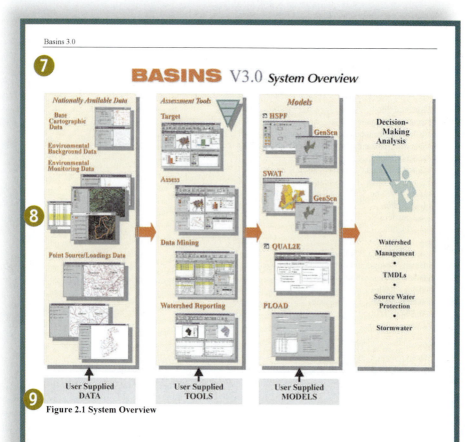

BASINS V3.0 *System Overview*

Figure 2.1 System Overview

Watershed Models:

- *WinHSPF* is an interface to the Hydrological Simulation Program Fortran (HSPF), version 12. HSPF is a watershed scale model for estimating instream concentrations resulting from loadings from point and nonpoint sources.

- *SWAT* is a physical based, watershed scale model that was developed to predict the impacts of land management practices on water, sediment and agricultural chemical yields in large complex watersheds with varying soils, land uses and management conditions over long periods of time. SWAT2000 is the underlying model that is run from the BASINS ArcView interface.

■ Page from a user manual for a software analysis tool

Features of a User Manual *(continued)*

7 After a previous page that uses text to describe the system overview, this page employs a visual to reinforce the way the BASINS system works.

8 Arrows illustrate how the different parts of the system are connected.

9 Each figure in this manual is labeled with a figure number.

2.4 Watershed Characterization Reports

The Watershed Characterization Reporting tools are designed to assist users in summarizing key watershed information in the form of standard and automated reports. These tools can be used to make an inventory and characterize both point and nonpoint sources at various watershed scales. The results are presented in table, chart and/or map layout formats. These reports allow users to quickly evaluate and define data availability for the selected watershed(s). Eight different types of watershed characterization reports are included in BASINS:

- Point Source Inventory Report

- Water Quality Summary Report

- Toxic Air Emission Report

- Land Use Distribution Report

- Land Use Distribution Report(Grid)

- State Soil Characteristics Report

- Watershed Topographic Report

- Watershed Topographic Report(Grid)

Point Source Inventory Report

Point Source Inventory Report provides a summary of discharge facilities in a given watershed. The report relies on the EPA Permit Compliance System (PCS) database to identify permitted facilities in a selected study area and summarizes their discharge loading for a given pollutant. Application of this report tool provides rapid identification of permitted sources, the receiving water body segment (Reach File Versions 1or 3), and a mapping function to display the geographical distribution of point sources in the study area.

Water Quality Summary Report

Water Quality Summary Report provides a summary of water quality monitoring stations within the selected watershed that monitored a particular pollutant during a given time period. The water quality data are presented as statistical summaries of the mean and selected percentiles of the observed data. The data is based on USEPA's Storage and Retrieval System (STORET). The information generated in this report can be summarized in tables and maps.

Toxic Air Emission Report

Toxic Air Emission Report provides a summary of facilities within the selected watershed(s) with air releases of selected pollutants. This data is based on USEPA's Toxics Release Inventory(TRI). Tabular summaries of TRI facilities are generated with their corresponding estimates of pollutant air releases and other pertinent information such as facility identification name, city location, status (active or inactive

---

■ Page from a user manual for a software analysis tool

Features of a User Manual *(continued)*

10 First level heading is numbered to correspond with the table of contents.

11 First paragraph is short and concise.

12 Types of reports are listed using a bulleted list, which makes the content easier for readers to skim through.

13 Each report type is then described in brief. Second level headings break up the text.

(14)

3 Hardware and Software Requirements

(15) BASINS Version 3.0 is a customized ArcView GIS application that integrates environmental data, analysis tools, and modeling systems. Therefore, BASINS' hardware and software requirements are, at a minimum, similar to those of the PC-based ArcView 3.1 or 3.2 system. BASINS can be installed and operated on IBM-compatible personal computers (PCs) equipped with the software, random access memory (RAM), virtual memory, and hard disk space presented in Table 3.1.

Because the performance (response time) under the minimum requirements option might be too slow for some users especially when dealing with large data sets, a preferred set of requirements is also included in Table 3.1. For some advanced features of BASINS 3.0, such as the automatic watershed delineator or the SWAT model, the ArcView Spatial Analyst software is required. This software is not included with ArcView and must be obtained separately.

(16) Table 3.1. **BASINS Hardware and Software Requirements**

Hardware/Software	Minimum Requirements	Preferred Requirements
Processor	166-MHz Pentium processor	400-Mhz Pentium II processor or higher
Available hard disk space	For a single 8-digit watershed (cataloging unit), allow for 300 Mb (50 Mb for BASINS system, 150 Mb for BASINS Environmental Data, and 100 Mb of free operating space for storage of generated themes and tables).	2.0 Gb (50 Mb for BASINS system, 750 Mb for BASINS environmental data for approximately 1 state, and 1.2 Gb of free operating space for storage of generated themes and tables).
Random access memory (RAM)	64 Mb of RAM plus 128 Mb of permanent virtual memory swap space	128 Mb of RAM plus 256 Mb of permanent virtual memory swap space
Compact disc reader	Quad speed reader (one-time use)	24X reader (one-time use)
Color monitor	Configured for 16 colors, Resolution 1024x768	Configured for 256 colors
Operating system	Windows 95, Windows 98, NT*	Windows 95, Windows 98, NT*
ArcView	ArcView Version 3.1 or ArcView version 3.2, Spatial Analyst Version 1.1 optional	ArcView Version 3.1 or ArcView version 3.2, Spatial Analyst Version 1.1

(17) *QUAL2E* cannot operate on NT.

Microsoft Excel 97/2000 is recommended for use with the PLoad extension. Internet Explorer 5.0 or newer is required to view help files.

■ Page from a user manual for a software analysis tool

Features of a User Manual *(continued)*

(14) First level heading is numbered to correspond with the table of contents.

(15) Requirements are described in an overview in the first two paragraphs.

(16) A table makes it easy for users to skim the columns and rows looking for needed information, rather than wading through lots of text.

(17) Green rules provide a simple yet effective way to organize the table.

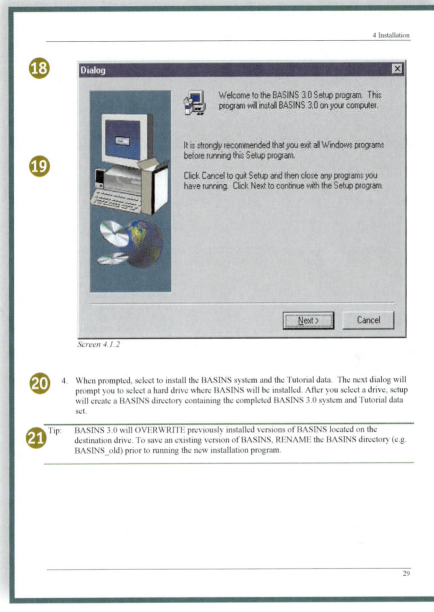

18

19

Screen 4.1.2

20 4. When prompted, select to install the BASINS system and the Tutorial data. The next dialog will prompt you to select a hard drive where BASINS will be installed. After you select a drive, setup will create a BASINS directory containing the completed BASINS 3.0 system and Tutorial data set.

21 Tip: BASINS 3.0 will OVERWRITE previously installed versions of BASINS located on the destination drive. To save an existing version of BASINS, RENAME the BASINS directory (e.g. BASINS_old) prior to running the new installation program.

29

■ Page from a user manual for a software analysis tool

Features of a User Manual *(continued)*

18 On this installation page, a simple screen capture lets users follow along with each step.

19 Screen captures also provide users with a visual reference, so that the visual reinforces what is in the text.

20 Each step in the process is numbered (this page is preceded by steps number 1-3).

21 Tips are set off with green bars. Consistent use of this feature makes the installation process easier for users.

Procedure Manuals

Also called a user's guide, the following manual functions as a procedure manual, in the sense that its primary purpose is to provide users with the steps and procedures needed to operate a sprinkler system timer.

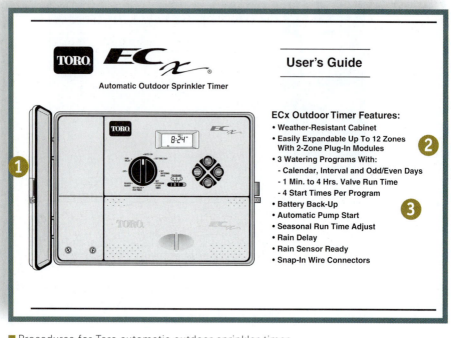

■ Procedures for Toro automatic outdoor sprinkler timer

Features of a Procedure Manual

1 A simple, yet well-detailed illustration helps orient users to all functions and features.

2 Placing text next to the illustration helps make the text more relevant.

3 The page layout is clear, with bulleted text items and plenty of white space.

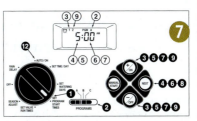

Setting Program Start Times

❶ Turn the control dial to **SET PROGRAM START TIMES** position.

❷ Check the **PROGRAMS** switch setting. If necessary, reposition the switch to select the desired program.

❸ Program start time number 1 will begin flashing. The current program start time or OFF will be displayed for start time number 1. To select a different program start time number, press the **+/ON** or the **–/OFF** button until the desired number is flashing.

Note: The numbers (1–4) shown at the top of the display designate program start times and should not be confused with zone numbers. The zone numbers will be shown at the bottom of the display when setting zone run time.

❹ Press the **NEXT** button. The hour digit(s) or OFF will begin flashing.

Note: To remove the start time, select **OFF** by pressing the **+/ON** and **–/OFF** buttons **at the same time**, and continue at step **❽**.

❺ To set the hour (and AM/PM), press the **+/ON** or the **–/OFF** button until the desired hour is flashing.

❻ Press the **NEXT** button. The minute digits will begin flashing.

❼ To set the minutes, press the **+/ON** or **–/OFF** button until the desired minute is flashing.

❽ Press the **NEXT** button. The next program start time number will begin flashing.

❾ To select another start time number, press the **+/ON** or the **–/OFF** button until the desired start time number is flashing.

10. To set, change or remove a program start time for the start time number selected, repeat all of the steps starting at step **❹**.

11. To set program start times for another program, repeat all of the steps starting at step **❷**.

⑫ Return the control dial to the **AUTO/ON** position.

■ Procedures for Toro automatic outdoor sprinkler timer

Features of a Procedure Manual *(continued)*

❹ These instructions are clear and easy to follow.

❺ Numbered items are easy to see due to the use of reverse type (white number inside a black circle).

❻ White space and indentations help users follow steps and not get lost.

❼ A labeled diagram is always useful when located near the actual instructions.

Assembly Manuals

Assembly manuals are usually longer and more comprehensive than procedures or instructions. These manuals provide complete information on how to assemble a product or device. They usually contain photographs, illustrated diagrams, or other visuals, which help provide an overview of what the device will look like, once assembled. Visuals can also help users connect text-based information with the parts of the device. The following assembly manual provides instructions for the assembly of a model spacecraft. These three pages demonstrate the key features of an assembly manual.

Galileo 1/45 Scale Model Assembly Instructions

Version 1.6

This is a detailed scale model of the Galileo Spacecraft, one of the most complex robotic spacecraft ever flown. Assembly is a project which is probably not appropriate for people younger than about ten years of age, depending on skill and motivation. The image above shows a completed scale model. Click on it for more views of the model.

Six Parts Sheets are available to download from this web site. ②

A. YOU'LL NEED THE FOLLOWING:

③

- A clear transparency sheet which your computer's printer will accept. These (as well as card stock sheets) are available at stationery stores and office supply stores. This is needed only for printing Parts Sheet 1.

- White paper card stock (also called "cover" stock, about the thickness of a postcard) which your computer's printer can accept. Parts Sheets 2 through 6 should be printed on this heavy white paper.

- Transparent adhesive tape, such as clear 3M Scotch Brand tape. Frosty tape will suffice if clear tape is not available.

- Scissors, to cut some parts from the parts sheets.

- An art knife, such as X-Acto #11, with a sharp new blade, and a proper pad on which to cut. This will be needed to cut some parts from the parts sheets. Adult supervision is required for children using sharp tools. Caution, one can injure oneself, as well as the furniture, with an art knife.

- Wooden toothpicks for applying glue.

- Glues. Use regular white glue (Elmer's Glue-All or equivalent). You might also try a thick white glue, sold in art and fabric stores, called "TACKY GLUE" (Aleen's or equivalent). For one part (the High-Gain Antenna) you'll need a different glue: BOND 527 Multi-Purpose Cement works very well. It's available in many art stores and craft stores, or from Bond Adhesives Company, Newark, NJ 07114 U.S.A. If BOND 527 isn't available, a rubber cement may be used for that part. An artist's spray glue, such as 3M Spray Mount, would be convenient for a few parts, but it is not absolutely necessary.

- A dense black felt-tip pen.

- A 35mm film can, and a beverage cup, would be convenient to support some parts while gluing.

■ Assembly manual for a model spacecraft

Features of an Assembly Manual

❶ Photograph illustrates the final product when assembled.

❷ Web site is provided if users wish to access these instructions online.

❸ A detailed list lets users know what items they will need in advance of getting started.

● 4. Get your bearings: Examine the sketch of the spacecraft below. For the purpose of these instructions, "Up" will be defined as toward the HGA SUNSHADE shown in the image. "Down" will be defined as toward the ATMOSPHERIC PROBE in the drawing. This also includes references to "Top" and bottom" in the instructions. "Inboard" is defined as toward the center of the spacecraft, and "Outboard" is defined as away from the center. You might also like to view a much larger line drawing of the spacecraft before getting started.

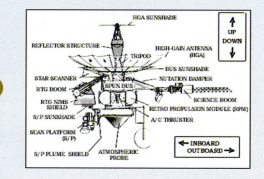

■ Assembly manual for a model spacecraft

Features of an Assembly Manual (continued)

● 4 Instructions are written in the imperative mood ("Get your bearings; Examine the sketch").

● 5 A carefully illustrated diagram shows how the parts connect to the whole.

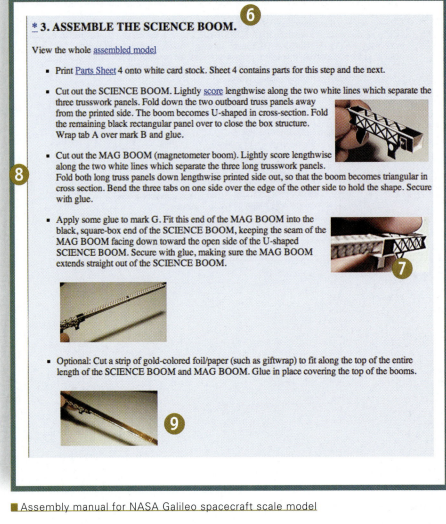

* 3. ASSEMBLE THE SCIENCE BOOM. **6**

View the whole assembled model

- Print Parts Sheet 4 onto white card stock. Sheet 4 contains parts for this step and the next.

- Cut out the SCIENCE BOOM. Lightly score lengthwise along the two white lines which separate the three trusswork panels. Fold down the two outboard truss panels away from the printed side. The boom becomes U-shaped in cross-section. Fold the remaining black rectangular panel over to close the box structure. Wrap tab A over mark B and glue.

- Cut out the MAG BOOM (magnetometer boom). Lightly score lengthwise along the two white lines which separate the three long trusswork panels. Fold both long truss panels down lengthwise printed side out, so that the boom becomes triangular in cross section. Bend the three tabs on one side over the edge of the other side to hold the shape. Secure with glue.

- Apply some glue to mark G. Fit this end of the MAG BOOM into the black, square-box end of the SCIENCE BOOM, keeping the seam of the MAG BOOM facing down toward the open side of the U-shaped SCIENCE BOOM. Secure with glue, making sure the MAG BOOM extends straight out of the SCIENCE BOOM.

- Optional: Cut a strip of gold-colored foil/paper (such as giftwrap) to fit along the top of the entire length of the SCIENCE BOOM and MAG BOOM. Glue in place covering the top of the booms.

■ Assembly manual for NASA Galileo spacecraft scale model

Features of an Assembly Manual *(continued)*

6 Each step is numbered and corresponds to a page. This page is for Step 3.

7 Unlike a simple sheet of instructions, an assembly manual offers more detailed information and illustrations.

8 Each major step is grouped into its own paragraph.

9 Photographs provide a clear view of what each device or part looks like.

Product Manuals

Product manuals are comprehensive manuals that provide a wide range of information, from instructions to operating procedures to safety information to ideas for using the product. Product manuals usually are provided in print format when you purchase the product, although for some products, these manuals are provided on a CD. Almost all companies now place copies of product manuals in

PDF file format on their company web site. This feature allows customers to find the manual for their specific product, long after the print copy or CD has gotten lost. The following example for a juicer illustrates many of these features. It is available on the web site but also comes with the product.

■ Product manual of the Champion juicer

Features of a Product Manual

1 Product manual starts with a colorful cover page.

2 A photograph illustrates the actual product.

Table of Contents

■ Product manual of the Champion juicer

Features of a Product Manual *(continued)*

❸ Detailed table of contents allows users to search easily for the topic of interest.

❹ Topics include Safety, Trouble-shooting, Assembly, Cleaning, and even recipes: sections typical for a comprehensive product manual.

WELCOME TO YOUR CHAMPION

YOU CAN SEE AND TASTE THE DIFFERENCE

The Champion is classified as a slow-speed, masticating type machine. It chews the fibers and breaks up the cells of vegetables and fruits. This gives you more fiber, enzymes, vitamins and trace materials. All this results in the darker, richer color of the juice and a sweeter, richer, more full-bodied flavor.

Your Champion is simple and easy to use. Assembly doesn't require nuts, bolts, screws or clamps. Just slide the floating cutter on the shaft, make a quick half turn, the body locks on and the machine is ready to use. Remove the screen and insert the blank and it is now a homogenizer. It is so simple and all the parts are easy to clean.

Assembly, disassembly, juicing and homogenizing instructions, plus a wide variety of recipes follow. Read the instructions completely to receive maximum efficiency and use from your new Champion Juicer.

Your Champion is powered by a full 1/3-horse-power, heavy-duty motor. All parts are made from 100percent FDA accepted nylon and stainless steel. The floating cutter has been designed to separate the juice from the pulp in a continuous operation. No intermittent cleaning is required.

MAKING HEALTHIER EATING CHOICES IS EASY

Research studies have repeatedly shown a direct relationship between diet and health. Recently, more doctors are discussing the ability of individuals to decrease their risk of certain cancers and heart disease through proper diet. Current recommendations are to increase the intake of fiber and complex carbohydrates, and decrease the intake of fat, cholesterol, sugar and sodium.

Our diet has, over the years, become filled with highly processed convenience foods. Many of these get their calories from fats and sugars, and contain large amounts of sodium to enhance flavor.

You have an opportunity too make changes in your diet and increase your intake of fiber and complex carbohydrates. The Champion Juicer and Grain Mill are designed to make preparation of fruits, vegetables in the prime and quickly prepare them to fit your family's needs and menu plans. You can control the addition of salt and sugar and make healthy choices in the types of fat you use.

From an early age, children love the flavor of fresh fruits and vegetables. Presentation of two or more vegetables tastefully prepared rounds out a dinner menu without large servings of animal protein, which are commonly high in fat and cholesterol.

Fresh fruit desserts and toppings are satisfying and good for you. Whole grains, used for breakfast as cereal and throughout the day in baked products, are a wonderful source of fiber and energy. The Grain Mill allows you to prepare your family's favorites.

We have chose to use safflower, corn or olive oil, wherever possible. All are without cholesterol and high in poly and mono unsaturated fats.

Begin using your Champion Juicer and Grain Mill at every meal. The more you use it, the easier it becomes to make healthier eating choices.

USING YOUR FREEZER

Fruits and vegetables should be processed in the Champion Juicer while still crisp and fresh. Old woody vegetables or soft mushy fruit will not produce good results or good flavor.
If you have more produce than you can eat while it is fresh, you have two options. One is to wash

---

■ Product manual of the Champion juicer

Features of a Product Manual *(continued)*

5 An overview page provides background information about the product, its uses, assembly and disassembly, and more.

6 Information is spread over two columns for ease of reading.

7 Subheadings in all caps correspond to the TOC sections and break up text.

and cut the produce to fit the feeding throat and quick-freeze to process later. Remember, the freezer will not stop the aging process, only slow it down. It gives you a few more weeks to use the food and still have good quality.

Your second opinion is to juice or homogenize the food and package it in freezing containers for use later.

Unless fruits and vegetables have been bleached, the aging process continues slowly in the freezer. For maximum flavor, uncooked fruits and vegetables should be used in 2 to 3 months. Always wrap fruits carefully to exclude all exchange of moisture of air between food and the cold air in the freezer. Always label with food type, amount and date.

 ASSEMBLY INSTRUCTIONS

Step 1

Slide the cutter onto the shaft. Make sure the shaft is greased with olive oil or coconut oil. Do not use other liquid oils, butter, margarine or petroleum jelly. About 1/4 inch onto the shaft, the cutter may stop; turn and jiggle the cutter slightly in either direction so that the flat edge of the shaft will match the flat edge of the cutter hole. It will then slide on easily.

Special Instructions on Champion Juicer Cutters:

At certain times while removing the juicing parts of your Champion Juicer you may find that the cutter has become sticky or vacuum locked. To prevent this we recommend the following:

Apply a thin film of coconut oil or olive oil to the motor shaft. Fill the cavity of the cutter with cold water. Empty the water out of the cavity and place the cutter on the motor shaft.

After juicing, remove parts and clean them thoroughly with cold water. The cutter should not be placed on the motor shaft until you are ready to shaft juicing once again.

Caution: To prevent injury from the exposed cutter when it is not in use, we recommend that you wrap the cutter in paper towels or place it in a paper bag.

Note: If the cutter becomes vacuum locked, place a small screwdriver behind the cutter and the stainless steel hub and pry the cutter forward.

Step 2

Slide the body over the cutter, holding it in a horizontal position with the feeder throat down.

Step 3

Place the juicer screen into the recessed grooves and hold it into position with one hand. Now slide the juicer screen holder over the screen. Hold the edge of the screen down for easier starting. Note: The screen holder will slide over the screen only one way.

The raised lip indicates front and the flat portion is the starting end. Hold the corners of the screen down for easier starting and, once started over the screen, slide the screen holder completely forward until the raised lip contacts the body slides. The screen holder should be level for easier starting. The nylon bank is inserted in the same manner.

Step 4

Pull the body forward slightly, so it is completely clear of the prongs on the hub. Turn the body one notch to the left, counter clock-wise, and match the openings in the back of the body with the prongs on the hub. Slide the body all the way back and turn it to the left until it stops. It is now locked in proper position for use.

■ Product manual of the Champion juicer

Features of a Product Manual *(continued)*

8 Assembly instructions are clearly numbered as Step 1, Step 2, and so on.

9 Instructions are written in imperative mood ("Slide," "Place," "Pull…").

Step 5

The Champion is now in proper position to juice. To homogenize, replace the screen with the blank.

Step 6

Add the funnel when using small feeding materials such as berries and nuts. Never use the funnel when juicing.

Note: If the Champion doesn't operate at this point, all the parts have not been assembled properly. Unplug and check to see that all parts are completely attached. If it still doesn't operate, take apart and reassemble carefully following complete instructions.

TO DISASSEMBLE THE CHAMPION

After shutting off the juicer, disconnect the electric cord. Turn the body one notch to the right, clock-wise. Jiggle the body slightly and remove. The screen, blank, screen cutter and cutter will slide off easily. However, if the cutter becomes vacuum locked, place a small screw driver at the back of the cutter and push forward slowly.

Note: A buildup of pulp behind the cutter and against the stainless steel hub is normal.

Note: Never switch on the juicer before the parts are properly assembled. Always turn off the juicer and make sure it is completely stopped before removing any parts.
Avoid dropping a hard object, such as a spoon or knife, down the feeder throat; it may damage the tempered stainless cutting blades.

CLEANING INSTRUCTIONS

Wash the nylon juicer parts immediately after juicing. Use only cold water and soap for cleaning the parts. Never wash nylon parts in hot water. Do not place in dishwasher.

The juicer body, screen, blank and screen holder may be soaked or submerged in soapy water. A foam-sponge brush, with a long handle, is handy for cleaning these parts. Sprinkle the parts with any cleanser containing bleach; clean well and rinse the parts thoroughly.

All food particles must be removed from the screen pores. Do not allow food to harden on the screen. Tap the screen gently on a table edge or sink to jar loose any food particle stuck in the pores. Scrub the screen with a stiff bristle brush.

■ Product manual of the Champion juicer

Features of a Product Manual *(continued)*

10 Photographs, showing a person performing the task, accompany the written instructional text.

11 The "Note" sections help users to troubleshoot during a complicated process.

Preheat oven to 375 degrees F. Spray muffin pans with non-stick spray. Set aside.

In a medium bowl, combine flour, baking powder, baking soda and cinnamon. Mix well.

In a large bowl, beat egg lightly. Stir in brown sugar, buttermilk, oil, orange peel and cranberries. Blend well. Add flour mixture and gently fold together until dry ingredients are moistened.

Fill muffin cups 3/4 full. Bake 20-25 minutes or until a toothpick inserted in center comes out clean. Serve warm.

Makes 18 muffins.

Note: Cranberries homogenize easily fresh or frozen.

ENERGY BARS
An easy alternative to candy bars for quick energy.

1-1/2 cups whole wheat flour
3/4 cup all purpose flour
1/2 cup brown sugar, firmly packed
1/4 cup wheat germ
1 teaspoon each baking powder and cinnamon
1/2 teaspoon salt
2 eggs
1/3 cup corn, safflower, or light olive oil
1/4 cup molasses
1/4 cup raw sugar
1 tablespoon finely grated orange peel
1 cup orange juice
1 cup chopped dried figs
1/2 cup golden raisins
1/2 cup chopped almonds

Combine flours, sugar, wheat germ, baking powder, cinnamon, and salt. In smaller bowl, blend eggs, butter, honey molasses, orange peel, vanilla, and orange juice with wire whip. Add liquid to dry ingredients; whip until smooth. Add figs, raisins, and almonds. Spread in a greased 9x13 inch baking pan. Bake in a 350 degree F oven 35 minutes, until it tests done.

Makes about 24 bars.

TROUBLESHOOTING

Motor will not start.

   A. Machine will not start unless all parts are in position.
   B. Check screenholder sensor located on screenholder closest to machine for possible damage.

Food backing up feeder throat:

   A. This is normal since machine works on a back-pressure system.

Motor section running warm:

   A. The motor is designed for a 40 degree heat rise so, depending on amount juiced, motor will get very warm to touch.

Juice hot:

   A. Feeding machine too slowly.
   B. Possible dull cutter.

Wet pulp:

   A. First two handfuls of pulp should always be refed.
   B. Possible clogged screen. Note: Refer to page 10.

Stuck cutter:

   A. Lack of lubrication
   B. Cutter needs to be cleaned. If cutter will not come off with help of screwdriver contact the Service Department. Note: Refer to page 10.

Leakage:

   A. Some juice leakage is normal on juicing

■ Product manual of the Champion juicer

Features of a Product Manual *(continued)*

12 End of manual includes recipes to generate user interest.

13 Complete troubleshooting information is clear and easy to read.

14 White space helps users scan the troubleshooting main points and the possible reasons (A and B).

(15) **IMPORTANT SAFEGUARDS**

When using electrical appliances, basic safety precautions should always be followed including the following:

1. Read all instructions prior to use.

2. To protect against electrical hazards, do not immerse motor or base in water or other liquid.

3. Close supervision is necessary when any appliance is used by children.

(16) 4. Unplug appliance when not in use and before cleaning.

5. Avoid contacting moving parts.

6. Do not operate any appliance with a damaged cord set or after the appliance has been dropped or damaged in any other manner. Return the appliance to the nearest authorized service facility for examination and repair.

7. The use of accessories not recommended by the appliance manufacturer may cause hazards.

8. Do not use outdoors (unless the product is specifically designed for outdoor use).

9. Cutter blades are sharp. Handle carefully.

10. Never feed food by hand. Always use food pusher.

11. Be sure to turn switch to "OFF" position after each use of your juicer. Make sure the motor stops completely before disassembling.

12. Do not put your fingers or other objects into the juicer opening while it is in operation. If food becomes lodged into the opening, use another piece of fruit or vegetable to push it down. When this method is not possible, turn the motor off, unplug juicer cord and disassemble juicer to remove the remaining food.

13. Do not let cord hang over edge of table or counter or touch hot surfaces.

14. Do not place on or near a hot gas or electric burner or in a heated oven.

SAVE THESE INSTRUCTIONS

■ Product manual of the Champion juicer

Features of a Product Manual *(continued)*

(15) Safety features are printed on a separate page for emphasis.

(16) Numbered items and lots of white space make this information easy to read.

Quick Reference Guides

Quick reference guides function like very brief user manuals and are appropriate for situations where users don't need details, they just need a summary of essential tasks to help them get started or remember how to do something. Quick reference guides may be delivered in print or electronically and can range from a small card to a short brochure. The following page is from a quick reference guide for video production software. Note that the cover of this guide also provides users with a link (techsmith.com/learn) in case they need more information.

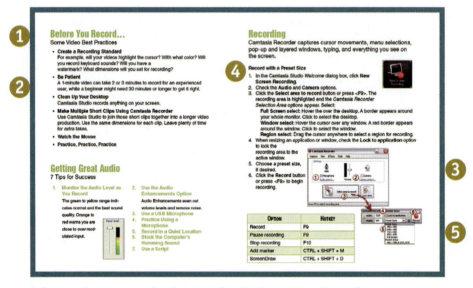

■ Quick reference page for Camtasia Studio video production software

Features of Quick Reference Guide

1 Headings help group material into main categories.

2 Sub-headings use action verbs (create; clean up; make) to help users identify tasks.

3 Annotated screen capture, using red arrows, provides visual reference.

4 Numbered list provides exact steps that users need to perform.

5 Plenty of white space makes the page easy to read.

Audience Considerations

As the name implies, all user manuals must be written and designed with specific users in mind. But people do not read user manuals unless they have a specific question or problem while using their new cell phone or software program. As a technical communicator, you need to put yourself in the position of being a user of the product or application. The questions, categories, and topics you use should imply what these users may be thinking or looking for in the manual. Test your user manuals on others to see whether there are places where users fail to understand your message and collect troubleshooting information based on their questions.

Your organization may even have a set format for procedures or instructions dictated by a regulatory agency or the company's legal department. Manuals and reference guides are typically used by global audiences, so recognizable or universal symbols can greatly increase the manual's usability.

Design Considerations

Like all forms of instructions and other task-based documents, the best manuals include accurate illustrations with clear careful labels, steps, or other brief text. Provide good quality illustrations or photographs for all parts and processes. The links between these illustrations, steps, and other items should be clear and easy to follow.

User manuals include much material written the imperative mood: "Get your bearings" or "Examine the sketch." Avoid passive voice because it makes the doer of each step unclear. Instead use the active voice or the imperative form of a verb. Written text should generally be brief and concise. Keep the words to a minimum. Write short instruction sequences and procedures in numbered lists. When the user needs to know about several items, chunk your information into logical sections. For example, when you purchase a new digital video recorder, the first section should be on setting it up, while the next set should be on using it to record, and so on. Make sure chapter and section names are also clear and obvious in their meanings for users of the product.

While many user manuals are still printed in hard copy, manuals may also come packaged with the product on a CD or DVD. Many products still come with a short user's guide, a quick start guide, or a troubleshooting guide. However, most companies realize that they can keep their manuals up to date if they also direct readers to a web site. Give users a logical online location for finding lost manuals or newer information on the product. Also, there are web sites that collect hundreds of different product and procedural manuals.

FOR CREATING USER MANUALS

- Decide what kind of manuals you need to help your users get started right away.

- Manuals and guides should be self-explanatory and self-contained.

- Use descriptive headings or short phrases to introduce procedures.

- Keep the words to a minimum, and be clear and concise.

- Chunk your information into logical sections.

- Describe each step in the process with an action verb (press, click on, insert, place) and describe the result of this action.

- Use photos and illustrations that correspond to the numbered steps, especially if your audience is global or multilingual.

- When possible, use labeled illustrations of a product that require no additional text.

- Test your manual on others to find places where users fail to understand your message.

- Deliver the instructions in whatever is the most appropriate medium—print, Web, or PDF file, or in multiple media.

See also

Web Sites

Web sites are collections of web pages and associated files that create a unified set of documents for an organization or individual to distribute over the Internet—a worldwide network of electronic documents connected by file servers, otherwise known as "the Web." Publicly available documents published on the Web become part of a world-wide network, which connects thousands of computers from across the planet. If you think of the Internet as the world's electronic office, each web site is an entire file cabinet, and each individual web page is like one document in a file. A simple web site may consist of just a single page, while a complex web site can consist of a large collection of web pages, which are linked together in some type of hierarchical organization structure.

Designing, building, updating, and maintaining the content and hierarchy of a web site is the job of web site designers, content managers, and web administrators. After a web site is built and posted to the Web, maintenance is ongoing as the information that the owners of the web site need to display grows and changes. Thus, beneath their home pages, the structures of the many web sites out on the Web vary widely, and they change frequently.

Many software tools are available for creating web sites. Word processors and other software programs usually provide an option to save files as a web page. If you work for a large company, they may have an entire department or consulting firm that works on web site design and creation. If you want to make a web site on your own, try software such as Adobe's Dreamweaver or Netscape's Composer. See the web page design entry in Part 3 for a detailed explanation of how to design specific pages for your site.

Home Pages, Addresses, and Links

The main entry point to a web site is its home page, which serves as the front page to view the site. The home page identifies the owner or subject of the site, its general purpose, and the types of information and activities that the site offers. The home page also provides *links* to associated visual media and files within the site, as well as other organizational identity markers, policies, products, and even legal information. The physical space on the screen (called "screen real estate") on the home page organizes complex, up-to-date information into sections and links that users can easily follow. The links take users to other web pages on the web site, which are linked together in a structure that may be flat, deep, interwoven, cross-referenced, and indexed in a variety of ways: this web-like structure of links within and among pages is called *hypertext*. Links can also connect users to other web sites or to file attachments, including text documents, photographs and other static visuals, video clips, audio recordings, and other types of multimedia.

Web sites are accessed via their uniform resource locator, or URL (more commonly called the "web address"). Web site addresses are chosen based on the organization's overall purpose: commercial, educational, governmental, community, and organizations are the most common. The web address ending indicates the *domain name* (.com, .edu, .gov, .net, .org), and can provide a key to the type of web site, its purpose, and the kind content you will find there. It can even indicate a region of the world, as with .asia, and the European Union (.eu), along with specialized purposes such as sites for pocket technologies that include web mobility (.mobi). New domains and domain names are added regularly. You can request almost any domain name you like; the domain name provider (companies available on the Internet) will let you know if that name is already taken.

Common Types of Web Sites

■ Education site—University of Minnesota front page

Features of Education Sites

1 School symbols, logos, and colors that many audiences will recognize dominate the site's visual presentation.

2 Large picture acts as a focal point appealing to prospective students and also alumni—the targeted users and school supporters.

3 Provides complex information for users at the university—faculty, staff, and students—as well as identity markers for visitors.

4 Represents organizational structure with links to departments, colleges, and people or other topic areas like the arts or student life.

5 Other campuses within the state system appear prominently as graphical links, as do the popular sports teams.

6 A prominent search field helps users navigate the dense information of this site.

■ Government site—Centers for Disease Control and Prevention (CDC) front page

Features of Government Sites

① Front page is filled with news, public events, publications, and technical information that both the public and medical experts will find useful.

② Governmental information and identifiers include a simple logo and a top banner that locates the CDC

within Health and Human Services federal government structure.

③ Key links that the public will be most interested in are listed on the left side of the page.

④ Site design divides information (based on the general divisions along the left and right side of the

screen) into clear paths for public health topics and events, publications, and data sets.

⑤ Provides a simple search feature.

⑥ Offers a translation into Spanish.

■ Organizational site home page—American Society for Nutrition

Features of Organizational Sites

1 Professional organization is made clear by a visible logo and organization name at the top of the page. Plenty of white space around the logo and name allow these to pop out.

2 A visible (white on blue) heading and brief paragraph make clear the organization's mission and purpose.

3 Concise photographs with short headings indicate three areas (research, clinical practice, global) where the organization has an impact. Readers can click on these for more information.

4 Publications and news feeds make it easy for readers to learn more about this organization.

5 Contact information is on the bottom of the page, where readers would expect to find it.

Commercial site—Allmusic.com

Features of Commercial Sites

1 Front page features new music releases and offers numerous categories for browsing and searching.

2 Large photographs display information (albums and recognizable artists) that change frequently.

3 Registration feature allows users to log in and receive specialized information.

4 Sophisticated but easy-to-use searching and browsing capability allows users to find what they are

looking for with as little work as possible.

5 Like most commercial sites, this site sells advertising space and collects information about its market.

Audience Considerations

Web sites are complex, multipage documents that are often aimed at multiple audiences and for multiple purposes. A web site about a new prescription drug, for example, will be read by patients and also by medical professionals. Audiences may include the public, technical experts, and inside members of the company or organization. Sites can generate attention to a product, company, or organization in ways that tie into its other organizational publications and advertisements. While the pages should be as correct and timely as possible, the overall site must be clearly organized, subdivided, accessible, and usable for these different audiences. This process is complex and multilayered: most web sites, unless they are password protected or limited to those with a subscription, can be viewed by many kind of readers, from specialists in the field to lay audiences.

Important audience considerations for web sites include the following:

Audience and purpose assessment (see Part One), including an audience needs analysis, is essential as you begin to plan the details. Focus your needs analysis on the core audience but plan for other site visitors as well.

Demographics: With an aging population and governmental accessibility requirements, web pages need to let users change the font size or otherwise make the information larger and easier to read. Other visual and audio features may need to be designed such that they can be accessed by special software that make the material accessible to people with certain physical challenges. See Accessibility in Part 1.

Global reach: Web sites are often translated into numerous world languages. Page design can influence how the automatic translation software works. See International Communication (Part 1) for more information.

Speed of reading: People do not read web pages as slowly or in the same linear format as they read print-based materials. Instead, they move quickly between chunks of text, visuals, and links to other information.

People who visit web sites want to find what they need quickly and easily. Create a site that does not overwhelm with too much information. Keep in mind that your web site may be visited by many kinds of users, from the general public to technical specialists. Be sure to provide links and information for different levels of users. Don't assume that the visitors to the site will know about the organization or company; use a link such as "about us" to provide contact information for specific audiences and methods (email, phone, mailing address).

Design Considerations

Creating web sites involves constant attention to the audience and purpose as well as the design of the site. It makes no sense to write large sections of beautifully written prose when, in the end, the text needs to be created in short chunks, linked across the entire web site. Especially on the front page, write in short, crisp sentences, using active voice, and limit the size of text to one short paragraph on one topic. Save longer text for secondary web pages. Create storyboards for your entire site structure using pencil and paper (one sheet per screen) or a software tool.

You can use nearly any word processor and many software programs to create web sites in a point-and-click environment. Web sites are actually created with a markup language called Hypertext Markup Language (HTML). Other software code, such as Flash or Java, is also employed to create certain interactive effects. When you use a software package such as Dreamweaver to create your web site, you never actually set the HTML or other code. But for sophisticated, complex web sites, you or a computer programmer may be required to know some of the more technical aspects of creating web site. It is fair to say that professional web sites are rarely created by one person. Usually, web site development requires a teamwork approach. See Web Page Design, in Part 3 of this book, for a detailed discussion about specific web page layout.

Guidelines

FOR CREATING WEB SITES

- Identify primary and secondary audiences as well as the key purposes of your site.

- Interview perspective users early in this process so that you understand how they wish to use the site.

- If you are creating a large, complex site, work with a team.

- Determine which domain name (.org, .com, or .net) is most suited to the purpose of the site and purchase the name, if necessary.

- Make an inventory of assets, including existing media, logos, and organizational documents that you can use.

- Sketch out a general structure for your site's content. You can use pencil and paper, a flow chart, or any other planning tool.

- Storyboard your entire site structure using pencil and paper or a software tool.

- Use standard linking and consistent navigational features that users expect from your type of web site.

- Build accessibility into your web site design.

- Use web creation software to build your site.

- Edit your content for grammar, spelling, and consistency in style.

- Test the first version of the web site with members of the organization and of the general public audience before publishing the site on your server.

- Create a plan for storing the files and for regular updates to the site.

See also

Accessibility (Part 1)

Blogs and Wikis (Part 2)

Multimedia (Part 3)

Storyboards (Part 3)

Web Page Design (Part 3)

Usability and User Testing (Part 4)

White Papers

White papers are reports that present information and advocate for certain standards, recommendations, or approaches. Usually, the topic is one that has economic, technical, or policy implications. White papers vary widely in scope and length. A good white paper will pull together information from a variety of research-based sources and present that information in a high-level analysis. White papers are often authored by a specific company, by a consortium of companies and organizations, or by a government agency. Unlike scientific and technical journal articles, white papers are self-published and produced in house or by writing consultants. They focus on a position, a summary of information and research, and other overview statements. Readers use white papers as important sources of information when conducting research or considering technical standards or policy positions. White papers can become marketing tools for businesses that want to persuade readers about products, and thus educate and influence readers who want information.

The audience for a white paper often includes non-specialists. Whatever their purpose, white papers should balance relevant historical background with current trends and solutions. Avoid a heavy-handed marketing approach if you want to establish credibility and maintain reader interest. This white paper on from the Environmental Protection Agency (EPA) on nanotechnology offers 136 pages of expert research, including a formal introduction and several appendices.

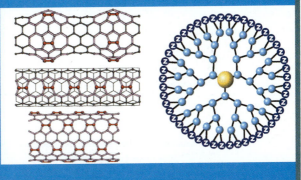
■ White paper on nanotechnology from the EPA: Cover page

Features of a White Paper

① Professional cover page identifies the organizational author and the topic of the paper.

② The EPA's organizational logo and name are prominent in the upper left corner.

③ The date of this publication, its location on the EPA web site, and the document's unique number lend credibility and currency.

④ The title of the white paper incorporates several technical illustrations to generate interest.

Nanotechnology White Paper

Workgroup Co-Chairs

⑤

Jeff Morris
Office of Research and Development

Jim Willis
Office of Prevention, Pesticides and
Toxic Substances

⑥

Science Policy Council Staff

Kathryn Gallagher
Office of the Science Advisor

Subgroup Co-Chairs

⑦

External Coordination Steve Lingle, ORD Dennis Utterback, ORD	**Ecological Effects** Anne Fairbrother, ORD Tala Henry, OPPTS Vince Nabholz, OPPTS	**Risk Management** Flora Chow, OPPT
EPA Research Strategy Barbara Karn, ORD Nora Savage, ORD	**Human Exposures** Scott Prothero, OPPT	**Converging Technologies** Nora Savage, ORD
Risk Assessment Phil Sayre, OPPTS	**Environmental Fate** Bob Boethling, OPPTS Laurence Libelo, OPPTS John Scalera, OEI	**Pollution Prevention** Walter Schoepf, Region 2
Physical-Chemical Properties Tracy Williamson, OPPTS	**Environmental Detection and Analysis** John Scalera, OEI Richard Zepp, ORD	**Sustainability and Society** Diana Bauer, ORD Michael Brody, OCFO
Health Effects Deborah Burgin, OEI Kevin Dreher, ORD	**Statutes, Regulations, and Policies** Jim Alwood, OPPT	**Public Communications and Outreach** Anita Street, ORD

■ White paper on nanotechnology from the EPA: Authors and credentials

Features of a White Paper *(continued)*

⑤ White papers are generally written by a group of technical experts. This page lists names and provides credibility for the paper.

⑥ The key workgroups and sub-group co-chairs indicate the EPA's Science Policy Council structure (much like a organizational chart).

⑦ This structure indicates a significant amount of teamwork used in producing the paper.

⑧ **Table of Contents**

■ White paper on nanotechnology from the EPA: Table of contents

Features of a White Paper *(continued)*

⑧ A detailed table of contents provides a quick overview of the sections.

⑨ Subsections are clearly named and numbered.

⑩ Numbering format of sections provides a scientific look to the paper.

⑪ A running head with a line and page number keeps readers oriented to their location in the document.

⑫ **EXECUTIVE SUMMARY**

Nanotechnology has potential applications in many sectors of the American economy, including consumer products, health care, transportation, energy and agriculture. In addition, nanotechnology presents new opportunities to improve how we measure, monitor, manage, and minimize contaminants in the environment. While the U.S. Environmental Protection Agency (EPA, or "the Agency") is interested in researching and developing the possible benefits of nanotechnology, EPA also has the obligation and mandate to protect human health and safeguard the environment by better understanding and addressing potential risks from exposure to nanoscale materials and products containing nanoscale materials (both referred to here as "nanomaterials").

Since 2001, EPA has played a leading role in funding research and setting research directions to develop environmental applications for, and understand the potential human health and environmental implications of, nanotechnology. That research has already borne fruit, particularly in the use of nanomaterials for environmental clean-up and in beginning to understand the disposition of nanomaterials in biological systems. Some environmental applications using nanotechnology have progressed beyond the research stage. Also, a number of specific nanomaterials have come to the Agency's attention, whether as novel products intended to promote the reduction or remediation of pollution or because they have entered one of EPA's regulatory review processes. For EPA, nanotechnology has evolved from a futuristic idea to watch, to a current issue to address.

⑬ In December 2004, EPA's Science Policy Council created a cross-Agency workgroup charged with describing key science issues EPA should consider to ensure that society accrues the important benefits to environmental protection that nanotechnology may offer, as well as to better understand any potential risks from exposure to nanomaterials in the environment. This paper is the product of that workgroup.

The purpose of this paper is to inform EPA management of the science needs associated with nanotechnology, to support related EPA program office needs, and to communicate these nanotechnology science issues to stakeholders and the public. The paper begins with an introduction that describes what nanotechnology is, why EPA is interested in it, and what opportunities and challenges exist regarding nanotechnology and the environment. It then moves to a discussion of the potential environmental benefits of nanotechnology, describing ⑭ environmental technologies as well as other applications that can foster sustainable use of resources. The paper next provides an overview of existing information on nanomaterials regarding components needed to conduct a risk assessment. Following that there is a brief section on responsible development and the Agency's statutory mandates. The paper then provides an extensive review of research needs for both environmental applications and implications of nanotechnology. To help EPA focus on priorities for the near term, the paper concludes with staff recommendations for addressing science issues and research needs, and includes prioritized research needs within most risk assessment topic areas (e.g., human health effects research, fate and transport research). In a separate follow-up effort to this White Paper,

---

■ White paper on nanotechnology from the EPA: Executive summary

Features of a White Paper *(continued)*

⑫ This 136-page white paper is condensed into a short executive summary, similar to one used with long technical reports.

⑬ The second and third paragraphs condense all the agency's relevant background activities for non-specialist readers.

⑭ The fourth paragraph focuses on the purpose of this paper and gives a brief summary of the entire contents.

2.0 Environmental Benefits of Nanotechnology

⑮ 2.1 Introduction

As applications of nanotechnology develop over time, they have the potential to help shrink the human footprint on the environment. This is important, because over the next 50 years the world's population is expected to grow 50%, global economic activity is expected to grow 500%, and global energy and materials use is expected to grow 300% (World Resources Institute, 2000). So far, increased levels of production and consumption have offset our gains in cleaner and more-efficient technologies. This has been true for municipal waste generation, as well as for environmental impacts associated with vehicle travel, groundwater pollution, and agricultural runoff (OECD, 2001). This chapter will describe how nanotechnology can create materials and products that will not only directly advance our ability to detect, monitor, and clean-up environmental contaminants, but also help us avoid creating pollution in the first place. By more effectively using materials and energy throughout a product lifecycle, nanotechnology may contribute to reducing pollution or energy intensity per unit of economic output, reducing the "volume effect" described by the OECD.

⑯ 2.2 Benefits Through Environmental Technology Applications

2.2.1 Remediation/Treatment

Environmental remediation includes the degradation, sequestration, or other related approaches that result in reduced risks to human and environmental receptors posed by chemical and radiological contaminants such as those found at Comprehensive Environmental Response, Compensation and Liability Act (CERCLA), Resource Conservation and Recovery Act (RCRA), the Oil Pollution Act (OPA) or other state and local hazardous waste sites. The benefits from use of nanomaterials for remediation could include more rapid or cost-effective cleanup of wastes relative to current conventional approaches. Such benefits may derive from the enhanced reactivity, surface area, subsurface transport, and/or sequestration characteristics of nanomaterials.

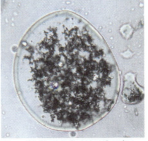

Figure 13. Nanoscale zero-valent iron encapsulated in an emulsion droplet. These nanoparticles have been used for remdiation of sites contaminated with variuos organic pollutants. (Image cortesy of Dr. Jacqueline W. Quinn, Kennedy Space Center, NASA) ⑰

Chloro-organics are a major class of contaminants at U.S. waste sites, and several nanomaterials have been applied to aid in their remediation. Zero-valent iron (Fig. 13) has been used successfully in the past to remediate groundwater by construction of a permeable reactive barrier (iron wall) of zero-valent iron to intercept and dechlorinate chlorinated hydrocarbons such as trichloroethylene in groundwater plumes. Laboratory studies indicate that a wider range of chlorinated hydrocarbons may be dechlorinated using various nanoscale iron particles

---

■ White paper on nanotechnology from the EPA: Major section on environmental benefits

Features of a White Paper *(continued)*

⑮ Each major section presents an overview of the topic in its own "Introduction." This section offers an overview of the environmental benefits.

⑯ Headings and subheadings help break up long sections of technical information.

⑰ Photographs provide visual references for the adjacent textual material.

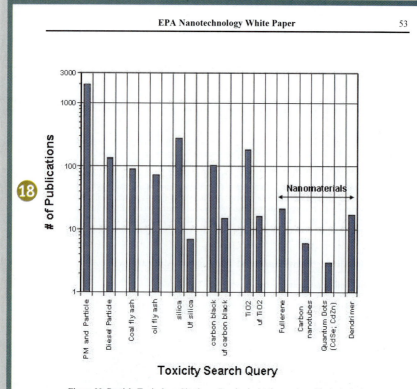

Toxicity Search Query

Figure 20. Particle Toxicology Citations. Results depict the number of toxicological publications for each type of particle obtained from a PubMed search of the literature up to 2005 using the indicated descriptors and the term "toxicity." Uf denotes ultrafine size (<100nm) particles.

These initial findings indicate a high degree of uncertainty in the ability of current particle toxicological databases to assess or predict the toxicity of intentionally produced carbon-based nanomaterials displaying novel physicochemical properties. Additional comparative toxicological studies are needed to assess the utility of the current particle toxicological databases in assessing the toxicity of other classes or types of intentionally produced nanomaterials, as well as to relate their health effects to natural or anthropogenic ultrafine particles.

■ White paper on nanotechnology from the EPA: Research summary on toxicity

Features of a White Paper *(continued)*

⑱ Bar chart provides an overview of research citations, by topic. A good white paper will synthesize research in a way that is easy for the reader to grasp.

⑲ The figure caption gives the methods used to general these search results.

⑳ A summary of the meaning of this research is very clear in this first sentence ("These initial findings indicate a high degree of uncertainty . . .").

6.0 Recommendations

This section provides staff recommendations for Agency actions related to nanotechnology. These staff recommendations are based on the discussion of nanotechnology environmental applications and implications discussed in this paper, and are presented to the Agency as proposals for EPA actions for science and regulatory policy, research and development, collaboration and communication, and other Agency initiatives. Included below are staff recommendations for research that EPA should conduct or otherwise fund to address the Agency's decision-making needs. When possible, relative priorities have been given to these needs. Clearly, the ability of EPA to address these research needs will depend on available resources and competing priorities. Potential lead offices in the Agency have been identified for each recommendation. It may be appropriate for other EPA offices to collaborate with the identified leads for specific recommendations. EPA should also collaborate with outside groups to avoid duplication and leverage research by others. Identified research recommendations were used as a point of departure for Agency discussion and development of the EPA Nanotechnology Research Framework, attached as Appendix C.

6.1 Research Recommendations for Environmental Applications

6.1.1 Research Recommendations for Green Manufacturing

- ORD and OPPT should take the lead in investigating and promoting ways to apply nanotechnology to reduce waste products generated, and energy used, during manufacturing of conventional materials as well as nanomaterials.

6.1.2 Research Recommendations for Green Energy

- ORD and OPPT should promote research into applications of nanomaterials green energy approaches, including solar energy, hydrogen, power transmission, diesel, pollution control devices, and lighting.

6.1.3 Environmental Remediation/Treatment Research Needs

- ORD should support research on improving pollutant capture or destruction by exploiting novel nanoscale structure-property relations for nanomaterials used in environmental control and remediation applications.

6.1.4 Research Needs for Sensors

- ORD should support development of nanotechnology-enabled devices for measuring and monitoring contaminants and other compounds of interest, including nanomaterials. For example, ORD should lead development of new nanoscale sensors for the rapid detection of virulent bacteria, viruses, and protozoa in aquatic environments

---

■ White paper on nanotechnology from the EPA: Recommendations

Features of a White Paper *(continued)*

21 Most white papers end with sections on conclusions or recommendations for new standards, research, or industry guidelines.

22 This section emphasizes specific areas and needs for further research, which fits the overall purpose of the paper.

Audience Considerations

Even when white papers discuss technical subjects, their readers can be non-specialists. It is important to write for these more general audiences and to avoid too much technical complexity and mundane detail about the topic. The paper should balance accurate, relevant historical background with current trends and solutions in a way that captures readers' attention and makes readers want to read on. Although not always peer reviewed, white papers like this one have been vetted within the organization by a number of specialists and government administrators. These kinds of indicators establish credibility and will help you maintain reader interest. In general, try to avoid a marketing approach and focus instead on providing the best information to convince your readers with subtle persuasion.

Design Considerations

White papers need to follow all of the elements of document design and visual communication, as well as data display. Depending on its length, a white paper can include many or all the elements of a report. You should aim to write objectively, and reference only credible third-party sources. Use clearly named headings and subheadings to break up long sections of technical information. Clear and accurate charts, graphs, and other detailed visual information are especially important in a technical or research-based white paper. High-quality photographs and illustrations can provide visual references for the textual material and also sustain reader interest. Be sure to introduce all figures and tables carefully and include sequenced figure numbers, brief captions, research methods, and image credits, as appropriate.

You need to decide whether your white paper will appear on a web site, in print, or in multiple media. White papers available on the Web have the benefits of digital publication such as PDF file format, which allows for color text and visuals without adding cost. Some white papers are also printed using color on high-quality paper. Either type of media allows you to design and include many helpful graphics. Photographs and charts will probably be more useful to general readers, while tables of data help you condense more complex research or provide comparative information. Anything that lets you summarize and visualize the trends in technologies, processes, and other information will help readers.

Guidelines

FOR CREATING WHITE PAPERS

- Determine the purpose and scope for your white paper.

- Know your audience; conduct an audience and purpose assessment to determine whether the paper is intended primarily for a technical audience, an audience of experts, or an audience of nonexperts.

- Decide whether you will need multiple sections, appendices, or other additional content material.

- Provide a historical overview or other relevant background early in the paper.

- Write clearly and objectively, condensing research and technical information whenever possible.

- Avoid a heavy-handed marketing approach and use subtle persuasion through information instead.

- Use visuals, charts, tables, and illustrations as appropriate.

- Offer a section of conclusions and recommendations for further research or development.

- Add credibility by including all your team's credentials.

- Provide references to credible third-party sources throughout the paper or include a separate bibliography.

See also

Visuals and Other Media

Audio

Audio refers to audible sound and the sound elements included or recorded with various technologies. Audio can be an effective medium for technical communication, for example, when you need to describe a process in a training video and you want users to hear spoken instructions while an animation or video plays on the screen (called *voice-over narration*). Companies and industries that use computer-based training (CBT) or tutorials may employ other kinds of sound as part of a training video—from the actual sounds of machines and work environments to music that helps create a mood. Digital audio files are easily created and distributed with computer programs and equipment, and these files may contain recorded music, voice, or other sound effects that can then be included on a web page, in a blog, in a PowerPoint presentation, or played on pocket technologies like MP3 players and cellular phones. Because video usually includes audio tracks with the sounds of the person talking and other location sounds, it depends heavily on this recorded sound track to make sense. Sound files are also frequently used alone to distribute music samples on the Web, offer live radio or online broadcasts, and send or receive digital voice mail. Audio broadcasts on iPods, called podcasts, let users record and publish their own programs for others to download and play on their MP3 players. You can also use a tool called RSS (really simple syndication) to let users subscribe to a wide range of audio broadcasts from your site or from a collection of sites based on similar information.

Audio File Types

Sound is recorded, sampled, stored, and can be visualized as an acoustic sound wave using audio technologies and software applications. Like video clips, audio clips for the Web must be short and compressed to download quickly and hear using appropriate software. Audio files use the frequency, sample rate (KHz), and size (e.g., 16 bit) in an encoder to determine the quality of the sound—the higher the sound quality, the larger the file size.

Some common audio file formats include:

- AIFF and WAV files—large files for commercial music and other high-quality recordings
- MP3 files—the Web standard, where quality and size can be widely adjusted to maximize disk space
- MPEG audio files—the standard for audio broadcasts, highly compressed files

- AAC and other proprietary formats—developed for specific applications like Apple iTunes
- Streaming files such as RealAudio—play while downloading, but with limited size and audio quality

Choices for sound quality and file format are determined in each computer's sound control panels and also in the control settings for saving and importing files in all audio software programs.

Music

Technical communicators who create instructional videos, presentations, or podcasts may find themselves functioning as writers, producers, and sound engineers. Programs included with personal computers make creating and including music in your presentations relatively easy. The widely used audio program iTunes, for example, stores, categorizes, and copies music files that you own or purchased from an online store onto other media or into an MP3 player. Music tracks can help set a mood or support the action in a project.

If you plan to use an existing song in a published project, you may need to pay record company permission or licensing fees. Free music composition software like Apple's GarageBand offers high-quality instrument samples and mixing tools for creating original music even if you are a novice. Other software programs such as Audacity allow you to edit audio files in multiple tracks, mixing and adding effects as you would in a recording studio. Music files are then easily imported into another media program and used as a sound track. Even PowerPoint offers simple ways to record and integrate music and other sounds into a presentation.

Voice-over Narration

For podcasts or training videos, technical communicators and specialists may need to provide a script to be recorded as a voice-over narration. When you decide to use audio voice-over versus on-screen written directions depends on the amount of information and its purpose. On-screen text works well for summarizing and repeating main points or steps in a process. Audio voice-overs can provide the depth and presence of vocal inflection that on-screen text cannot. Voice-overs also provide details that are too long to read on a screen. Studies suggest that a combination of text on the screen and spoken narration of that same text offers the best retention. The steps or concepts need to be covered in the same order in both the audio and the visual media. If a voice-over provides a sequence of steps using numbers—1, 2, 3— and the slides use letters—a, b, c, for example—users may become confused. Write a complete and concise script for your voice-over narration. You may consider hiring a voice-over talent for your application if you want your project to have wide commercial appeal.

Equipment and Software

For recording audio, most desktop computers have built-in microphones and basic recording software in the operating system. The sound quality from a built-in microphone varies, so using an external microphone with an audio interface is usually a good idea. Video cameras can also include high-quality microphones that will record sound with the video portion turned off. When using external input sources of any kind, take care to keep the volume low and check the levels so as not distort the sound. Be aware of the environment and surrounding noise when you record. Whenever possible, record sound in a quiet location wearing headphones to monitor the sound and with the microphone close to the source.

Video editing programs such as Windows Movie Maker also include tools for recording a voice-over narration. You can use professional quality computer-based audio programs such as Sony Sound Forge and Digidesign Pro Tools to capture sounds with a microphone, from within a video, or from analog media (conventional audiotape or videotape). These tools save the sounds as digital file formats for use with other computer applications. Note that sound editing interfaces can be quite complicated, since they offer extensive options for sound mixing and special effects. Remember too that the audio portion and the visual stills, the videos with sound, and the text messages all need to be carefully integrated and timed to reinforce each other in your project or presentation.

Guidelines

FOR CREATING AUDIO

- Determine your application, purpose, and context for using audio: a sound clip for the Web, a voice-over for a video, a presentation, or a podcast.

- Decide whether the final application will be used as a stand-alone module, distributed via a web site, or saved onto on high-capacity storage media.

- Decide what audio editing software program you will use, the file quality needed, and storage requirements.

- Write a script or plan your recording.

- Record your audio in a quiet, controlled environment.

- Use a good quality external microphone if possible and headphones to monitor sound.

- Check sound input levels regularly to avoid distortion (too loud) or low input (too soft).

- Make sure the audio and visuals synch in the final project so you don't confuse users.

- Save only small audio files to your web site and provide links and explanations for clips.

Cartoons and Comics

Cartoons and comics are drawn artwork including visuals and words that engage an audience, usually through humor. Traditional print comic strips and single-frame cartoons may provide effective visuals in a technical document or presentation to tell a story, to illustrate a specific subject, or to make fun of human foibles. Depending on your audience and purpose, cartoons and comics can be used in reports and other documents to introduce a general topic or even to make a difficult topic more accessible with a funny example. You might also use comics in a presentation to connect with your audience through laughter, perhaps even defusing tension while also engaging the group. Humor, when used appropriately, offers an effective emotional appeal for persuasion. Showing that you have a sense of humor can also build rapport with your audience.

Dilbert Comic Strip: 07/25/06

■ This *Dilbert* comic strip uses humor and cynicism to critique frustrating experiences with teamwork in the workplace.

Because they often use satire and other humorous techniques, comics sometimes drawn on familiar stereotypes, like the nerdy and observant Dilbert character and his demanding, wisecracking boss. These characters can be a bit crass, but readers of this strip enjoy this kind of humor because it reveals the absurd realities of many workplaces. However, you do need to keep in mind that what one person finds funny, another person may not. Be careful when using humor in a professional setting.

Uses of Cartoons and Comics

The success of any comic depends on audience members recognizing themselves or a situation they can understand without being offended. Comics artist and expert Scott McCloud defines comics as two or more "juxtaposed pictorial and other images in deliberate sequence" that "convey information" or evoke "an aesthetic response in the viewer" (1994, p. 9). Thus, while their purpose is usually to evoke laughter, comics sometimes draw on serious current or historical events. The editorial cartoon, for example, can use a single frame to illustrate an opinion about current news events quite pointedly and often appears in an editorial section of a newspaper or magazine. Such openly opinioned pieces may be difficult to use in a diverse professional setting or with any audience that you don't know well. Humor does not translate well across cultural differences, so it's best to avoid using cartoons and comics in intercultural settings.

Knowing your audience well is the key to using a comic or cartoon appropriately. The targeted audience must be receptive to the message and also to the types of humor used. The main character in *Betty*, illustrated below, is a smart, ordinary professional woman with a family who regularly confronts current work issues.

***Betty* Comic Strip**. 07/21/06

In this conversation, Betty and her friend debate the privacy issues about blogs, a format where people sometimes include private details but also make that private information public. This comic points out confusing issues of online privacy and permissions by having each character play a personal role in the exchange. Many readers, not just women, who are confronting new tools and technologies may identify with these characters and their issues. Even when comics target a particular group, they can still have a wide appeal if they effectively use humor to comment on current issues.

Where to Find Comics and Cartoons

In additional to newspaper and magazine web sites, online sites like *Comics.com* offer huge collections of comic strips and editorials that are searchable by author, date, topic, or other details. The online magazine *Slate.com* includes "Today's Cartoons," where hundreds of political cartoons appear along with useful links by topic or artist to large archives. If you decide to use a commercial artist's work, remember to check terms of use and whether you need to get permission. Always cite the comic or cartoon's source and date in your document.

You can also create your own comics. Comic template programs like Plasq.com's *Comic Life* allow users to construct comic strips with their own content (pictures, etc.) along with provided bubbles and clip art. The suitability of this or any type of comic depends on the situation.

Guidelines

FOR USING CARTOONS AND COMICS

- Determine whether humor is appropriate for your audience and context.

- Be sure your audience understands the subject and type of humor you plan to use. Choose something neutral and simple enough so that most people identify with the cartoon.

- If you do not know an audience, it may be best to avoid using comics.

- Be aware that many cartoons have political overtones that may offend some audiences.

- Check your final version to see that the cartoon will display accurately on paper, on the Web, or from an overhead projector.

- Cite the comic or cartoon's source and date in your document.

- Always check terms of use. Get permission if necessary.

Charts

The terms *chart* and *graph* are sometimes used interchangeably, but there are important differences between these two types of visuals. Charts are used to represent relationships of parts to a whole or to show visual relationships that are not plotted on the *x-* and *y-axis* used in most graphs. For example, one common type of chart is a pie chart, which displays the relationship of various parts to the whole or total. The example below shows a pie chart of expenses for an engineering company. Pie charts are often accompanied by numerical data presented as a spreadsheet or table to allow readers to explore the displayed information in more detail. This pie chart allows readers to see a visual snapshot of the relative amount each category of expenses contributed to the company's total yearly expenses.

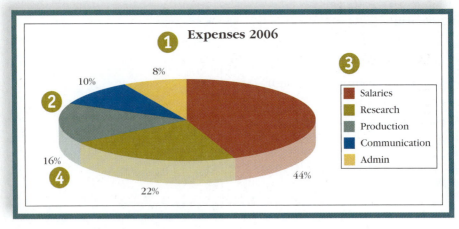

■ Pie chart created using Microsoft Excel

Features of a Pie Chart

1 The title clearly identifies the information the chart is intended to display.

2 Colors are used to visually distinguish each category of information.

3 Each category on the chart is labeled in the legend at the right.

4 A numerical percentage adds quantitative detail to the visual information.

Other types of charts include flow charts, organizational charts, and Gantt charts. This detail from a flow chart illustrates a series of events and decisions involved in generating drinking water.

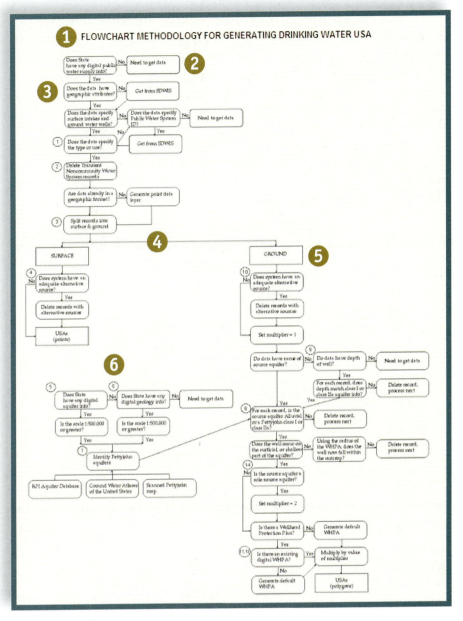

1 FLOWCHART METHODOLOGY FOR GENERATING DRINKING WATER USA

■ Sample flow chart

Features of a Flow Chart

1 The title identifies exactly the process that is described in the chart.

2 Each step in the process is represented by an oblong box.

3 In a decision chart like this example, each step is labeled in the form of a question.

4 Different paths may be taken depending on answers (yes or no) to each question.

5 Key categories or divisions in the process are represented by different boxes ("Surface" and "Ground").

6 Numbered items are used to divide major components of the process.

The next example shows a typical organizational chart. Organizational charts are useful for displaying the structure and internal relationships of units or individuals within an organization. In this case, the chart displays the relationships among the various divisions of the Federal Communications Commission (FCC), an agency of the U.S. government. Government and corporate organizational charts are often quite large and complex. Notice that it is set up with a top-down structure, which best represents the structure of this particular organization. The commissioners who oversee the FCC are shown at the top of the chart, and the lines display how each division reports to the commissioners. Notice in this example that all of the divisions report directly to the commissioners at the top. This chart tells us that the FCC is a highly centralized organization.

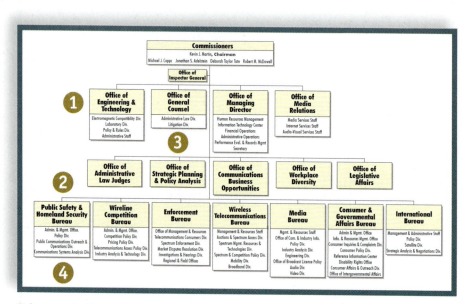

■ An organizational chart

Features of an Organizational Chart

1 Each position or department is represented by a shaded box (or other visually similar element).

2 The hierarchical structure of the organization is represented by parallel vertical layers or levels on the chart.

3 Vertical lines or arrows clearly show relationships of oversight and accountability within the organization.

4 Descriptive text is written so that each division or unit is described in similar terms, making it easy to compare units across all levels on the chart.

Another chart commonly used in the workplace is called a Gantt chart (named for engineer Henry L. Gantt who developed the chart format in the early 1900s). Gantt charts are useful for planning and tracking projects. A Gantt chart is a special type of bar chart that shows the timing of tasks as they occur and helps to schedule resources and events needed in a collaborative project. (See Project Management Visuals for more about Gantt charts.)

Designing Charts

Often, people can read and comprehend charts more easily than the information on which the chart is based. Readers scan charts for a general impression and do not read the fine print to evaluate the chart's accuracy. It is important that the information delivered in a chart is true to the data from which it came. Look at your chart carefully to be sure that it represents your data or information clearly and accurately. Charts can be used, intentionally or not, to distort information, or to make exaggerated claims based on partial or misinterpreted data.

Charts should be designed to meet the needs and expectations of their intended audience. Effective chart design can be achieved through a number of computer programs, which help you create charts. Using a program like Microsoft Excel will help you build a professional looking chart or graph. With a few clicks, for example, a data set can be displayed in the form of a pie chart. Be careful about using too many features (colors, fonts, and so on): your graph or chart can easily be "overtaken" by design features that detract from your message or purpose, creating what visual communication expert Edward Tufte (1983) calls "chart junk." Simplicity and clarity should never be sacrificed in the name of technology. Sometimes what looks great on your computer screen looks terrible in print.

Guidelines

FOR CREATING CHARTS

- Think about how your audience will view the data when creating a chart.

- Design each chart to show one primary visual idea or a specific relationship.

- Make your chart clear; do not include too much information.

- Use labels and titles effectively; do not use too much text.

- Do not create charts that could be distorted or are purposely misleading.

- Take advantage of Microsoft Excel and other software but be careful about creating a chart with too many bells and whistles.

See also

Data Display (Part 3)

Graphs (Part 3)

Project Management Visuals (Part 3)

Spreadsheets (Part 3)

Tables (Part 3)

Clip Art

Clip art describes the simple artwork or illustrations commonly included with a software program or available on the Web or in clip art books. While not always humorous like cartoons and comics, clip art can add also visual interest and a visual appeal to reports or presentations that engages the audience and builds rapport. Clip art tends to look informal and to represent objects as simple rather than realistic pictures or drawings. A frequent application of clip art in technical writing is to use symbols or icons that signal particular sections or highlights in a text. In presentations, clip art should be used sparingly so as not to give an unprofessional or cluttered look to the visual presentation.

The iconic clip art shown here is a widely recognized symbol indicating wheelchair accessible facilities.

Other types of clip art include the numerous line drawings, illustrations, and other visual representations that can be inserted into a word processing

■ The International Symbol of Access

document, slide show, or web page. Most software comes with a choice of clip art.

This type of clip art can be useful if you are writing instructions or procedures and need to refer users to a visual reference for a piece of hardware or other item. Clip art can also be a helpful way to guide readers through a lot of textual material. For instance, in this training manual from the National Library of Medicine, clip art of a notebook and pencil is used selectively in the left margin, as a way to cue readers to take notes.

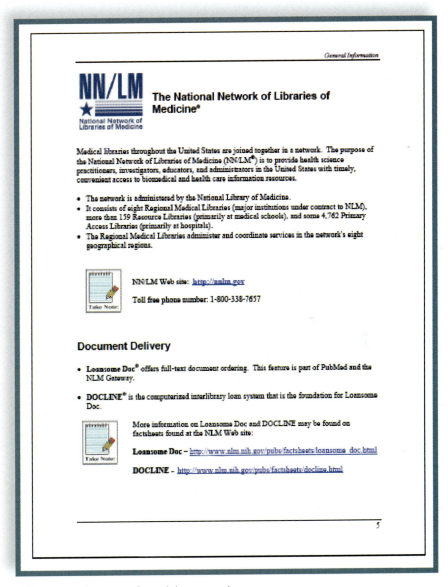

■ Clip art used as part of a training manual

FOR USING CLIP ART

- Decide whether clip art–style visuals are appropriate for your audience and context.

- Ensure that your audience will understand the meaning of the image or icon.

- Use the visual consistently throughout your document or presentation.

- Use clip art sparingly so as not to clutter the design of the document.

- Situate the clip art as close as possible to the text that it is supporting.

- Check your final document to ensure the clip art prints clearly or displays clearly on the screen.

See also

Cartoons and Comics (Part 3)

Presentations (Part 2)

Reports (Part 2)

Symbols (Part 3)

Data Display

Displaying data in an easily understandable, accessible format is an important feature of clear technical communication. By using visual display techniques, you can help focus reader attention on key aspects of the data or give readers visual cues to help them quickly find what they need. Readers understand the visual display of data in a different way than they do by just looking at the statistical or numeric data in a chart or table. By revealing relationships and trends in the data, data visualizations engage your reader in the material and help them make decisions or take action. Different types of research lead to different kinds of numerical and other data. Both quantitative and qualitative data sets suggest strategies for thinking through how you want readers to view the results you present in a document.

Quantitative Data Visuals

For any group of numerical data, you will have many potential choices for how to display it. The context of your communication, the needs of your audience, and the data itself all influence visual design choices. The type of display you choose depends on what story you want the data tell to tell, and that story often depends on relationships between data sets.

Emphasizing Specifics

Tables are preferable when grouping many different related data sets into one place and when your reader needs to focus in on the specifics. The sample table below, for example, tracks more than five separate variables (worker characteristics, mean hourly earnings, relative error, mean weekly hours, and type of industry). A transportation worker can easily hone in on just the type of information needed by scanning across one row's worth of information. This one table helps readers who have a wide range of interests focus in on very specific, relevant information.

Table 1-1. Summary: Mean hourly earnings[1] and weekly hours by selected characteristics, private industry and State and local government, National Compensation Survey, Minneapolis-St. Paul, MN-WI, May 2004

Worker and establishment characteristics	Total			Private industry			State and local government		
	Hourly earnings		Mean weekly hours[3]	Hourly earnings		Mean weekly hours[3]	Hourly earnings		Mean weekly hours[3]
	Mean	Relative error[2] (percent)		Mean	Relative error[2] (percent)		Mean	Relative error[2] (percent)	
Total	$21.80	2.2	35.2	$21.18	2.7	34.8	$24.31	2.7	37.1
Worker characteristics:[4]									
White-collar occupations[5]	25.40	2.5	37.4	25.22	3.1	37.3	26.00	3.0	37.8
Professional specialty and technical	31.21	2.1	37.4	31.60	3.0	37.3	30.43	2.4	37.7
Executive, administrative, and managerial	32.90	5.7	40.1	32.96	6.4	40.1	32.61	10.7	40.0
Sales	21.29	16.4	31.5	21.32	16.4	31.5	–	–	–
Administrative support	15.95	2.3	38.1	15.87	3.0	38.4	16.22	1.5	37.2
Blue-collar occupations[5]	18.14	3.4	36.0	17.93	3.6	35.7	20.57	2.9	38.9
Precision production, craft, and repair	21.96	5.2	40.0	21.99	5.8	40.0	21.78	3.8	40.0
Machine operators, assemblers, and inspectors	16.31	5.6	37.9	16.31	5.6	38.0	–	–	–
Transportation and material moving	16.25	3.2	31.3	15.90	3.9	30.6	18.45	4.3	36.4
Handlers, equipment cleaners, helpers, and laborers	13.98	5.5	30.8	13.59	5.7	30.3	19.49	6.2	40.0
Service occupations[5]	12.51	5.0	27.0	10.29	3.0	25.4	19.37	5.6	33.5

■Table providing breakdown of jobs and pay scales

Highlighting Trends and Relationships

Unlike tables, graphs can generally only handle two or three different variables effectively. These visual devices help tell a story about the relationships, comparisons, and trends in your data. Graphs are a way for your reader to quickly grasp a particular message. While graphs are a great way to present complex information, in doing so they sacrifice the precision of a table. In this way, graphs work to present the big picture, leaving more specific presentations of data to tables and written discussion. At a glance, the graph below shows, for example, the sharp increase in rice yields over time.

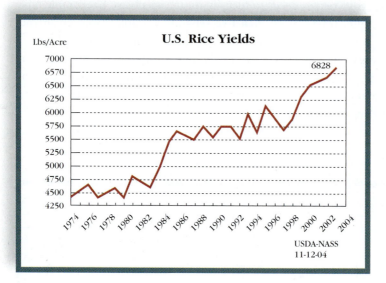

■ Line graph showing a trend

By tracking multiple trends in a single graph or chart, you can help your reader more easily draw comparisons than if the same data were in a table or paragraph format. For example, notice the contrast between trends in the graph versus the table below. While the table format provides all of the data, the bar graph makes the data visually apparent and easy to understand.

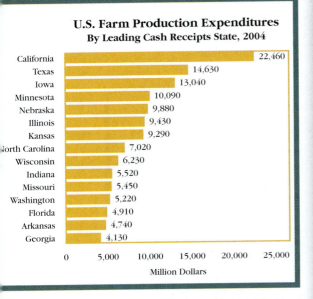

U.S. Farm Production Expenditures
By Leading Cash Receipts State, 2004

State	Million Dollars
California	22,460
Texas	14,630
Iowa	13,040
Minnesota	10,090
Nebraska	9,880
Illinois	9,430
Kansas	9,290
North Carolina	7,020
Wisconsin	6,230
Indiana	5,520
Missouri	5,450
Washington	5,220
Florida	4,910
Arkansas	4,740
Georgia	4,130

Bar graph showing farm production data with multiple trends

US Farm Production Expenditures (2004)	
State	**$ (millions)**
California	22460
Texas	14630
Iowa	13040
Minnesota	10090
Nebraska	9880
Illinois	9430
Kansas	9290
North Carolina	7020
Wisconsin	6230
Indiana	5520
Missouri	5450
Washington	5220
Florida	4910
Arkansas	4740
Georgia	4130

■ Multiple trends in a table format

Showing contrasting trends in one line graph can also be a valuable strategy. For example, the difference between the adult and juvenile trends shown in the next graph may lead the reader to ask different questions about crime causality than an adult-only or juvenile-only line graph might.

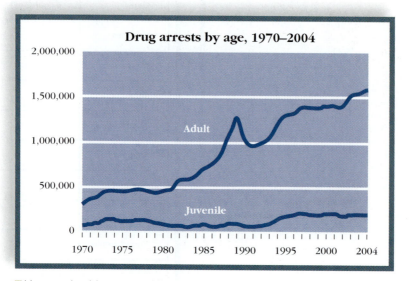

■ Line graph with two trend lines

Visualizing Complex Numerical Data

Science, engineering, and medical fields use various visual techniques to create images that represent complex arrays of numerical data. In mathematics, three-dimensional (3D) representations of calculus formulations are now commonly used to help conceptualize otherwise abstract information. Meteorologists use mapping software to visualize weather modeling results that represent literally thousands of calculations. Through maps and the use of color for symbolic representation, the next example below shows predicted temperature over a large section of the country. The second image converts raw numerical data into a visually powerful 3D representation showing the temperature and structure of a vortex (Jupiter's red spot).

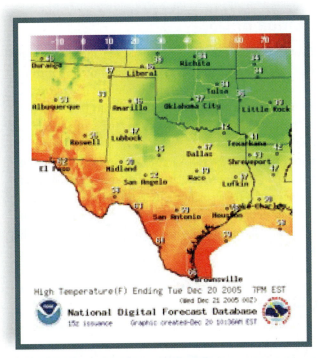

■ Map depicting weather modeling results

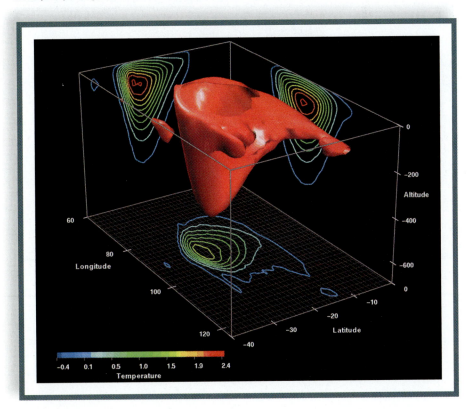

■ 3D map depicting temperature and structure of Jupiter's red spot

Individual Customization of Data

Increasingly, the scientific community is moving beyond static paper documents to share information online. Digitally exchanged information allows opportunities to design visuals so that users can choose how and when to view particular parts of a complex data set. Designing for user interactivity can be more time consuming, but it can also dramatically increase the effectiveness of your visual display for audiences in different contexts.

Geographical information system (GIS) software, for example, can combine geographical and other spatial data with statistical data that the user can selectively view or hide, depending on the need. GIS also allows users to zoom in to particular geographical areas for greater detail. Through one interactive visual, literally thousands of bits of data and hundreds of variables are available to understand the context and make comparisons. As shown below, for example, California's Digital Atlas offers a range of variables for a user to choose from in order to generate a map of the state or part of the state. The result of these selections color codes the state according to wetland classifications.

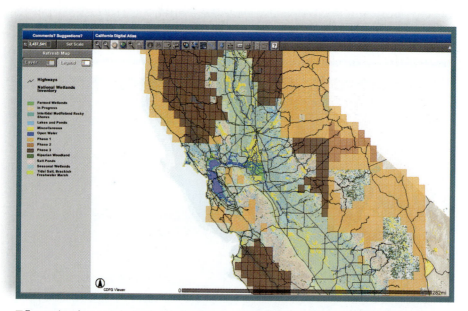

■ Example of user-selected visualization

Using the same section of the map as an example, but with Impaired Water Bodies selected instead of National Wetlands Inventory, the image changes (see below).

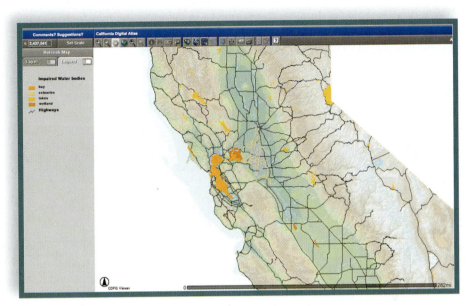

■ Example of variable data selection options and their visualization

This digital interactive visual gives users the widest range of options for displaying and interpreting complex data sets.

Qualitative Data Visuals

Qualitative data includes information that is not numerical. Qualitative data can be displayed in a variety of ways. By using visual strategies, you can help organize the information to show patterns and relationships that emerge from the data.

When sharing qualitative data, you can use themes that emerge from the information in order to organize and display it in manageable pieces. Using a document design with headings and subheadings, you can help highlight overall themes. Often, the table format is an effective visual tool for qualitative data. One of the simplest and most ubiquitous examples is a two-column glossary or definition table. Planning documents, which track tasks, resources, and timeline information, also can use tables, such as this example below.

Table 13 - Erosion Control Implementation Plan

Activity Steps	RMP Unique ID No.	Resources	Measurement	Target Date
1. Construction erosion and sediment control ordinance.	4-02-R	• City Staff	• Review ordinance language and requirements. • Complete process and Council approval.	2004 2005
2. Development plan review process.	4-04-R	• City Staff	• Site developed.	Annual
3. Construction site inspection and street sweeping follow-up.	4-05-R	• City Staff • Contractor	• Completed inspections and follow up actions.	Annual

■ Planning document data displayed as a table

As the qualitative data categories become more complex, you might also consider including symbols to represent some of the categories in a table. Using this type of strategy, the following table from a water manual uses column headings to show categories of pollutants for different types of activities, and then includes symbols to indicate the four levels of pollutant contribution.

Table 10.10 Stormwater Pollutants Associated With Common Operations at Potential Stormwater Hotspots (Schueler et al., 2004)

Operation or Activity	Nutrients	Metals	Oil / Hydrocarbons	Toxics	Others
Vehicle Repair	◔	●	●	●	
Vehicle Fueling	◔	●	●	●	(MTBE not used in MN)
Vehicle Washing	●	◐	◐	●	Water Volume
Vehicle Storage	○	◐	●	◔	Trash
Outdoor Loading	◐	◐	◔	◔	Organic Matter
Outdoor Storage	◐	◐	◐	◐	
Liquid Spills	◐	◐	●	●	
Dumpsters	◐	◐	◐	●	Trash
Building Repair	◔	◐	◐	◐	Trash
Building Maintenance	○	●	◔	◐	
Parking Lot Maintenance	◔	◐	●	◐	Chloride
Turf Management	●	○	○	●	Pesticides
Landscaping	●	○	○	●	Pesticides
Swimming Pool Discharges	○	○	○	○	Chlorine
Golf Courses	●	◔	○	●	Pesticides
Hobby Farms/Race Tracks	◐	○	○	○	Bacteria
Construction	◐	◔	◔	◐	Trash, Sanitary Waste, Sediment
Marinas	◐	◐	◐	●	Bacteria
Restaurants	◐	○	●	○	Grease

Key:
- ● major contributor
- ◐ moderate contributor
- ◔ minor contributor
- ○ not a pollutant source

Table from a water manual that uses data symbols

Technical illustrations can also depict relationships of categorical information. In the simple graphic on the next page one image quickly shows the increasing breadth of water protection, based on the various categories.

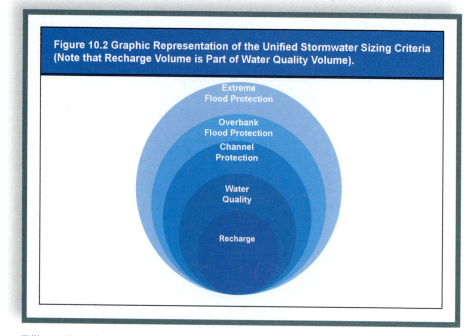

Figure 10.2 Graphic Representation of the Unified Stormwater Sizing Criteria (Note that Recharge Volume is Part of Water Quality Volume).

Extreme
Flood Protection

Overbank
Flood Protection

Channel
Protection

Water
Quality

Recharge

Illustration of stormwater size criteria

Reporting Qualitative Research Results

Qualitative data is common in market research, usability testing, and some academic research environments. Such data may begin in written format, such as notes regarding personal observations about people or phenomena. Sometimes, the source data are in audio or video format, which then gets translated into a textual format for analysis, as with transcripts of interviews or focus group feedback. Other qualitative data sources may be what others have written, such as internal corporate memos, published articles, or historical archives. The challenge, then, becomes how to organize the plethora of words into usable categories and formats.

In the following below, the table provides a succinct summary of some major themes emerging from the transcripts of a series of interviews. The table allows the reader to compare quickly some of the views of the major categories of interviewees.

Exhibit 10.
Data matrix for Campus A: What was done to share knowledge

Respondent group	(a) Activities named	(b) Which most effective	(c) Why
Participants	• Structured seminars • E-mail • Informal interchanges • Lunchtime meetings	• Structured seminars • E-mail	• Concise way of communicating a lot of information
Nonparticipants	• Structured seminars • Informal interchanges • Lunchtime meetings	• Informal interchanges • Structured seminars	• Easier to assimilate information in less formal settings • Smaller bits of information at a time
Department chair	• Structured seminars • Lunch time meetings	• Structured seminars	• Highest attendance by nonparticipants • Most comments (positive) to chair

■ Table representation of interview data

Researchers who work with large volumes of complex qualitative data often use software to analyze the results and help illustrate the themes or trends in the data. Such software can also make visible the relationship between ideas that emerged during data analysis.

Combining Quantitative and Qualitative Data

You may choose to combine the precision of quantitative data with the descriptive aspects of qualitative data to give an even clearer picture of a situation. This dual approach is common in usability studies, which collect a range of data from both categories. For example, a usability test of software may include quantitative data, such as how much time or how many keystrokes it takes to accomplish a task. It might also include qualitative data, such as audio or video recordings of what users say as they run the software or what a group of people says in a conversation about their experiences.

Note that the distinction between qualitative and quantitative data can dissolve. A common approach to data analysis turns qualitative information into quantitative information by assigning numbers to the various trends in the data. For example, a series of interviews with 25 researchers might include an open-ended question about common ethical choices they face in their professions. A review of the transcripts might reveal five main ethical conflicts, with a certain percentage of researchers mentioning each one. Reporting these percentages provides a quantitative approach to reporting the data. A qualitative approach to reporting the same data might use these main ethical conflicts as headings but include quotes or summaries from the interviews that provide concrete, specific examples revealing much more about the context and dynamics of ethical conflicts than mere percentages might capture. Both types of data might be important to give the most complete and most understandable picture of what the interviews tell readers about the ethics of research.

Guidelines

FOR DATA DISPLAY

- Decide what readers most need to understand and create visual displays of data that help readers make decisions or take action.

- For statistical or numeric data that emphasize the details, consider a chart or table.

- To reveal relationships and trends in the data, use bar or line graphs with multiple variables.

- For complex data sets, consider using three-dimensional maps or providing variable data options that users can select.

- For qualitative (non-numerical) data, use tables or technical illustrations to organize information and show patterns and relationships that emerge from the data.

- To combine qualitative and quantitative results, convert findings into usable categories and visual formats, as appropriate.

See also

Charts (Part 3)

Visual Communication (Part 3)

Document Design (Part 3)

Graphs (Part 3)

Maps (Part 3)

Symbols (Part 3)

Tables (Part 3)

Technical Illustrations (Part 3)

Types of Research (Part 4)

Usability and User Testing (Part 4)

Document Design

Every document you help produce as a technical communicator requires elements of design, from the most basic margins of a single printed page to the many visual features of book-length reports. This entry covers the basic elements of page layout and other key concepts for print document design:

Basics of page layout

Captions

Headings

Labels and callouts

Lists

Tables of contents

Typefaces and fonts

Basics of Page Layout

Page layout is the arrangement of all design elements on the page. Visual appearance gives readers their first impression, and a document's design largely determines whether that document is usable. If information is difficult to find or read, the document does not serve its purpose well.

In effective page layouts, form follows function; the layout of the page must help readers find and grasp information quickly. What constitutes effective page design varies based on the document's purpose, usage, scope, and intended audience. Centered headings may be effective design in the case of a short product brochure where style and elegance matter most, but those headings would not work well in a 500-page reference manual, which requires fast scanning of vast material. Likewise, a pamphlet on health care written for the elderly could use an inviting, spacious layout and larger type, while legal contracts aimed at corporate attorneys might emphasize efficiency and conserving space with small type and less white space.

An effective page layout should be

Readable: easy to read and understand the purpose and scope of the document.

Consistent: the document makes it easy to determine how information is organized. Visual and layout features stay the same throughout the document.

Easy to navigate: readers can find the specific information they need.

Preliminaries to Page Layout

In most large organizations, the format for a particular document has already been determined, and you may not have many design questions to consider. For instance, most large companies have preset style sheets and templates or other tools that are used to create reports, strategic plans, proposals, and so on. In fact, it is rare in any workplace setting to create a document design from scratch. Templates from programs like Microsoft Word, or in-house custom publishing tools often specify the headings, layout, page size, and so on.

However, it is still important when thinking about page layout to assess the kinds of information the document will contain. The answers to these questions will be important from choosing page size to producing a high-quality layout:

Does the document consist mostly of body text? Or, will the document contain lists, illustrations, charts and graphs?

How many levels of heading does the content require?

What are the dimensions of the page? Is the orientation portrait (vertical) or landscape (horizontal)?

Will a single-column or a multicolumn layout be optimal?

How many chapters, pages, and sections will the document have?

Will different types of page layouts be needed, such as chapter pages, tables of contents, right pages and left pages, or special appendixes?

Will software templates or style sheets provide these layouts?

How will the document be bound or produced?

Once these questions are considered, page layout can begin.

Principles of Page Layout

Page layout consists of creating visual "chunks" of information that emphasize the hierarchy of information and are easy for the reader to process. It is important to balance text and images with white space and to use headings that make information easy to identify. Pages should not be cluttered with so much information that they overwhelm the reader. The placement of elements such as headings, body text, and page numbers should be consistent from page to page. In addition, each element can be made visually distinct by varying the fonts, spacing, and alignment for different elements. The key to an effective page layout is to achieve the right combination of certain basic principles of visual arrangement.

The two sample pages below each contain the same information, but it is easy to see that the second example is easier to read, clearer, and professional looking. The first page of a sample resume displays many of the layout errors inexperienced writers commonly make.

NORISHA M. PRICE

1223 Maple Lane
Springtown MN 55555
(656) 222-1234
nprice@notasite.org

PROFILE

15 years experience in human factors and user interface design in the software industry. Including experience with medical devices and related Web sites. Academic background in science and art and a focus on visualization of multivariate quantitative data. Strength in combining creative and analytical approaches to conceptualize complex interfaces.

Experise in the following areas:

Human Factors

Implementing knowledge acquisition methods including user requirements identification, task analysis, and user profiles, and usage scenarios

Implementing usability evaluation methods including formal usability testing in a tab environment, contextual inquiry, heuristic evaluation, cognitive walkthrough, and field studies.

User Interface Design

Designing user interfaces for enterprise server/client software applications
Designing page layout and navigation structure for Web pages and Web application
Researching technologies that support Web-based UI architecture and display
Writing detailed GUI specifications for user interface designs

Concept Generation

Collaborated with a team to generate a method, a basis for implementation, and a conceptual graphical user interface to visualize large amounts of data for an expert decision support system

Filed (with team) several patent applications in 2004 based on the method: one patent applications received the 2004 Intellectual Property Award

Extensive experience with brainstorming activities, including experience with electronic brainstorming tools

Academic Research

Cognitive processes in visualization
Visual representative of multivariate quantitative data
Statistical analyses and usability research methods

PROFESSIONAL EXPERIENCE

STARTUP SOFTWARE   *Springtown. MN*
Enterprise Data Management Group
Senior Staff Usability Engineer 2000 - Present

Conduct a wide range of usability and system analysis activities for cross-platform enterprise backup and disaster recovery software. Acitivies include user and task analysis close collaboration with customers and partners, product definition. GUI design, heuristic evaluation, and usability testing. Led development of a program for disabled accessibility compliance that meets federal government standards. Designed and supervised construction of a digital state-of-art usability lab.

■ Ineffective resume layout

Features of Ineffective Page Layout

❶ Lack of consistency in fonts: The decorative fonts used to emphasize the author's name and contact information are difficult to read. The document uses too many different fonts rather than a consistent font.

❷ Lack of contrast between type elements: Italics and other styles work well for selective contrast, but should be limited to highlighting specific words. In longer texts like the author's profile, italics are hard to read.

❸ Lack of alignment: The page centers too many of the elements. The author has centered the headings while running the body text flush with the left margin. This lack of alignment between related elements makes it hard to see the connection between each heading (human factors, user interface design) and the descriptive detail.

❹ Lack of visual consistency and balance: The document suddenly shifts from centered section headings to flush left. The inconsistency in layout looks like a formatting error, or, worse, carelessness.

❺ Lack of emphasis and balance: The bottom of the page does not have a visual anchor, page number, or running footer to emphasize it as the bottom and to balance it with the top.

A second version of the resume page shows how the same information appears when principles of alignment, balance, emphasis, consistency, and contrast are applied and work together in an effective page layout.

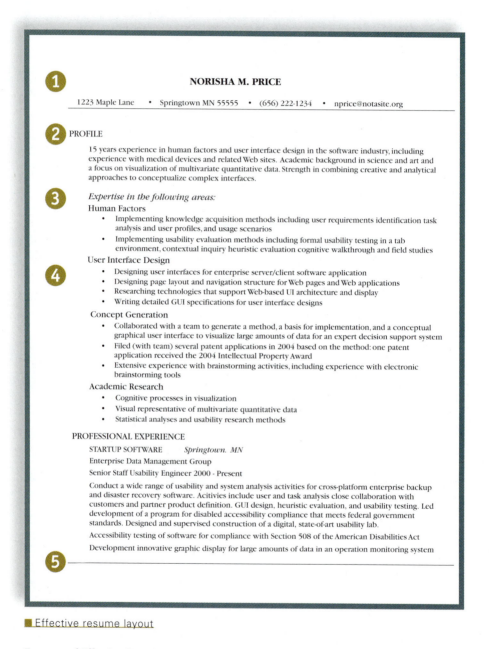

NORISHA M. PRICE

1223 Maple Lane   •   Springtown MN 55555   •   (656) 222-1234   •   nprice@notasite.org

PROFILE

15 years experience in human factors and user interface design in the software industry, including experience with medical devices and related Web sites. Academic background in science and art and a focus on visualization of multivariate quantitative data. Strength in combining creative and analytical approaches to conceptualize complex interfaces.

Expertise in the following areas:

Human Factors
- Implementing knowledge acquisition methods including user requirements identification task analysis and user profiles, and usage scenarios
- Implementing usability evaluation methods including formal usability testing in a tab environment, contextual inquiry heuristic evaluation cognitive walkthrough and field studies

User Interface Design
- Designing user interfaces for enterprise server/client software application
- Designing page layout and navigation structure for Web pages and Web applications
- Researching technologies that support Web-based UI architecture and display
- Writing detailed GUI specifications for user interface designs

Concept Generation
- Collaborated with a team to generate a method, a basis for implementation, and a conceptual graphical user interface to visualize large amounts of data for an expert decision support system
- Filed (with team) several patent applications in 2004 based on the method: one patent application received the 2004 Intellectual Property Award
- Extensive experience with brainstorming activities, including experience with electronic brainstorming tools

Academic Research
- Cognitive processes in visualization
- Visual representative of multivariate quantitative data
- Statistical analyses and usability research methods

PROFESSIONAL EXPERIENCE

STARTUP SOFTWARE   *Springtown. MN*

Enterprise Data Management Group

Senior Staff Usability Engineer 2000 - Present

Conduct a wide range of usability and system analysis activities for cross-platform enterprise backup and disaster recovery software. Acitivies include user and task analysis close collaboration with customers and partner product definition. GUI design, heuristic evaluation, and usability testing. Led development of a program for disabled accessibility compliance that meets federal government standards. Designed and supervised construction of a digital, state-of-art usability lab.

Accessibility testing of software for compliance with Section 508 of the American Disabilities Act

Development innovative graphic display for large amounts of data in an operation monitoring system

■ Effective resume layout

Features of Effective Page Layout

❶ Emphasis and alignment of name: The author's name uses an all caps, bold version of the same font used throughout the document (Arial). Because the author's name is the only text element on the page that is centered, it clearly stands out as the most important visual element. Spreading the elements of the author's contact information above a basic ruled line makes the contact information readable and easy to pick out on the page.

2 **Balance and alignment:** White space above and below the "profile" heading, coupled with an effective use of indenting in the text below, makes it easy to see how everything that follows, down to the heading "professional experience," profiles the author's core areas of expertise.

3 **Contrast by using italics sparingly:** These are the only italics on the page. Italic type works effectively here since this is a short phrase and serves as a unique element introducing a series of four main topics.

4 **Balance and consistency of elements:** Indented bullets help to reinforce the list of key areas of expertise and to distinguish between the items in a series.

5 **Emphasis and balance:** The ruled line and running footer balance the page visually with the top and also help readers navigate the second and third pages of the resume.

Applying even a couple basic principles to all visual elements in a layout can make a noticeable difference in clarity and readability, even in a simple document like a page from a resume.

Techniques of Page Layout

This section describes the basic techniques for the layout for a printed page. All of these features can easily be set up and even customized using any basic word processing software.

Vertical Spacing

Vertical spacing between lines of texts and paragraphs, lists, and images should always be consistent. Point sizes and spacing actually vary according to the font used, and you can define the space you want before or after specific elements. You should establish spacing guidelines first, and then apply them consistently throughout the document. For example, if you allow a certain amount of a space between an illustration and its caption, do the same for all other illustrations and their captions. Note that double-spaced body text seldom appears in printed technical communication. Double-spacing is primarily used for draft manuscripts to allow room for comments.

White Space

White space refers to all empty areas (even if not white) around the elements in your document. White space is crucial to a layout's impact and its readability. Some elements may need more space than others just to see them. For example, you might allow a bit more space between each item in a list than you would between lines of body text. Other elements, such as graphics, need a margin of space around them to set them off from text and also to draw the reader's attention. Good page layout does not break up all the white space on the page, but instead uses ample margins and alignments, as in the well-designed resume page.

Margins

The left, right, top, and bottom margins define the white space around the printable area of each page in your document. A 1-inch margin all around is standard for double-spaced drafts and manuscripts, but most typeset documents will require more space or different margins on various sides, depending on their purpose and the final printed format.

Gutters

Gutters are additional margins on the sides or tops of pages to leave space for binding the document. They are most commonly used with facing pages (see below) or in between columns of text (as on a newspaper page).

Alignment

Left, right, centered, and justified are standard horizontal alignments that determine where an element appears in relation to the margins of your document.

Left, the most common, is used for full lines of running text and leaves a "ragged" right edge of text.

Right, used for short lines of text, is primarily for headers, footers, and page numbers that "stay" at the right margin for emphasis.

Center places the midpoint of the word or line at the exact center of the page for emphasis of the primary element.

Justify, often used for columns, makes both right and left margins smooth and "stretches" the text in between to fill the space. This feature is often used on pages that have columns of text (such as newspaper or magazine pages).

Columns

Columns align elements horizontally spaced across the page in more detail than the standard alignments. A full page is actually one column. When creating multiple columns, each one can be equal or have its own width and spacing that contain the elements within it. Multicolumn layouts will help keep items separated across the page.

Tables

A table refers to a set of cells that align elements both horizontally and vertically. "Invisible tables," where the gridlines aren't apparent, are commonly used for alignment on printed pages and also for placing elements on web pages. A "floating object" indicates a graphic or object that is not in line with the rest of the text. Use tables or text boxes if your document requires precise alignment of text and graphics with each other.

Lines and Line Length

Lines that are too long or too short are difficult for readers to process. Odd lines breaks also can be distracting in a layout. For example, a single word should never stand alone on a separate page or in a new column. Likewise, lines should stay with the paragraphs to which they belong. Most programs have what is called "widow/orphan control," which keeps single words or single lines of text from being printed alone at the top or bottom of a page. Most programs also ensure that words are hyphenated properly.

Indentation and Tabs

Indentation is used to show hierarchical relationships between information. Indentation is often (but not always) used for bulleted and numbered lists. Remember that the bullets headings, and numbers themselves serve as a graphical signal, and indentation may be unnecessary or redundant.

The first lines of paragraphs are sometimes indented and sometimes not. In much of technical writing, paragraphs are usually are not indented because this interferes with the streamlined appearance of the information. The more common technique is to skip a line between paragraphs.

Tabs can be used to horizontally align columns of information, such as aligning definitions with key terms. If using tabs becomes too cumbersome, you can create a table with invisible grid lines and use the table cells to align the items.

Headers and Footers

Headers and footers are used to include consistent text at the margin of every page in your document. Headers appear on the top of every page, and footers on the bottom, within the printable margins and set off slightly from the rest of the page. The most common uses of headers and footers include dates, pages numbers, names, or shortened titles, especially in longer documents. A *running head* commonly includes this kind of information on every page of longer manuscript documents.

Page Numbers

Page numbers should always appear in the same location on every page. Centering the number or placing it in the lower right corner is fine for a draft manuscript, but professional manuscripts usually place page numbers in a running head for readability. Remember that if your document will be using right and left pages then the "outside" of the page is different for each page and must be defined separately. In some cases, page one is left unnumbered.

Complex Layouts

Complex projects can benefit from a page layout program that gives you much more precise control of alignment than the settings in a word processor.

Grids

Grids help to define specific areas of the page, including the margins and alignment of elements into columns of various widths. Grids determine the horizontal placement of elements along vertical grid lines. For example, the "hanging indent" is a simple grid technique where body text aligns with a vertical grid line indented just to the right of the vertical line used for headings. This technique gives the body text the appearance of "hanging" under the first level heading to which they belong. Grids are common in page layout programs.

Here is an example from a tutorial for Adobe InDesign, a page layout software program. The colored lines show the top and bottom margins (in pink) and both horizontal and vertical grid lines (in blue) that the user defines to create aligned elements on the page.

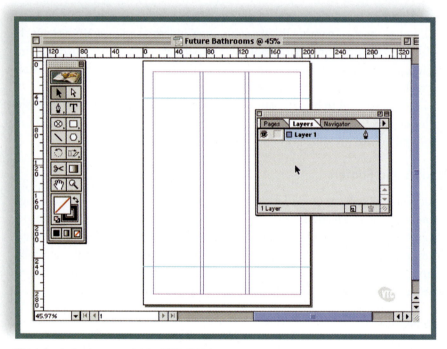

■ Adobe InDesign tutorial by VTC Training

Facing Pages

Facing pages must be used whenever a document will be printed and bound along the side. Right and left facing pages must use different layouts and margins to accommodate. Different odd and even numbered pages are defined for facing pages; generally, right pages are odd numbers and left pages are even.

Audio

Audio refers to audible sound and the sound elements included or recorded with various technologies. Audio can be an effective medium for technical communication, for example, when you need to describe a process in a training **video** and you want users to hear spoken instructions while an animation or video plays on the screen (called *voice-over narration*). Companies and industries that use computer-based training (CBT) or tutorials may employ other kinds of sound as part of a training video—from the actual sounds of machines and work environments to music that helps create a mood. Digital audio files are easily created and distributed with computer programs and equipment, and these files may contain recorded music, voice, or other sound effects that can then be included on a **web page**, in a **blog**, in a PowerPoint **presentation**, or played on **pocket technologies** like MP3 players and cellular phones. Because video usually includes audio tracks with the sounds of the person talking and other location sounds, it depends *heavily* on this recorded sound track to make sense. Sound files are also frequently used alone to distribute music or samples on the Web, offer live radio or online broadcasts, send or receive digital voice mail. Audio broadcasts on iPods, called podcasts, let users record and publish their own programs for others to download and play on their MP3 players. You can also use a tool called RSS (really simple syndication) to let users subscribe to a wide range of audio broadcasts from your site or from a collection of sites based on similar information.

Audio File Types

Sound is recorded, sampled, stored, and can be visualized as an acoustic sound wave using audio technologies and software applications. Like video clips, audio clips for the Web must be short and compressed to download and hear using appropriate software. Audio files use the frequency, sample rate (KHz) and size (8 or 16 bit) in an encoder to determine the quality of the sound—the higher the sound quality, the larger the file size.

Some common audio file formats include:

- AIFF and WAV files—large files for commercial music and other high-quality recordings
- MP3 files—the Web standard, where quality and size can be widely adjusted to maximize disk space

- MPEG audio files—the standard for audio broadcasts, highly compressed files
- AAC and other proprietary formats—developed for specific applications like Apple iTunes
- Streaming files such as RealAudio—play while downloading, but with limited size and audio quality

Choices for sound quality and file format are determined in each computer's sound control panels and also in the control settings for saving and importing files in all audio software programs.

Music

Technical communicators who create instructional videos or presentations may find themselves functioning as writers, producers, and sound engineers. Programs included with personal computers make creating and including audio in your presentations relatively easy. The widely used audio program iTunes, for example, stores, categorizes, and copies audio files that you own or purchased from an online store onto other media or into an MP3 player.

If you plan to use an existing song in a published project, you must pay record company permission fees. Free music composition software like Apple's GarageBand offers high-quality instrument samples and mixing tools for creating original music even if you are a novice. Other software programs exist that allow you to edit audio files in multiple tracks, mixing and adding effects as you would in a recording studio. Music files are then easily imported into another media program and used as a soundtrack. Even PowerPoint offers simple ways to record and integrate music and other sounds into a presentation.

Voice-over Narration

Technical communicators and specialists may need to provide a script to be recorded as a voice-over narration. When you decide to use audio voice-over versus on-screen written directions depends on the amount of information and its purpose. On-screen text works well for summarizing and repeating main points or steps in a process. Audio voice-overs can provide the depth and presence of vocal inflection that on-screen text cannot. Voice-overs also provide details that are too long to read on a screen. Studies suggest that a combination of text on the screen and spoken narration of that same text offers the best retention. In any of these instances, the steps or concepts need to be covered in the same order in both the

■ Facing pages layout example

Features of Facing Pages Layout

1 The heading and page number changes from left to right.

2 The gutter leaves a wider inside margin for binding.

Guidelines

FOR PAGE LAYOUT

- Identify the purpose and scope of your print project.

- Determine the hierarchy and organization of information that is best for readers.

- List all different types of design elements needed, such as chapter pages, tables of contents, and right and left pages.

- Consider the length of the document and where it requires lists, illustrations, charts, and body text.

- Design your layouts on paper or screen, using principles for effective page design.

- Maximize white space in your document by using margins and alignments of elements. Don't clutter your document.

- Consider using a template or grid, especially for longer or complex documents.

- Once you've laid out the pages, turn every page and check for consistent headings and other elements.

- Test the document on target readers using an open-ended approach to usability.

Captions

A caption is a short phrase or sentence that identifies a visual image within a document. Note that tables usually require a separate numbering scheme in technical writing and often have titles but not captions. Technical documents that contain multiple images require a caption for each labeled image. Captions for the figures in a document often begin with some kind of numbering or labeling system (for example, Figure 1, Figure 2, Figure 3). In addition to the label or code, the caption should contain some kind of description. Here is an example of a simple caption for a figure in a technical document:

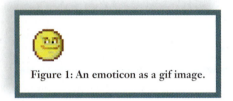

Figure 1: An emoticon as a gif image.

Designing Captions

Your first consideration in designing captions is to decide whether you need them or not. If your document requires photographs, charts, illustrations, or screen captures and if you need to refer to these images within your text, then you should label these graphical elements with captions that identify them as *figures*. However, if your document requires only a few visuals or if they always appear within the text that discusses them, you may include a descriptive caption beneath each image but choose not to use formal, numbered figure labels.

If you want to include photographs and illustrations that are not specifically discussed in your text, if they serve only to add general visual interest, or if they are self-explanatory, you can include them without any captions at all. If your document contains many figures that you think the reader will need to locate quickly, you might want to include a list of figures at the beginning of the document.

Once you have determined that you need captions, you can then consider their content and visual design. When designing captions, be consistent in the style of captions throughout a document. Consistency includes wording, level of detail, typography, and layout. For example, do not use bold 12-point type for some captions and italic 9-point type for others. Software programs often include automated figures and captions features that maintain this style and consistency. In terms of page layout, captions should be adjacent to (usually directly beneath) the images that they identify.

Headings

Headings are titles for the major sections and subsections of a document. Headings tell readers what each section is about. Experienced technical writers know that headings are one of the most basic yet most important of elements in technical writing. Headings serve a critical function for the reader: they make a document easy to scan for content. Effective headings must be both visually prominent and informative.

Headings can be used in all types and lengths of technical documents and correspondence, whether printed or displayed on screen, except in the case of a one-page document in which subsections of information are usually unnecessary. In that case, a title is sufficient. Very short documents, such as memos, may have only a few headings, but longer documents work best with multiple headings, and even multiple levels of headings, to indicate the various types of information and show how this information is organized.

The example below shows part of a chapter from a user's manual for accounting software:

What Are Accounts? **1**

Accounts are information repositories that collect, store, and identify financial data. Accounts can get their values from user input or from calculations generated by your chart of accounts…

2 Subaccounts

Subaccounts are input accounts that are automatically totaled into the major account to which they belong. Subaccounts are often used to capture line item detail, such as various types of expenses…

3 Account Types

Account types define how account values affect the subtotals and totals in the chart of accounts…

4 *Balance Accounts*

Balance type accounts store values as a snapshot; these types of accounts do not accumulate value over time, but rather are used to measure the current state of the balance sheet.

Flow Accounts

Flow type accounts store values that accumulate over time…

Adding Accounts

5 You can add accounts anywhere in the chart of accounts…

Editing Accounts

You can edit accounts names, descriptions, and attributes…

Deleting Accounts

You can delete accounts that have a zero balance…

■ Document using three heading levels

Features of Headings

1 Heading 1 is flush left and largest for easy scanning and emphasis.

2 Headings 1 and 2 use a larger type size and bold style to distinguish them from body text.

3 Headings 2 and 3 are left aligned with the body text.

4 Heading 3 is smaller and changes the style to italics for contrast.

5 The last three headings begin with the gerund (-*ing*) form of a verb to indicate that those sections explain how to perform a task.

This document contains three levels of headings; the highest level of heading is usually called Heading 1, the second highest is Heading 2, and so on. Notice that the font, size, style, and spacing are consistent at each heading level. All the headings are professional-looking, consistent, and can also stand alone.

Designing Headings

When designing a technical document, keep in mind that headings at each level should be visually consistent in wording, type style, placement, spacing, and other factors. Documents that contain multiple levels of headings must provide distinct visual cues to help readers distinguish and differentiate heading levels; each level should be visually apparent through use of type style and size, placement, spacing, color, and any other special effects, such as icons or images.

You can use an outline (see Part 4, Research) to help identify the types and levels of headings that you will need. Also, keep your readers in mind. What will they be looking for in the document? What kinds of headings will make it easier for them to find information? When the primary purpose of a section is to tell readers how to perform a task, headings should be verb phrases, beginning either with the gerund form (ending in *-ing*) or the imperative (command) form of the verb followed by the task to be performed, which, grammatically speaking, is the object and, operationally speaking, is the objective of the task.

Guidelines

FOR USING HEADINGS

Use effective wording for headings:

- **Informative**: Use terms that identify each section and work together to describe the scope and content of the document.

- **Consistent**: Use similar grammatical structures for similar types of information.

- **Easy to scan**: Put the words that distinguish the information at the beginning of the heading; eliminate leading articles.

Organize headings clearly:

- **Levels of headings**: Use one level for short documents; use a hierarchy of headings for longer documents.

- **Table of contents**: Make sure headings will look, sound, and appear coherent when viewed together.

Use a consistent visual style for headings:

- **Type style**: Choose a font type; a plain, bold, or italic style; and a size for each level.

- **Placement on the page**: Use alignment and indentation consistently to organize "white space" around the heading to give it emphasis.

- **Spacing**: Keep consistent space above and below each heading.

- **Color**: Use highlight color or color scheme to distinguish levels (when available).

- **Special effects**: Use special effects such as shadow, small caps, or underline in a limited way, if at all.

- **Icons or symbols**: Use sparingly, in conjunction with heading text.

Labels and Callouts

Labels and callouts are identifying text that is located directly on an image or connected to the image by lines. The purpose of labels and callouts is to identify specific elements or features within an image. Although the terms are used interchangeably, labels are text identifiers that are self-explanatory in an image, while callouts are labels that require further information outside the image to explain what they are identifying. Writers and designers use labels extensively for technical and explanatory visual displays.

Labels as Text

As text placed directly on the visual, the label simply becomes part of the picture. This example shows a map that uses labels to identify city parks.

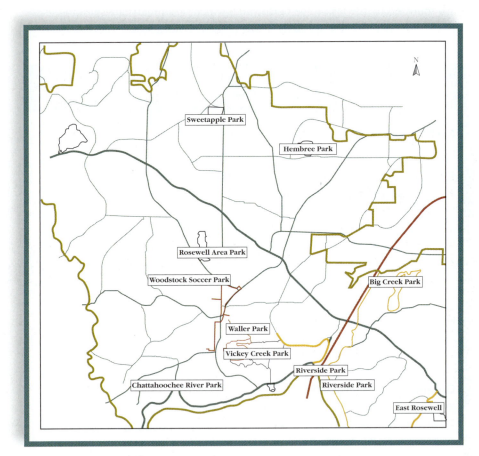

■ Labels used to identify areas on an image

The city map identifies and emphasizes the location of parks by placing each park name as a rectangular label directly over each park's location on the map. The use of green text further reinforces that these locations are parks, while the rectangular boxes with white backgrounds make the park labels stand out as the dominant images on the map.

Labels as Pointers

Labels can also use lines to point to each item on the visual. Labels in this example help explain how to read the label on a passenger car tire.

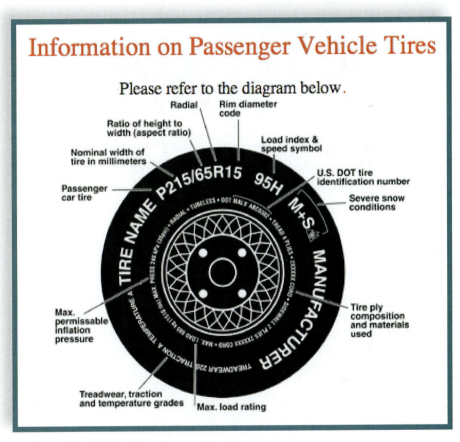

Labels with rule lines identify parts of a tire

Labels as Links

Labels can also add interactivity in a digital medium. "Anatomy of a Steam Locomotive" shows how labels can act as links or hotspots in a screen-based image to set up an interactive display of the parts of a train. Moving the mouse over these links allows users to learn more about this particular feature of the locomotive.

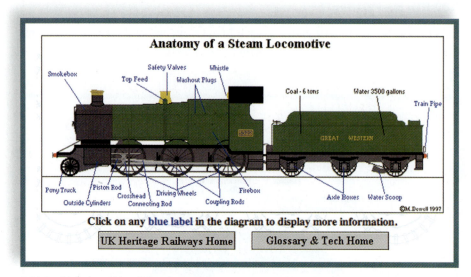

■ Linked labels to identify train parts

Callouts

When too many labels will obscure the image or simply not be clear, create a list of ordered callouts. The advantage of using coded callouts is that they allow more of the image to remain visible. The disadvantage is that coded callouts make image interpretation a two-step process: readers must first visually scan an image and then move to a key or an explanation outside of the image to identify the meaning of each callout.

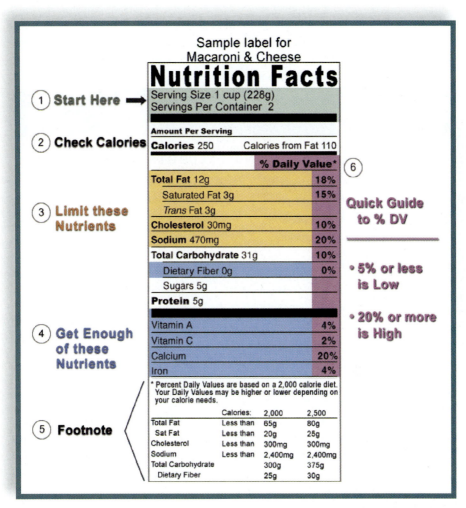

Example of the use of callouts

Designing Labels and Callouts

To begin, you need to decide several things including the number of items to identify, how much of the image you want visible, and the design and placement of identifying items. Sketch the design and include the rough placement for labels or callouts on the image. If considering callouts, also sketch out where the corresponding explanatory text will be located. Then, choose between labels and callouts, noting the differences discussed above.

Ideally, labels consist of entire words or short phrases that explain various areas of an image, which allows the reader to immediately gain a basic understanding of the parts of an image without having to look at supplementary explanations. However, too many labels can obscure the image that you are trying to display, and you will likely use callouts. Coded callouts are useful when many parts of an image need to be labeled and when each part requires a longer description. Coded callouts can use numeric (1, 2, 3 …) or alphabetic (A, B, C …) sequenced items to identify areas on an image. The codes must be explained in an adjacent key or legend, or right next to the callout itself.

Guidelines

FOR CREATING LABELS AND CALLOUTS

- Determine the number of items that you need to identify on your image.

- Make a quick list of the items, and estimate how much explanation each item requires.

- Write the text for your items to become each label or the list of callouts. If using callouts, include your explanatory text.

- Create a consistent visual style for your labels or callouts text, including font size, font color, and the use of text boxes, background colors, shapes, or other design elements.

- Lay out the labels or sequenced items on your image.

- Place the callout explanatory text in a separate key or next to the image.

Lists

A list is a simple, easy-to-scan format for presenting a set or series of related information. A list can consist of a bulleted series of items or a set of ordered instructions. While lists are popular and often useful in technical documents, they need to be used carefully and should be designed to be concise. See Parallelism (in lists), under "sentences" in Part 5, for additional discussion of proper list style, and Colon, Part 5, for advice on punctuating lists.

For example, you might use a simple bulleted list used to introduce one chapter of a user's guide.

1 This chapter explains how to complete the following tasks:
- Add individual contacts in your address book.
2 • Edit contact information.
3 • Set up group lists.

■ A bulleted list

1 The preceding sentence introduces the list to the reader using
the colon.

2 The list consists of three
items, indented and signaled by

bullets. Notice that the spacing
between each bullet and the text
that follows it is consistent
throughout the entire list.

3 The text is parallel in wording:
each of the three items begins
with an imperative verb (*add, edit,
set up*).

Numbered Lists

Use numbered lists when order matters, such as in step-by-step instructions.
Numbered lists can also be used when numeric labels add clarity, emphasis, or
precision to an explanation. For example, the web page of an insurance provider
might include a numbered list of the ten most frequently asked questions by its
customers.

A numbered list might be used, for example, to provide instructions for
working with contacts in an email program:

---

1 To set properties for a contact folder, follow these steps:

    1. Click the **Contacts** icon.

2   2. Select **File/Folder/Properties**. The Contacts Properties dialog box
      appears.

3   3. Edit the default values of the fields as needed.

    4. Click **OK** to save changes to the contact folder properties or click
      Cancel to discard them.

■ A numbered list used in instructions

Features of a Numbered List

1 The sentence preceding the
list introduces it as set of
instructions.

2 The instructions contain
four steps; the text following all
steps is left-aligned along the
same grid.

3 Each item in the list begins
with an imperative verb (*click,
select, edit, click*), and each is
written as a complete sentence,
to ensure that the list is parallel in
its style.

Unnumbered lists

An unnumbered list consists of a series of small paragraphs or sentence groups,
which each use a consistently formatted heading or label. When you need to present a set of descriptions or other relatively compact information about a related
series of things, this type of list is an informative and visually effective format.

The example below shows a list of the tasks users can do through the Contact Administration pages of their email program.

1 The Contacts Administration page lets you do the following:

Configure contacts folders—You set properties for each contacts folder to specify how it interacts with email and displays information.

2

Create contacts—You can create an online directory of people, companies, and organizations, their e-mail addresses, mailing addresses, phone numbers, and any notes that you want to file under a name.

Set call specifications—You can specify automatic dialing and call recording facilities for a contact through your keyboard interface.

■ Unnumbered list

Features of an Unnumbered List

1 The sentence preceding the list introduces it as a set of features.

2 Each list item begins with a word or noun phrase that is visually distinct, boldface, and set off from its associated text by a dash.

Designing Lists

If you need to present a series of related but separate items, steps, features, or concepts, a list may be a good way to present information. Effective lists should be integrated with the other parts of the document and should function to communicate efficiently and clearly to the reader. Lists are used to simplify complex information; therefore, the meaning and purpose of a list should be readily apparent to the reader. For example, if the bullet list will act as a checklist, use blank boxes or check marks. If your document is for a specialized audience or subject, you might select a bullet symbol whose meaning and style relate to the context.

One problem that can happen when writing lists is that they can become too long and complex, especially if each list item provides a detailed explanation or description. Long list items may indicate that a list needs to be edited or that lists are not a good format for the information.

FOR CREATING LISTS

- Introduce a list with an explanation of what the list contains and how it will help the reader.

- Use consistent symbols, numbering, sizing, tabs, and alignment for each list item.

- Avoid wordiness in lists. Begin each item on the list with a distinguishing word or words; this approach will make your list easy to scan.

- Try to keep the text for each list item verbally parallel; for example, avoid beginning some items with verb forms and other items with noun phrases.

For bulleted lists

- Select a clear and simple graphic symbol, such as a dot, box, or diamond. Avoid using symbols that are overly complex or cute.

- Try to keep each list item to two lines at most. If list items keep running more than two lines, consider changing the information into a nested list or paragraphs with headings.

For numbered lists

- Make sure the items are listed in the correct order. (Numbered lists imply a set of steps to be followed in sequence.)

- If your text includes a detailed discussion of each item in your list, be sure to address the steps in the same order you use to present them in your list (*first*, *second*, *third*, and so forth).

For unnumbered lists

- Begin with a word or phrase that is visually distinct and set off from its associated text by special formatting, such as colon or a dash, or by appearing on a separate line altogether.

- Use consistent formatting and, when possible, parallel wording to begin each list item.

Tables of Contents

A table of contents (TOC) is a navigational aid that lists the major headings in a document and their corresponding page numbers. Located at the beginning of a document, the table of contents helps readers quickly find the parts of the document they need. Readers usually expect documents that are longer than ten pages to have a table of contents.

Very long books and manuals sometimes have more than one table of contents. For example, a long manual may have a master table of contents at the beginning of the book and also have a chapter table of contents at the beginning of each chapter. The master table of contents may list only the highest level of headings (see "headings" above), while the chapter table of contents might list three or four levels of headings.

The examples below shows a master table of contents followed by a chapter table of contents:

TABLE OF CONTENTS

■ Master table of contents showing chapter names and heading level 1

Features of a Table of Contents

1 The master table of contents lists the chapter titles and each chapter's first level headings (Heading 1) with the corresponding page numbers.

2 Each chapter heading is flush left for easy scanning. Chapter names are parallel and visually consistent.

3 First level headings are indented under the chapter name. First level heads are parallel and visually consistent both within and across chapters.

■ Chapter table of contents showing heading levels 1 to 3

Features of a Chapter-Level TOC

❶ The chapter table of contents shows the chapter title at the top of the page then lists the top three heading levels with corresponding page numbers.

❷ Heading 1 levels are flush left.

❸ Heading 2 and 3 are each indented to show the visual hierarchy of the chapter.

Designing a Table of Contents

The table of contents serves as an outline to the major parts (chapters and sections) of a document. Most word processing software can automatically generate a table of contents based on headings already in the document. For example, Microsoft Word provides this dialog box for generating a table of contents.

■ Table of contents screen in Microsoft Word

Features of TOC Software

1 The dialog box provides a preview of how the table of contents will look in print or on the Web.

2 This table of contents will be generated for three heading levels.

3 Other format options (pages numbers, alignments, etc.) are selected here but can be modified later.

4 "Tab leader" indicates the dots or other graphic leading from the heading to the page number.

It makes the most sense to generate a table of contents based on heading styles. However, special types of tables of contents, such as a list of figures, might be generated using another style type.

Do not wait until the end of your writing process to generate a table of contents. Generating a table of contents every so often as you are writing is an excellent way to check for consistency in your heading usage across a large document and can also help you get a feeling for what you have accomplished and what you have left to do.

FOR CREATING TABLES OF CONTENTS

- Consider the look and effectiveness of your table of contents when designing headings for a document.

- Choose a layout for the table of contents that shows the hierarchy of your entries. Use consistent visual strategies for hierarchy, such as type size and style, indentation, and vertical spacing.

- Select a format that associates each table of contents entry with its corresponding page numbers.

- Consider how many heading levels will be most useful to the reader in the table of contents. While the table of contents needs to help the reader navigate a document, too many levels will make it unwieldy.

- In the case of a long multipart document, consider having a master table of contents just listing the major parts, and provide tables of contents with more detailed listings for each section.

- Consider the need for different types of table of contents, for example, a list of figures.

Typefaces and Fonts

Typefaces and fonts are two terms that describe the character sets that are used for designing and displaying text. Although *typeface* and *font* are often used synonymously, they are not exactly the same thing. *Typeface* refers to the visual appearance of a character set (most of which consist of the alphabet, numbers, punctuation marks, and other common typographic symbols); the characters of a typeface share a common look and feel. The art of creating typefaces is called typography and is performed by highly specialized graphic designers.

Technically, *font* refers to the computer-based information that defines and describes the typeface rather than the actual visual appearance of the typeface itself; a font includes information such as the shape, thickness, height, and width of each character, the horizontal distance between characters and words, and default vertical distance between lines of text. Font definitions are loaded onto computers to give writers and designers a selection of typefaces to apply to their texts.

Almost any kind of computer program, including word processors and page layout programs, as well as web page and help authoring tools, provide a choice of fonts, font sizes, and other style options, such as underlining, bold, and italic. More specialized fonts are often available on the Internet, either for free or for a cost, depending on the copyright terms and conditions. In most practical discussions, the terms *typeface* and *font* can be used interchangeably. *Font* has become the more commonly recognized term today.

Fonts convey more than just a literal presentation of characters. Like other graphic elements, fonts convey tone and mood. Fonts, along with layout, also largely determine the legibility of a document. A clean font like **Arial** is professional and readable for a report with equations, for example, while a decorative font like COPPERPLATE GOTHIC should be reserved for headings or special applications, such as business cards or brochures.

Fonts Types and Their Uses

The basic distinctions for fonts are whether they are *serif*, *sans serif*, or some type of decorative or scripted design. Serif and sans serif are by far the most commonly used and legible fonts while decorative and script fonts are much less legible and should only be used sparingly for special effects.

Serif fonts, often called formal or old style fonts like Baskerville have serifs: the tails, feet, and other flourishes on each letter. Serif fonts give a text a bookish or literary feel. Serif fonts work well for headings and particularly well for body text (the main text of your document).

Sans serif fonts like Arial are modern, more informal fonts without the serifs. Sans serif fonts give text a clean, scientific, and streamlined feel. They can work well for headings, as well as body text, especially text with many numbers because the characters are cleanly drawn with more space between them.

Display fonts like Optima and `Courier` are evenly spaced and used specifically on screen.

Decorative fonts like Chalkboard have a specific visual meaning or connotation and can be used sparingly for special effects or humor.

Script fonts like Lucinda Handwriting resemble handwritten text and can be used to personalize a message or for special effects. But they're hard to read, so use sparingly.

Some research suggests that paper documents are easier to read when the body text consists primarily of a serif font, such as Times Roman, while electronic documents and interfaces are easier to read when the body text consists primarily of a sans serif, such as Arial.

Font Size

Another element of text that is very important for the reader is its size. All fonts are available in a variety of sizes, which are measured in *points*. While actual sizes are relative to the font used, body text is displayed in 12-, 11-, or 10-point type, with 12 being the largest. Normally, fonts below 8 points are very difficult to read.

Vertical Space

The vertical space that a line of type takes up is called the *leading* (pronounced "ledding"). By default, most fonts define the leading to be two points larger than the point size of the text. For example, 12-point text would have a 14-point leading by default. Most word processors and other text layout programs allow the user to control the vertical space, or leading, even further by offering such options as double-space, 1.5 space, or even allowing the user to specify additional points of space above or below a line. The best way to determine what is most appealing in your document is to print out different versions of the document using various combinations of font size and spacing.

Combining Fonts in a Document

A document, whether printed or electronic, can use a combination of fonts, font sizes, and font styles. The heading of a document may use a larger, sans serif font, such as 18-point Arial, while the body text may use a smaller serif font, such as 12-point Garamond. The strongly contrasting type styles make the headings stand out and make the document more scannable for content. In addition, sidebars and captions can use yet a different contrasting font or font style than the body text and headings. Remember that textual style effects, such as bold, italic, and underline, as well as indentation, can be used in conjunction with fonts to create distinct typographical elements. Making different kinds of information typographically distinct is important, but too many differences will confuse the reader. Stick to two or three fonts at most, and achieve distinction of elements by varying the size and style of those fonts in a purposeful schema that will become intuitive for the reader.

Once you select the various fonts, font sizes, and other textual effects for your document, be sure to use them consistently. Probably one of the biggest indicators of a professional print document is consistent use of typography and spacing throughout. In other words, if your document uses major headings and subheadings, be sure that the font, font size, spacing above and below, and indentation are used consistently and distinctly for the two levels of heading. Finally, be sure that the spatial relationships between your body text and headings are always consistent—don't skip a line after the heading in some places and not in others. One way to ensure consistency and allow for easy global changing of a textual element is to define styles using the features provided by your software.

Guidelines

FOR USING TYPE

- Use one font or typeface family as a general rule for body text.

- Use strongly contrasting fonts between headings and body text, usually serif and sans serif.

- Use font sizes to differentiate between headings and subheading levels, as well as to create large, eye-catching headlines or small items like captions, addresses, or page numbers.

- Use only one font style at a time for emphasis and use it sparingly (bold, italics, all caps).

- Be consistent with your use of each font and font style throughout the document and for each level of heading.

- Be sure the font size and the amount of space allows your body text to be easily read.

See also

Audience (Part 1)

Colon (Part 5)

Outlines (Part 4)

Sentences (Part 5)

Purpose (Part 1)

Style Sheets and Templates (Part 1)

Tables (Part 3)

Technical Illustrations (Part 3)

Usability and User Testing (Part 4)

Visual Communication (Part 3)

Graphs

Graphs are used to represent and compare numerical data visually. Graphs make abstract information accessible to the reader at a quick glance. Most graphs plot points along a horizontal (x-) and vertical (y-) axes. Typically, the horizontal dimension is used to represent intervals of time while the vertical axis is used to show a value or quantity. For example, if you wanted to show how real estate values in your area have suddenly increased much more rapidly than in the past, a graph would be an effective way to show a sudden change in the property values. Some graphs include a third dimension or z-axis as well. Instead of many tables of numerical information, graphs allow your reader to see in one image a trend or pattern within a large data set. Technical communicators often need to use two types of graphs, bar graphs and line graphs, for data display.

Bar Graph

Bar graphs use shaded or colored bars to represent a specific theme or trend over time. Bar graphs are useful for showing comparisons among groups or classes of data. Bar graphs can look at a single relationship or many, can be horizontal or vertical, and may represent specific parts alone or the relationship between a part and its whole, as shown in this sample. Most readers are familiar with bar graphs and can quickly grasp quantitative relationships by comparing the heights or lengths of the bars.

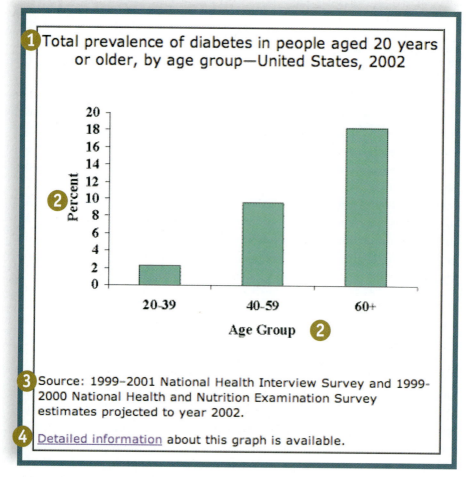

1 Total prevalence of diabetes in people aged 20 years or older, by age group—United States, 2002

2 Percent

2 Age Group

20-39   40-59   60+

3 Source: 1999–2001 National Health Interview Survey and 1999-2000 National Health and Nutrition Examination Survey estimates projected to year 2002.

4 Detailed information about this graph is available.

■ Single bar graph

Features of a Single Bar Graph

1 The title and heading of this graph provide the overview and context.

2 Each axis is clearly labeled.

3 The source of the data used to create this graph is provided in enough detail for readers to understand.

4 A hypertext link directs readers to more information about the graph.

Bar graphs can also be created using multiple bars within the same class of data. In the next example, the U.S. Securities and Exchange Commission tracked the type of queries received by its customer assistance center. For each year, queries broke down into two categories: complaints and questions or other queries. Each type of query, plus the total, is displayed by a bar representing years 2002–2006.

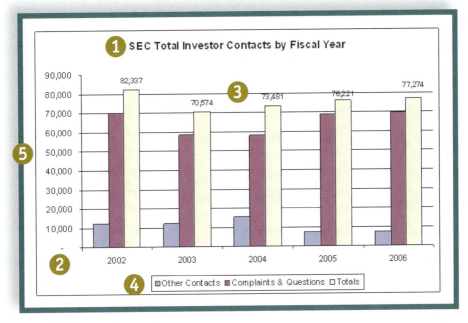

■ Multiple bar graph

Features of a Multiple Bar Graph

1 The title and heading of this graph provide the overview and context.

2 Each axis is clearly labeled.

3 For each year, data (type of query) is broken down into two categories and then totaled. Each bar represents one of the categories or the total.

4 Bars are color coded consistently.

5 Despite the amount of data represented, this graph is uncluttered and easy to comprehend.

Line Graphs

Line graphs also look at one or more events over a time period. They can be a more precise visual method for depicting trends. This graph presents an interesting example of how to use the surrounding text to explain the graph to the reader. In the actual EPA report, this graph is preceded by a paragraph explaining, for example, that "[t]he dotted line shows the trend in observed values at monitoring sites, while the solid line illustrates the underlying ozone trend after removing the effects of weather. The solid line serves as a more accurate ozone trend for assessing changes in emissions" (EPA 2007, p. 4). The graph provides a visual snapshot, and the text provides details.

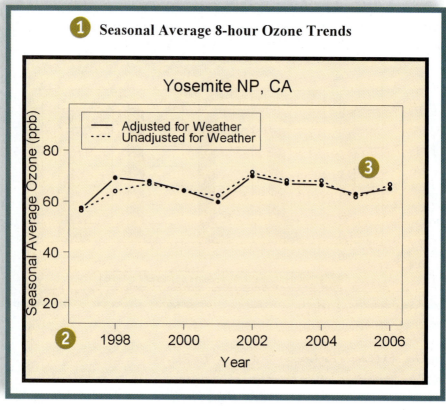

Seasonal Average 8-hour Ozone Trends

Yosemite NP, CA

— Adjusted for Weather
---- Unadjusted for Weather

Line graph

Features of a Line Graph

1 The title and heading of this graph provide the overview and context.

2 Each axis is clearly labeled. The "y" axis (ozone) is assumed to start at zero.

3 The two lines represent different data sets. A dotted and solid line are used to illustrate the difference.

Designing Graphs

Because graphs represent complex data in visual form, they can be powerful and persuasive. Readers who may not be willing to carefully read through a long table of numerical data may glance at a graph and come away with a general impression: "prices are up" or "oil is getting scarce" or "our profits are growing." The persuasive power of graphs means that technical communicators need to be aware of the possibilities of visual distortion and misrepresentation. The purpose of a graph is to show trends and comparisons, so take extra care that the data you are working from is accurate and that the final graph is clear and easy to read.

FOR CREATING GRAPHS

- Label each axis clearly and make sure the units of scale or measurement are identified.

- Readers assume that an axis begins at zero, so if your axis begins at a different value, be sure to make this clear.

- Be sure that your graph does not distort or exaggerate the change or trend.

- Always indicate the source of the data you are representing in your graph.

- Be sure to explain how your graph supports the claims you are making in the document's text.

- As with any visual, aim for a simple, minimalist style. Avoid using too many different colors or typefaces.

See also

Maps

Maps serve many purposes to help people connect with and use information. Most people are already familiar with maps as tools to find particular locations or understand distances and routes between locations. Maps also show boundaries, such as counties, voting precincts, or school districts. Maps can be very powerful tools for visualizing numerical data by illustrating how such data is distributed over different areas. The U.S. Census Bureau web site, for example, allows users to generate census data into map format, letting them compare regional patterns of data such as income, age, agricultural practices, demographics, and more. You can also access this site (Census.gov) as well as many other web sites to locate premade maps that are available for download.

The following map is from the U.S. Census Bureau's collection called "Census 2000 Special Reports (CENSR/01-1)" and illustrates the location of major U.S. cities. This map is part of a larger report on "The Geography of

U.S. Diversity." Notice that the map employs one primary color, green, but uses it in shades. Counties are visible as light green shapes in each state. Source information is clearly listed.

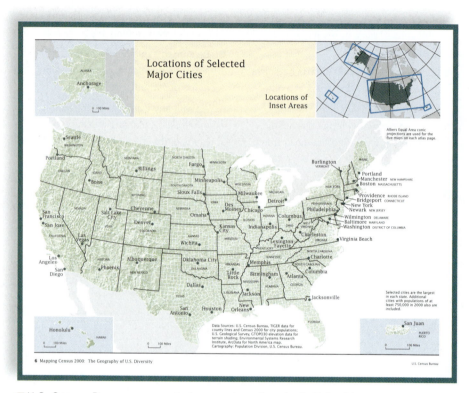

6 Mapping Census 2000: The Geography of U.S. Diversity

U.S. Census Bureau

■ U.S. Census Bureau map, part of a report on diversity in U.S. cities

Designing Maps

The design and content of a map depends on the purpose and type of map being constructed, the conventions for that type of map, and the audience using it. For example, a topographic map, commonly used by backpackers and developers, typically includes coordinate information, a scale showing how far an inch or centimeter represents, and lines and numbers representing elevation data, along with a range of other technical data. This type of information would be irrelevant and distracting in a road map, however. Keep some overall mapping strategies in mind, using visual strategies that fit the type of data (see Data Display for more information).

Many digital technologies provide users with quick, easy access to maps, especially street and road maps. Digital mapping tools such as Google maps or

Mapquest, mapping information over a cell phone, a global positioning system (GPS), or other pocket technologies, all allow users to create quick custom maps. If you are creating a map to give driving directions, use one of these tools: their data will be more accurate and updated than a map you create yourself. You can usually link from your web site directly to a custom map.

Mapping Sequential Data

In general, for sequential information in which you want to emphasize the progression from lower to higher amounts of some variable, it is more effective to use shades of one or two colors than a broad color range. As you can see from the figures above and below, by using color gradations rather than a full color spectrum to represent the data, you can help the user quickly glance at a map and identify the high and low areas.

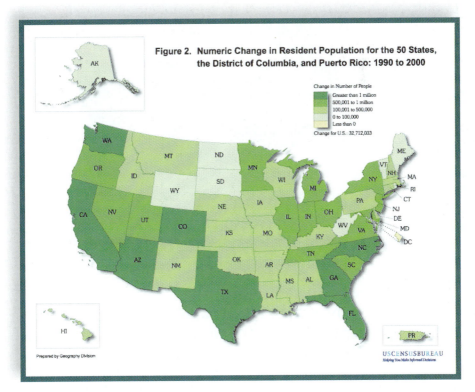

■ Use of color in map

Mapping Categorical Data

If you want to show differences in kind rather than differences in amount, a wider color range may make sense. You can color code your map using a range of easily distinguishable colors to indicate the different types of information. In the example below, the colors indicate the demarcations of different federal circuit courts. Note that in this example, no legend is necessary since the circuit court numbers are included as part of the map. In a different map without this internal labeling, a legend explaining the meaning of the colors would be necessary.

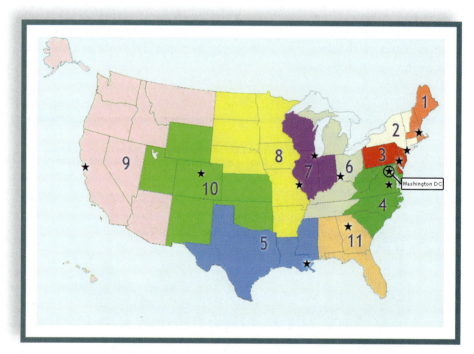

■ Map of federal circuit courts using multiple colors

Interactive Mapping Techniques with Geographical Information Systems (GIS)

A range of digital technologies has evolved to bring map visualization to new levels. Three-dimensional representations of data, such as the figure below showing elevation and water data for the Hoopa Valley Indian Reservation, help the viewer more easily grasp the meaning of the data than a two-dimensional map might.

■ Map of Hoopa Valley Indian Reservation

Geographical Information Systems (GIS)

The Hoopa Valley example was created using 3D options in geographical information system (GIS) software; readers can interact with the topography by using the mouse to hover over different areas to obtain elevation information or other data. By using GIS to design maps, you can combine data from a variety of databases about all kinds of variables. Data on species distribution, watershed areas, political districts, census data, weather data, streams, and a number of other variables are often available as free downloads from state and government web sites. Each of these data sets can become a layer you can add to your map.

Immersive Maps

An increasingly popular tool, called Google Earth, allows users to move through geographical areas at different angles and elevations. As the user zooms down toward a city, 3D buildings become visible, for example, as well as land cover and land use information. With the higher-end versions of this product, you gain access to a range of additional preset layers, including aerial and satellite images, and you can create layers of your own to add to the viewing experience, such as site plans, sketches, and all sorts of importable images. At this point, the technology requires high-speed, PC-based computer access, limiting its accessibility.

The challenge with having such rich data sets available all at once is that too much information at one time can overwhelm the user. The ability to select and deselect different layers is thus critical with these approaches to mapping. Consider using these kinds of engaging and interactive mapping techniques with user selected customization whenever possible.

Guidelines

FOR CREATING MAPS

- Select appropriate color schemes for your context and the kind of data.

- For sequentially oriented data, use gradations of one or two colors rather than the full color spectrum, choosing light colors for low values and dark colors for high values.

- For categorical data, use a wide color range with clear contrast.

- For complex data sets, use a 3D mapping tool to analyze different layers of information.

- Include a legend where you define the meaning of the different visual categories shown in the map.

- Include only the necessary details so that the user can focus on what's important.

- Use available digital mapping tools whenever possible.

See also

Data Display (Part 3)

Document Design (Part 3)

Pocket Technologies (Part 2)

Symbols (Part 3)

Visual Communication (Part 3)

Multimedia

Documents and presentations that incorporate any combination of different media types are often called multimedia. Most people understand multimedia to be the Web and other computer-based applications and documents, but there is no one single standard for what constitutes "multimedia." Software applications that produce presentations, web pages, and even word processing documents all offer multimedia because they can incorporate many types of digital and interactive media elements into one document. Media elements can be purchased or licensed from professional sources or created in house. Multimedia documents can also be read by many types of devices, including computers, cell phones, and personal digital assistants (PDAs). The basic kinds of multimedia files that are most commonly used in technical communication include the following:

Photographs: Still images, often acquired with a scanner, or taken with a still camera or cell phone.

Video: Moving images, usually recorded on a video camera or cell phone.

Audio: Audible sounds with music, voice, or other sound effects, usually recorded with a microphone.

Animation: Shapes, drawings, text, or illustrations that are designed to move on the screen.

Storyboards: A visual outline or overview of a media or multimedia project in sequenced drawings or screens.

Podcasts: Audio and video broadcasts about a particular topic that can be downloaded to a computer or portable MP3 player.

Each of these media types has its own entry in this section of the Handbook. Technical communicators are often asked to create a document or web site in a multimedia format. It is important to determine exactly what mix of sound, moving images, still images, text, interactivity, and other features are required for a particular project.

The following web site illustrates many features of multimedia.

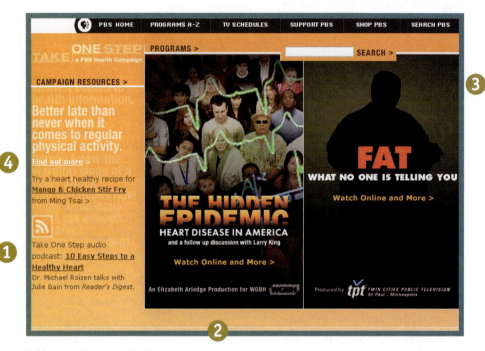

■ Multimedia-rich web site

Features of Multimedia on a Web Site

1 **Audio:** A podcast link allows users to download and listen to a podcast about heart health.

2 **Video:** These links let users watch the television program on their computer.

3 **Photographs:** Still images from the featured programs help draw users to video content.

4 **Text links:** Users can click to access a heart-healthy recipe or to read more about heart health.

Most digital media contains some combination of multimedia, such as photographs, music, audio, animation, video, podcasts, text, and illustrations. The key is not to overwhelm your audience with too many moving elements. Keep the use of multimedia straight forward and simple, with an eye toward what people really need.

FOR USING MULTIMEDIA

- Decide what mix of sound, moving images, still images, text, and interactivity are required for a particular project.

- Determine the types of devices where your project will be viewed, such as computers and cell phones.

- Start simple and do not overwhelm your audience with too many moving elements.

- Conduct usability testing on your document to make sure people actually need and use the multimedia elements.

Photographs

Photographs use light and exposure to create a reproduction or "picture" of what appears in the camera frame and then save this image onto film or a digital storage device. Scientists and technical communicators use photographs for many purposes:

To provide evidence for a claim made in the document (a picture of the polluted site for an EPA report)

To present an accurate picture of an object described in the document (a close shot of a drill in a training manual)

To present information or a location in a holistic, unified way (a workplace interior or an aerial view of a geologic site in a proposal or report)

To provide data or information that explains or illustrates something not easily said in words (a series of photos showing river erosion over time or a photograph of a screen in user documentation)

Printed and Digital Photograph Formats

Photographs are available in many print and digital formats that can serve various purposes:

Prints: Hard copy photo paper in matte or glossy finishes, made from film or digital files

Slides: Made from film or files to project or view on a light table

Posters: Large format prints made on oversized printers to create a display piece

Image files: For displaying on screen, projecting from a computer or inserting into documents; see various formats below

Photo CDs and DVDs: Developed film or digital files saved onto CD or DVD

Image Resolution

Before selecting a digital file format, it is important to consider the resolution in a digital image file. Resolution refers generally to the density and number of pixels (dots on the screen) in an image file. More pixels usually mean that you see a finer level of detail and clarity; more pixels also result in larger file sizes because they include more information. Images for web pages are usually measured in pixels, width × height. Images in printed publications are usually measured in dots per inch (dpi). In either case, the higher the numbers, the better the resolution.

Print quality photos and slides require images with the highest resolutions. Higher resolution is especially important in larger images and in printed publications, but is unnecessary if the image will only appear on screen. You may need to save several copies of an image using different resolutions for separate applications. Be sure to confirm how your images will be used and what devices will be used to print or display them before finalizing your documents. Most digital cameras allow you to choose the quality of the photograph you'll be taking. If you are unsure about the final use of the photo, take it at the highest quality setting. You can always save it as a smaller file or at a lower resolution, but the higher resolution file will ensure the best quality for either print or screen.

Photograph File Formats and Quality

Most digital cameras have a range of options that adjust how images are captured and stored. Your settings will depend on the purpose and function of the photographs you will be taking or acquiring and using in your documents. Digital photographs may be saved in several common file formats. Software, email readers, or different computer platforms may convert photo files into these different formats. It's useful to know the differences among these formats, because each has its own advantages and disadvantages.

Joint Photographic Expert Group (JPEG): JPEG files (pronounced "jay-peg") are the most common form of digital photograph. JPEG is the best choice for displaying photographs on the Web. JPEG files are "compressed," meaning the file sizes are much smaller than with other formats. Highly compressed files can suffer some loss in image quality (images can look blurry or "pixilated") but generally this is not noticeable in smaller images published on web pages.

Tagged Image File Format (TIFF): TIFF files are an older but still another common digital file format used for printing. At higher resolutions, TIFF files can be quite large, and they are rarely used on the Web for this reason.

Graphic Interchange Format (GIF): GIF files are compressed for use on the Internet. Unlike JPEG files, GIF files are not suitable for photographs, which require many more colors. GIF files are useful instead for graphics and illustrations with a small number of colors.

Photoshop (PSD): Adobe Photoshop is so widely used that its file format has become a fourth type of standard file format for saving and manipulating digital image files. Many cameras and most scanners will save files directly to PSD format, which makes it easy to open and edit the files in Photoshop. PSD files, however, cannot be opened or read by users who do not have Photoshop installed on their computers.

Scanned Images

A scanner works like a photocopier to save images onto a computer as a digital file. Printed photographs can be converted to digital ones by scanning the printed copy and saving the file. (If you use film, you can request a CD when you have the film developed.) Scanners come in several sizes and levels of quality, but even an inexpensive flatbed scanner can be used to acquire high-quality versions of printed photographs.

When scanning an image, you should first decide the following:

What type of digital file you need (JPEG, TIFF). Choose the format based on the final project output and also the easiest for your software to access.

Whether you need color or a black-and-white image, which can have one-third the file size of color.

If you want portrait (vertical) or landscape (horizontal) orientation for the image.

What resolution quality you need. Use the lowest resolution for the project at hand without compromising the image quality.

What the dpi setting should be (lowest for onscreen, higher for print, highest for photos).

These settings are all defined in advance in the scanner driver software program. Each setting affects the size of a digital photo file, so generally you want to minimize file sizes, which then increases the speed of handling the file.

Altered Photographs

The availability of editing software makes it easy to enhance and also to alter photographs for various purposes. Some changes in digital photos serve a clear scientific purpose. For example, the U.S. Department of Energy's Idaho National Engineering and Environmental Laboratory developed the Change Detection System (CDS) software, which highlights subtle photographic changes that are not easily visible to the human eye. These satellite images of Alaska's Hubbard Glacier from 1985 and 1986 were compared to show how the glaciers had been receding. Changes to the images are highlighted in yellow with the CDS software.

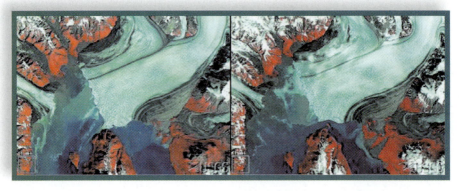

■ 1985 image   ■ 1986 image

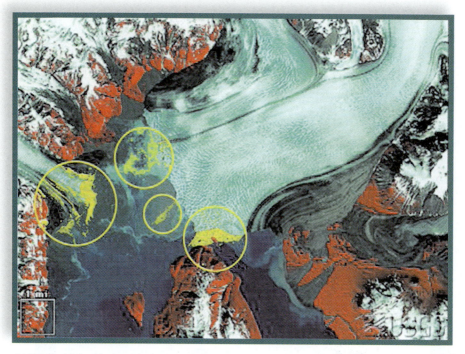

■ NASA's Idaho National Laboratory. CDS software compares photographs and makes information readily available for further analysis

Photo-Illustrations

A photo-illustration is a deliberately altered photograph that clearly shows readers a meaning that the photograph alone cannot convey. Many people expect photographs in the media to be "true," real representations that document events objectively and accurately. When newspapers alter photographs for illustrative purposes, they label the image a "photo-illustration." Technical publications may use photo-illustrations for a magazine or journal cover to enhance images and also to make explicit concepts or arguments more clear. This example from the Berkeley Lab's research review uses a photo-illustration to show the importance and also the elusive, science-fictional qualities of nanotechnology.

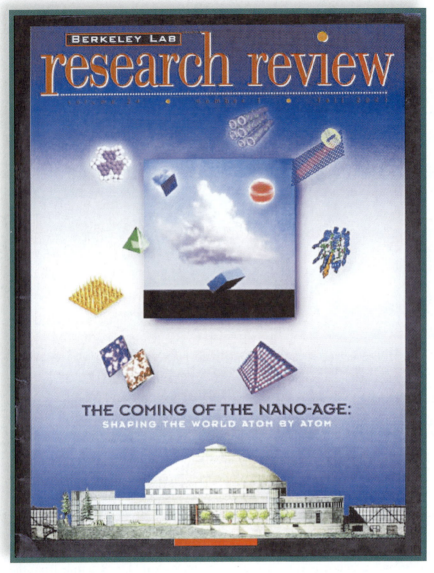

■ Photo-illustration

This photo-illustration contains photographic elements but is clearly distinct from a traditional photograph and shows how a photo-illustration can be used to represent an abstract or future-oriented concept.

Editing Digital Photographs

Commonly used software for photo editing includes the following:

Adobe Photoshop and Image ready (industry standard)

Photoshop Elements

GIMP (GNU Image Manipulation Program), an open-source tool available over the Internet

Apple iPhoto (consumer/home software with wide sorting and publishing capabilities and basic editing tools that comes with McIntosh computers)

Picasa, a similar program for Windows users (provides photo editing and storage software, offered free from Google at http:/ /picasa. google. com/)

Basic paint programs and accessories that come with computers also do minor editing and sizing of photos

Word processors and PowerPoint provide minor editing and sizing of inserted photos

What follows is an example in Photoshop where tools let a user select and then edit the center of the photo-illustration with brightness and contrast to change its contrast significantly.

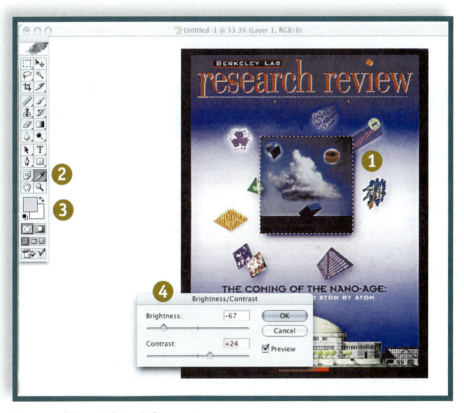

Features of Photo Editing Software

❶ Dashed lines, called "marching ants," define the selected area.

❷ Tool bar along left shows that the color picker tool (eyedropper) is selected (highlighted in gray) from the palette of tools. The eyedropper tool allows the exact colors from the existing image to be selected and then used elsewhere.

❸ The color squares show foreground (black) and background (yellow) colors pulled from the image with the eyedropper.

❹ The brightness and contrast dialog box shows sliders and percentages for changes and a preview of the results within the selection.

Ethics of Using Photographs and of Photo Alteration

When you use a photograph in a document or on a web page, you should always include a photo credit. The owner of the photograph will usually specify how he or she wishes to have this credit listed. For example, the photo credit may be as simple as "Photograph by John Smith." You may need to request permission or pay a licensing fee depending on the copyright requirements. See Copyright and Intellectual Property (Part 4) and Ethics (Part 1) for more information.

If you find a photograph that is right for your document or web site, but you want to alter it, the ethical issues can be tricky. A deliberately altered photograph can make a powerful statement. When you alter another's original photograph significantly for illustrative or artistic purposes, it requires both a photo credit and a clear indication of the photo's altered status as an illustration. Unacknowledged alterations can result in biased meanings, stereotypes, or unethical statements. And as noted above, you may also need to seek permission to use the photo, depending on the copyright situation.

Guidelines

FOR USING PHOTOGRAPHS

- Decide on your purpose for using a photograph in your document.

- Determine what image file format and quality best complements your document delivery.

- Purchase or find a suitable photograph, or take the photo yourself with a digital camera.

- If necessary, select a basic image editing software program such as Picasa or a professional quality program like Photoshop and enhance your photograph.

- If you alter a photograph significantly, make sure the changes are ethical and in good taste. You should also label the photo as altered or as a photo-illustration.

- If a large format photo or an elaborate photo-illustration best suits your needs, hire a professional artist to make sure the results look professional and of high quality.

See also

Copyright and Intellectual Property (Part 4)

Ethics (Part 1)

Image Searches on the Web (Part 4)

Technical Illustrations (Part 3)

Visual Communication (Part 3)

Pocket Technologies

Personal digital assistants (such as Palm Pilots, BlackBerries), smart digital phones, MP3 players, global positioning systems (GPS), e-book viewers, iPhones, and other mobile technologies allow on-the-go access to information in an increasingly mobile world. These "pocket technologies" provide instant, portable access to the Web, email, audio and video programming, text documents, spreadsheets, maps, photographs, and other multimedia and information formats.

Designing Text and Visual Information for Pocket Technologies

Compared with a desktop computer, pocket technologies have very small screens with limited memory and computing capacity. If you are helping create information that will be used on a pocket technology, you face a wide range of design considerations. Text needs to be brief and to the point. Graphics need to be limited so they do not crowd out the text.

Some search engines and news sites have special web addresses that are for use with mobile devices. The sites offer a more brief, minimal screen design that is easier to read on small screens.

■ Google's interface for mobile devices

If you know that the document or information you are working on will be displayed on a pocket device, you may need to create one document for a regular web page and an alternative document, one with shorter amounts of text and fewer graphics that is specifically created for a pocket technology. It will not work simply to transport the web page for a computer onto a PDA's small screen.

When dealing with large amounts of information (user documentation or news), the more common approach is to create content once, in small chunks, independent of the platform, and store it in a database. Then, various chunks of information are pulled together in different formats, depending on whether the final form will be delivered on a PDA, on a regular computer, on a cell phone, or in print. (See Content Management Systems (Part 1) for more information on this approach to managing large sets of information.)

Many scientific and technical experts use pocket technologies for their everyday work. For instance, medical doctors access the *Physician's Desk Reference*, which contains detailed information on prescription medications, via their PDAs. A wealth of medical, pharmaceutical, engineering, and other technical reference material is available for PDAs, allowing doctors and others to have this information readily available in their pocket and not in a large volume book on a bookshelf somewhere else. This information must be designed much differently for pocket technology display than for a large reference textbook. A layered approach usually works best: display the high-level headings and summary information first, and let users click through the document for more detail.

Guidelines

FOR CREATING INFORMATION FOR POCKET TECHNOLOGIES

- Design for the small screen using a layered approach: display headings and summary information first.
- Don't import a web page design directly onto a smaller screen.
- Give users control over the font size.
- Use color to help users navigate on the small screen.

See also

Content Management Systems (Part 1)

Email (Part 2)

Podcasts (Part 3)

Visual Communication (Part 3)

Podcasts

Podcasts are audio files about a particular topic that can be downloaded and listened to on a computer or portable MP3 player. Users can subscribe to podcasts and have the audio files downloaded automatically using one of several software tools. Because MP3 players are portable and easy to use, technical communicators have begun to use podcasting as a way to distribute product and technical information to customers, clients, and employees.

Many MP3 players also include video playback capability, and video podcasts (or V-casts) are becoming commonplace in tech-related industries. Video podcasts can include presentation slides or even video tutorials ("screencasts"). Technical communicators have been quick to embrace podcasting as a means of information delivery because it offers the immediacy of oral presentation with the easy distribution of the MP3 audio file format. Popular audio and screen recording programs frequently offer the ability to format recorded presentations in compressed files suitable for podcasting, making it relatively easy and inexpensive to produce and distribute high-quality presentations through pocket technologies or computers.

Subscribing to Podcasts

Podcasts are available on almost any topic of interest, from news and information to hobbies to scientific and technical topics. Web sites and blogs often provide annotated links to current podcasts related to a specific industry or technology. Fields where information changes and needs to be frequently updated, like computer security, for example, rely heavily on podcasts to distribute critical information quickly.

■ Web magazine for chief information officers and other information technology professionals. Offers traditional written articles as well as podcasts about computer security

A number of software tools, including Apple iTunes, allow users to locate and subscribe to podcasts. Podcast directories, like Podcast Alley, also allow easy searching and downloading from among the expanding universe of download-able, and usually free, podcasts.

Creating Podcasts

Podcasts can be created on a computer using audio and video recording and edit-ing tools. Audio-only podcasts require a microphone, headphones, and audio ed-iting software (such as Audacity), and video podcasts also require a video camera or screen capture tool (like Camtasia Studio). Once the recording equipment and software are installed and tested, most podcasts can be created using a series of basic steps:

1. Plan the podcast (determine the audience, purpose, style, format, and frequency needed) and draft a script or outline to follow during recording.
2. Record the audio (voice, music, sound effects) and video (images and screen captures) files.
3. Edit the audio and video tracks.
4. Convert the edited files to MP3 or similar format.
5. Publish the podcast and promote it so users know where to find it.

Guidelines

FOR CREATING PODCASTS

- Speak in an informal tone and connect to your listeners as individuals with an interest in your topic.
- Practice your delivery to be sure that your voice is clear, direct, and audible.
- Edit audio and video tracks carefully to eliminate noise, repetition, pauses, and gaps, and to create an effective, concise listening experience for your audience.
- Test your files to be sure that your podcast can be downloaded and played back on an MP3 device.
- Post your podcast in a place where your audience can easily find it, and include a brief annotation or note to clarify what each episode includes or covers.

See also

Audio (Part 3)

Pocket Technologies (Part 3)

Presentations (Part 3)

Video and Animation (Part 3)

Project Management Visuals

Project management tools help you manage large projects by displaying benchmarks and deliverables based on when the work must be completed. With complex projects or large teams, a visual chart that schedules work based on other work can be essential.

For instance, when renovating a bath, tile cannot be completed until the drywall is finished. Wiring should be done before the drywall begins but after the demolition. Plumbing must be roughed in before the appliances are placed. Fixtures must be selected before installation can be completed. A *Gantt chart* for this remodel project is in essence a bar chart with multiple horizontal bars that schedules and tracks all these interdependent activities by working backward from the completion date.

Gantt chart used to plan and schedule a bathroom renovation project

Another type of visual used for project management is called the Program Evaluation and Review Technique, or *PERT chart*. PERT charts also use benchmarks but instead of creating a bar-type chart, they produce a flow chart: a network diagram with each task illustrated as a node. PERT charts illustrate task dependencies better than Gantt charts but at the same time may be hard to interpret for the most complex projects.

A work process method, on the other hand, describes tasks in detail and how the work will be completed. A wide range of work styles, technologies, equipment, and other factors may complicate the project. If a programmer works all night authoring code in a lab, for example, while the graphic designer works 9–5 at his home office and the software development team leader attends her meetings in various locations, how will this group exchange their work and communicate about their progress? Scheduling becomes a critical issue. Using a software tool like Microsoft Project, you can define and schedule work for a project in sequence.

Work process chart for a team project

These task-based plans reside on the Internet so that all team members can access and modify the schedule. A chart can also be printed and displayed at the work site for a quick common reference.

Guidelines

FOR CREATING PROJECT MANAGEMENT VISUALS

- Make a list of steps and deliverables for a large or complex project.

- Decide on the scheduling needs for your project.

- Choose the appropriate software tool and chart type (PERT, Gantt, or work process).

- Set a date for completion, define tasks, and add them to the project list.

- Print charts for reference, if desired.

See also

Charts (Part 3)

Project Management (Part 1)

Teamwork and Collaboration (Part 1)

Spreadsheets

Spreadsheets are electronic collections of data that allow you to track, sort, calculate, and display data in a variety of ways. Spreadsheets allow you to track information, perform calculations (including some types of statistical analyses), create tables, charts, and graphs, and much more. Because of these various functions, spreadsheet programs such as Microsoft Excel serve many critical functions in the workplace.

Using the Internet, you can find many examples of data that are downloadable in a spreadsheet format. For instance, the U.S. Department of Agriculture makes available this spreadsheet on the per capita consumption of red meat and poultry:

	A	B	C	D	E	F	G	H	I	J	K
1											
2									Filename:	MTPCC	
3	Red meat (carcass weight) and poultry (ready-to-cook weight): Per capita availability[1]							Poultry (ready-to-cook)[4]			
4	Year	U.S. population, July 1[2]	Red meat (carcass)[3]								Total[5]
5			Beef	Veal	Pork	Lamb	Total[5]	Chicken	Turkey	Total[5]	
6							SUM(C.F)			SUM(H.I)	SUM(G,J)
7											
8		Millions					Pounds				
9											
10	1909	90.490	74.2	7.3	67.0	6.7	155.2	15.3	1.0	16.3	171.4
11	1910	92.407	70.4	7.2	62.3	6.4	146.4	16.0	1.0	17.0	163.4
12	1911	93.863	68.5	7.1	69.1	7.4	152.0	16.2	1.1	17.3	169.3
13	1912	95.335	64.5	6.9	66.7	7.6	145.8	15.5	1.1	16.6	162.4
14	1913	97.225	63.3	6.3	66.9	7.2	143.7	15.1	1.1	16.2	159.9
15	1914	99.111	62.0	5.8	65.1	7.1	140.0	15.0	1.1	16.1	156.1
16	1915	100.546	56.4	5.9	66.5	6.1	134.9	14.9	1.2	16.1	151.0
17	1916	101.961	58.9	6.4	69.0	5.8	140.2	14.1	1.2	15.3	155.4
18	1917	103.414	64.7	7.2	58.9	4.5	135.3	13.7	1.2	14.9	150.2
19	1918	104.550	68.6	7.3	61.1	4.8	141.7	13.7	1.2	14.9	156.6
20	1919	105.063	61.5	7.8	63.9	5.7	138.9	14.8	1.3	16.1	155.1
21	1920	106.461	59.1	8.0	63.6	5.4	136.1	14.2	1.3	15.5	151.6
22	1921	108.538	55.5	7.6	64.8	6.1	134.0	13.8	1.3	15.1	149.1
23	1922	110.049	59.1	7.8	65.8	5.1	137.8	14.7	1.3	16.0	153.8
24	1923	111.947	59.6	8.2	74.2	5.3	147.3	15.1	1.3	16.4	163.8
25	1924	114.109	59.5	8.6	74.1	5.2	147.3	14.2	1.3	15.5	162.8
26	1925	115.829	59.5	8.6	66.8	5.2	140.0	14.8	1.3	16.1	156.2
27	1926	117.397	60.3	8.2	64.1	5.4	138.0	14.7	1.3	16.0	154.0
28	1927	119.035	54.5	7.4	67.7	5.3	134.8	15.8	1.4	17.2	152.0
29	1928	120.509	48.7	6.5	70.9	5.5	131.6	15.2	1.4	16.6	148.2
30	1929	121.767	49.7	6.3	69.7	5.6	131.3	14.8	1.4	16.3	147.5
31	1930	123.188	48.9	6.4	66.9	6.7	128.9	16.3	1.5	17.8	146.7
32	1931	124.149	48.5	6.6	68.3	7.1	130.6	14.6	1.4	16.0	146.6
33	1932	124.949	46.7	6.6	70.6	7.1	130.9	14.9	1.7	16.6	147.6
34	1933	125.690	51.5	7.1	70.7	6.8	136.0	15.3	1.9	17.2	153.2
35	1934	126.485	63.8	9.3	64.4	6.3	143.8	14.0	1.8	15.8	159.6
36	1935	127.362	53.2	8.5	48.3	7.2	117.3	13.6	1.7	15.3	132.6

■ U.S. Department of Agriculture spreadsheet

Once you have downloaded this publicly accessible data, you can use any spreadsheet software to generate a graph, such as the following graph illustrating the different meats consumed in 1999, a single entry (row) on the spreadsheet:

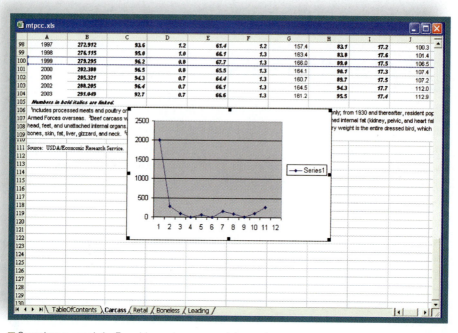

	A	B	C	D	E	F	G	H	I	J
98	1997	272.912	93.6	1.2	61.4	1.2	157.4	83.1	17.2	100.3
99	1998	276.115	95.0	1.0	66.1	1.3	163.4	83.8	17.6	101.4
100	1999	279.295	96.2	0.8	67.7	1.3	166.0	83.0	17.5	106.5
101	2000	282.308	96.5	0.8	65.5	1.3	164.1	90.1	17.3	107.4
102	2001	285.321	94.3	0.7	64.4	1.3	160.7	89.7	17.5	107.2
103	2002	288.205	96.4	0.7	66.1	1.3	164.5	94.3	17.7	112.0
104	2003	291.049	92.7	0.7	66.6	1.3	161.2	95.5	17.4	112.9

Numbers in bold italics are linked.

106 Includes processed meats and poultry on...

107 Armed Forces overseas. Beef carcass w...

108 head, feet, and unattached internal organs...

109 bones, skin, fat, liver, gizzard, and neck.

111 Source: USDA/Economic Research Service.

...only; from 1930 and thereafter, resident pop...

...hed internal fat (kidney, pelvic, and heart fat...

...ry weight is the entire dressed bird, which...

— Series1

TableOfContents | Carcass | Retail | Boneless | Leading

■ Creating a graph in Excel based on spreadsheet data

Entering data in a spreadsheet is typically the first step for making a chart or graph, which is then normally a simple task. Spreadsheets are also used for statistical analysis. For instance, if you want to calculate the average from a column of data or any other range of data, you can easily create an Excel formula. In the example above, you could calculate the average of items in column C with the formula AVERAGE(C1:C104). Other common Excel functions include SUM (adds all numbers in the specified range), SIN (used for calculating the sine of an angle), and PMT (used for calculating loan payments).

You may need to create a spreadsheet from scratch, but increasingly, you can locate templates and prewritten formulas to help with your calculations. For instance, you can locate formulas for calculating loan amortizations, sales figures, student grades, and so on. The best way to find out about formulas and statistical calculations using Excel is to take the online tutorial or a short course.

Guidelines

FOR CREATING SPREADSHEETS

Working with data and formulas

- When possible, look for data from government and other public sources.

- Verify when the data was last updated.

- Find templates and formulas from reliable sources instead of creating formulas from scratch.

- Use spreadsheets instead of complicated statistical software when you can.

Creating visual displays using spreadsheets

- Keep visual displays simple and easy to read.

- Create the appropriate visual for that particular data set—don't create a pie chart if what you want to show is trends over time.

- Remember that your visual will only be as accurate as the data from which it is created.

- Follow other guidelines in this handbook for charts and graphs.

See also

Charts (Part 3)

Data Display (Part 3)

Graphs (Part 3)

Tables (Part 3)

Storyboards

Storyboards are a visual outline or overview for a project or presentation that will be created in a multimedia format. A digital media project, a presentation, a video, or a complex web site might all use storyboards as part of the planning and development phase, especially with a team of writers and designers working together. Each storyboard shows all the authors what the planned presentation or project will include by offering a visual representation of each section, whether it be a screen design, a thumbnail picture, or a still frame of the video. Professionals use storyboards in film and video as well as in broadcasting to line up spoken scripts with the visuals that appear during that portion of the text. Storyboards can be used in many technical projects to illustrate a general outline, early navigation schemes, or simply change over time, along with how the content will be organized.

The example below shows a simple slideshow of photographs from Mexico in storyboard view, using a timeline in a simple movie editing program which organizes all media elements for a project.

■ Storyboard view of project in Windows Movie Maker

Features of a Storyboard

1 Begin by selecting "Import Pictures" from your collection.

2 Drag pictures onto the storyboard to create a sequence in any order.

3 Each storyboard box defines a specific still photo that will appear in the main video window.

4 The images change over time from left to right, creating the timeline for this sequence.

Designing Storyboards

Storyboards can also be created by drawing on pieces of paper that will appear on each screen, by working with slides and notes in presentation software programs such as PowerPoint, or by using commercially available storyboarding software such as StoryBoard Artist.

Like project planning visuals, a single storyboard is shared by the group and helps define the scope and structure of the project. When designing a technical project, remember that the storyboard is just a place to design and experiment with your project structure in a focused way. You can use your storyboards to help identify the types of visuals and content that you will need before you build the project.

Remember to keep your audience in mind as you storyboard. What will they be expecting in the final presentation? What order helps them best understand the information? How will you correlate the written text or spoken script with the visuals you have selected? As you decide these issues, storyboards give you the opportunity to conduct usability research early on with members of the intended audience.

Storyboards on Paper

■ Storyboard for Out of Bounds multimedia kiosk

To storyboard by hand, simply create sheets of paper that include a screen box and several lines for the written or spoken text and other notes at the bottom of the page. Use simple forms, even stick figures, to indicate images on the screen, with notes that describe any visual and interactive elements that will appear on each screen. Any links and other screen actions should be clearly marked so that when you lay out the paper storyboards, a user can "navigate" from one screen to another using those hand-drawn links.

Storyboarding with Software

In the example below, a storyboard is being used to plan an interactive web site on state and local insect populations. The image illustrates what will appear on the screen. The text below the image outlines what will happen when users get to this screen.

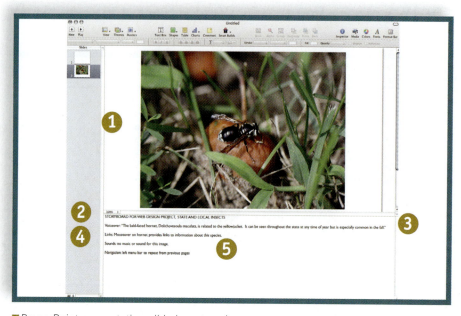

■ PowerPoint presentation slide in notes view

Features of a Storyboard

1 The slide displays the graphic and shows what this screen of the storyboard will look like.

2 Notes window gives space to list all the features of this screen.

3 Voiceover text (to be recorded later) is included in quotation marks.

4 Links explain all actions that the user can do from this screen (here, a mouseover on the hornet).

5 Navigation indicates a standard left menu bar that will be used on all pages.

To storyboard using presentation software such as PowerPoint, set the view display to speaker's notes and view this notes page and the slide together on your screen. You then have a space for working with visuals on the slide and a space for the text or a voice-over script on the notes page. You can insert links, pictures, and video clips directly from your own files onto the PowerPoint slide. The finalized storyboard can be printed as individual screens, saved as a presentation, or saved on a web page.

To storyboard with movie editing programs like iMovie, you must first import the media files—the pictures, video clips, and audio files that you want to use. The timeline lets you select between editing mode and the storyboarding mode. Change your view to storyboard (as shown above). Once you place each item on the timeline in the order you want it displayed, you can see in storyboard format how your project will look and change over time. The finalized storyboard can be saved as a movie or imported into another software program.

FOR CREATING STORYBOARDS

- Decide whether you will create storyboards by hand or using a software tool.

- Create a list for your sequence of screens or images.

- Design a flow chart of the navigational structure of your project.

- Create each individual storyboard, including all links and media elements.

- Do a walk-through to test usability using paper storyboards or your digital storyboard and have users work through the sequence of images.

- Make changes and rearrange items on the storyboards based on user feedback.

See also

Symbols

Symbols are graphical images that are meant to represent an item, concept, process, or some other type of information. Symbols are used in technical documents to convey a message quickly and efficiently or to help enhance the meaning of textual information. Symbols within a document might be large or small. They may be used within the context of a map or graph to illustrate a category of information, in a chart to illustrate some part of a process or hierarchy, in a technical illustration to show various components in a schematic diagram, or alongside paragraphs or other text to signal a particular type of information.

Technical documents often contain their own specific sets of symbols whose meanings are defined within the document or interface through a key or legend, interactive help, or some type of explanatory text. Sometimes symbols might even have short labels associated with them to clarify their meaning instantly.

A good example of symbols are those used on maps. In order to make information clear and consistent for the reader, maps must use the same symbol to represent similar locations or features. For instance, on a topographic map, the following symbols might be used to represent features of a coastline. Maps also use a key to define symbols for readers.

COASTAL FEATURES

| Foreshore flat |
| Coral or rock reef |
| Rock, bare or awash; dangerous to navigation |
| Group of rocks, bare or awash |
| Exposed wreck |
| Depth curve; sounding |
| Breakwater, pier, jetty, or wharf |
| Seawall |
| Oil or gas well; platform |

Symbols used on a topographic map

Certain symbols, such as a question mark or a stop sign, are easily recognizable to an audience within national or even international communities. These universal symbols usually require little to no explanation. Images can have very different connotations, however, depending on the cultural context. Finding ways of communicating across cultures without misunderstandings or offenses can be challenging. It can be easy to blunder if you're not careful—examples often cited as misunderstood include the A-OK sign and thumbs up sign, which have positive connotations in the United States, but which are obscene or offensive gestures in several Middle Eastern cultures.

Icons

Icons are a type of symbol that is used to signify a specific object or group of objects. Universal icons can provide critical instructions for international communication. In the following examples, the icons address food handling. In a world of international trade of goods, finding quick, effective communication methods is critical to public health as well as business profits. Notice how the middle icon indicates temperature in both Fahrenheit and Celsius to reach a global audience that might use either measurement standard. The instructions regarding hand covering on the left and hand washing on the right both use close-ups of the hands rather than a full-body view, preventing the viewer from becoming distracted by extraneous details and making the icon more universal.

■ Sample icons from the International Association for Food Protection

The International Organization for Standardization (ISO) and the International Electrotechnical Commission (IEC) have developed symbols and icons that are available for purchase and use in software, signage, and other applications. These standardized symbols incorporate the standards IEC 60417 and ISO 7000 for use on any equipment, from automobiles to office and home entertainment equipment to textile and earth-moving machinery. Most people are familiar with the ISO symbols used in airports worldwide.

Context-specific Symbols to Facilitate Reader Navigation

A printed training manual might use symbols and icons in the margin to give the reader a quick and easy signal for the purpose of each section, providing a symbol for instructions, then another symbol for exercises, and so on. The example below shows a symbol used in a training manual.

Budgets

Budgets are the different sets of financial data that you set up to track. You can set up any number of budgets in your application. The Budgets window lets you add, remove, and modify budget data sets and define their attributes. Periods are units of time for keeping financial data. When setting up a budget, you assign a budget type to determine the number of periods per year.

Learn to create a budget

Workshop: Creating a Budget

In this workshop, you will create a Budget called FY2009.

1. From the desktop, select **Budgets**.

2. In the Categories box, type **FY2009**.

3. In the Category type field, select **Monthly**.

4. In the Start Period box, select the month and year.

5. Select the **Setup** button and preview how your budget will be formatted.

Try creating a budget

■ Symbols within a text

Features of symbols

1 The symbols are simple generic items.

2 The symbol for the first section indicates explanatory information.

3 The symbols for the second section indicate a hands-on exercise.

4 The words "Learn to create a budget!" and "Try creating a budget!" elaborate on each symbol's meaning.

Notice that in the sample the symbols are accompanied with text so that their meaning is clear. Once established, the icons automatically represent two document sections with different purposes to the reader. While these icons might seem unnecessary in a one-page excerpt, they actually serve as effective markers in real documents in which each section is considerably longer than in this sample. Use principles of document design to place symbols effectively within documents. Be sure to test your symbol, and to understand the cultural filters of those who will view your figure. Avoid elements in your figures that might offend your audience or seem unnecessarily exclusionary.

FOR CREATING AND USING SYMBOLS

- When appropriate, use standard international symbols (such as ISO).

- Use crisp, clear graphic designs that will reproduce well in either print or digital formats.

- Use images that will look good in small or large sizes.

- When symbols are used as part of a text-based document, use them sparingly.

- Place ample white space around a symbol to increase its visibility.

- Pay attention to cues that your symbol may offend or mislead particular groups of people.

- Use gender-neutral figures unless the message needs to specify male or female (e.g., restroom designations).

- Avoid hand and arm gestures, since an innocuous expression in one culture could have a negative or offensive meaning to another culture.

See also

Audience (Part 1)

Charts (Part 3)

Clip Art (Part 3)

Graphs (Part 3)

International Communication (Part 1)

Visual Communication (Part 3)

Tables

Tables present a visual display of information within a two-dimensional grid of rows and columns. A table is an effective display for two-dimensional data, usually when one dimension is a collection or series of items and the second dimension consists of attributes or characteristics that all or most of the items have in common, such as description, type, size, and color.

There are many applications and uses for tables in technical communication. Particularly when you need to describe highly detailed information, both textual and numerical, a table can communicate simply the many details that would be lost in a paragraph or difficult to comprehend in a list. One advantage of a table is that it is easy to scan for information. Another advantage is that the commonalities and variations across items are quickly apparent.

The benefits of using a table are easily seen in this example of labor statistics about the midwest region of the United States.

Data Series	Back Data	Sept 2007	Oct 2007	Nov 2007	Dec 2007	Jan 2008	Feb 2008
Labor Market							
Civilian Labor Force [1]	🦕	34,921.1 [6]	34,930.7 [6]	34,914.9 [6]	34,933.7 [6]	35,011.3	
Employment [1]	🦕	33,084.3 [6]	33,084.5 [6]	33,085.7 [6]	33,088.1 [6]	33,186.1	
Unemployment [1]	🦕	1,836.8 [6]	1,846.2 [6]	1,829.2 [6]	1,845.5 [6]	1,825.2	
Unemployment Rate [2]	🦕	5.3 [6]	5.3 [6]	5.2 [6]	5.3 [6]	5.2	
Consumer Price Index							
CPI-U, All Items [3]	🦕	199.714	199.455	200.762	200.227	201.427	201.896
CPI-U, All Items, 12-month % change	🦕	3.1	3.7	4.1	3.8	4.3	3.8
CPI-W, All Items [4]	🦕	194.828	194.384	196.056	195.493	196.617	197.110
CPI-W, All Items, 12-month % change	🦕	3.2	3.9	4.6	4.1	4.7	4.2
Employment Cost Index, Private Industry							
Total Compensation [5] [7]	🦕	104.6		105.3			
Compensation, 12-month % change [7]	🦕	2.2		2.4			
Wages & Salaries [5]	🦕	105.0		105.6			
Wages & Salaries, 12-month % change	🦕	2.9		2.9			

■ Table of labor statistics from the U.S. midwest

Features of Tables

1 Colors make the table easy to navigate.

2 Major categories of information are separated visually.

3 Comparable data from each month can be compared easily across each row.

4 Data from a specific month can be viewed in each column.

5 Subcategories are created as hyperlinks, so readers can click for more information.

6 The dinosaur symbol provides a visual cue that readers can locate back data in any category.

Designing Tables

Tables are easy to design if you think about information in terms of categories and options. You might begin by listing or sketching out the various options or major categories that you want to describe. Then beside each major category, list the various characteristics that need to be pointed out about that category. Try to find the commonalities of your data—to design a table effectively, the types of data across major categories need to be similar.

Once you have identified your categories, you need to determine which items will go in the rows and which will go in the columns. To make the most effective arrangement of rows and columns, consider your readers and what you are trying to communicate to them in the table. Design the table so that the important information will be easily understandable and accessible:

To make information understandable, write informative row headings. If you abbreviate, be sure that the abbreviations will be understandable to the reader.

Avoid wordiness within each cell. When possible, use phrases rather than sentences and remove extraneous words while being clear.

Use bold text and color coding to highlight important information.

Tables can be created using a variety of tools. The most common forms of table are the types created within word processing software, such as Microsoft Word's "table" function, or spreadsheets created with programs such as Microsoft Excel. You can also create tables for web pages using HTML. These programs offer many options for table styles, including invisible grid lines that effectively hide the table. All of these formats have the added benefit of alignment: the information placed within cells will appear formatted on the page or screen according the table and cell style features. Be aware, though, that overuse of tables can lead to a boxy and rigid-looking document.

Guidelines

FOR CREATING TABLES

- Introduce a table with an explanation of what it contains and how it will help the reader.

- Give the table either a title or a caption (depending on what is appropriate for your document).

- Write informative, understandable, and visually distinct heading labels.

- Make rows distinct, either with row headings or by displaying the major categories prominently in the first cell of each row.

- Avoid wordiness in tables. Try to keep textual descriptions in cells to a few words.

- Use color coding and symbols (and provide a key if necessary) to create tables that allow for the fastest scanning and comparisons of data.

- Make sure the table will be readable in the format where it will be displayed (paper, screen, overhead projection).

See also

Data Display (Part 3)

Document Design (Part 3)

Spreadsheets (Part 3)

Symbols (Part 3)

Web Page Design (Part 3)

Technical Illustrations

Technical illustrations are used to communicate information and concepts visually in technical documents. Illustrations are most effective when they bring increased clarity to the topic or subject matter being described. In technical writing, illustrations are particularly valuable to show equipment and devices, spatial relationships between objects, site development plans, and other physical concepts, such as perspective, change, and growth. In addition, illustrations can convey cause-and-effect relationships and other important elements in task-based documents.

Types of Technical Illustrations

Technical illustrations can include drawings, illustrations, plans, and exploded view diagrams. Technical illustrations usually cannot stand alone; they require supporting text, including titles, captions, labels, and sometimes callouts to explain their significance in conveying technical concepts. The look, layout, and accompanying text depend on the purpose and type of illustration. Technical illustrations should be placed as near as possible to the text they illustrate or embellish.

Drawings

The following example shows a handmade drawing, the oldest type of technical illustration. This illustration is a hand-drawn diagram of some basic lighting

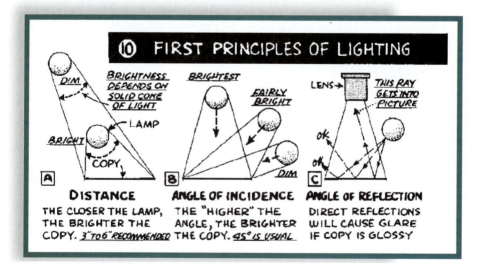

■ Technical drawing: Lighting an object to be projected

principles to consider when designing and building an opaque overhead projection system.

This drawing shows that a neat hand-drawn and labeled representation can impart technical concepts with a customized look.

Illustrations

Illustrations usually contain more detail than simple line drawings. Because people comprehend pictures more quickly than words, illustrations can have an immediate impact. Thus, your choice in illustrations will influence the overall tone of your document. Be sure to select illustrations that are appropriate for your subject matter, audience, and context.

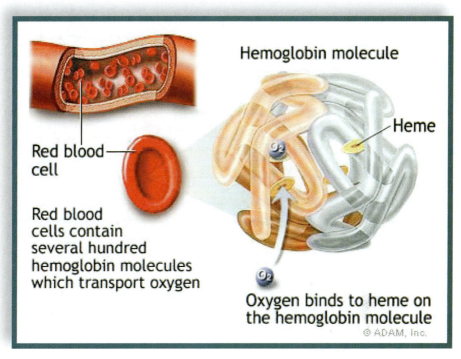

■Example illustration of red blood cell hemoglobin

This illustration of a red blood cell and hemoglobin molecule is clear, easy to understand, and more detailed than a simple drawing. The use of color helps distinguish the details. Most technical illustrations such as this are created using drawing software or computer-aided design tools.

Plans

Architects, landscape designers, construction project managers, and other professionals use site plans and blueprints for designing to scale and illustrating project specifications for a particular location. While traditionally drawn by hand, most plans are produced using programs like Softdesk's AutoCAD (computer-aided design) drafting tool. This example shows the layout of roads, buildings, landscape objects, measurements, and other details for a new construction development.

■ Site plan for new construction

Features of Plans

1 Objects are drawn to scale and identified by name or descriptions.

2 Doors, entrances, and windows marked similar to blueprints.

3 Labels drawn with leaders pointing to specific objects and features such as "stone wall" on the left.

Exploded Views

Technical illustrations do what photographs and other realistic images cannot: show different views and layers of an object. The exploded view is an extremely helpful illustration for instructions, specifications, and other documents. This exploded view shows part of the Hubble Space Telescope (HST).

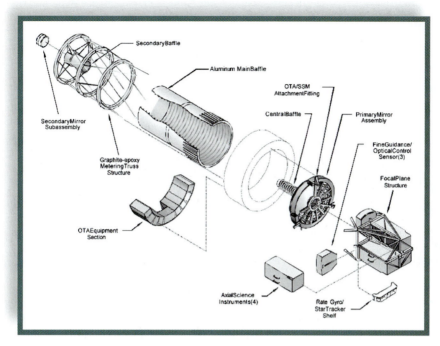

■HST optical telescope assembly: Exploded view

While these examples do not cover the many possible types of technical illustrations, they give a sense of the purpose and range for using them in your documents. Keep in mind that, if you plan to create a new illustration, you will probably need a graphic designer or illustrator to help you produce the image.

Guidelines

FOR CREATING TECHNICAL ILLUSTRATIONS

■ Use professional quality illustrations to elaborate, not decorate, your document.

■ Provide supporting text to introduce and explain the illustration.

■ Include a title, caption, and, if needed, labels and callouts to make the illustration clear.

■ Always check accuracy of your drawings, labels, and details.

■ Do not distort or misrepresent the subject matter.

See also

Data Display (Part 3)

Document Design (Part 3)

Instructions (Part 2)

Specifications (Part 2)

Visual Communication (Part 3)

Video and Animation

Video, film, and animation are all moving images that are recorded onto videotape, a computer disk, or other media formats. While film has always been an expensive, professional industry, video has been more accessible and less expensive for users. Film and video are separate media, yet with the popularity of digital video (DV) cameras and nonlinear video editing software, many of the differences in quality and usage are becoming obsolete. Video can be recorded with a video camera or captured by a screen recording software tool like Camtasia Studio to display moving images to the user.

Animations were originally sequenced drawings that, when viewed in quick succession, gave the appearance of movement. Computer-based animation allows you to create movement of shapes, text, and other illustrations within web pages, videos, and other media. An animation is designed and then drawn from the data, or "rendered," in a software program to illustrate full-motion moving images or three-dimensional concepts not possible in two-dimensional video. Animation can be effective when you need to demonstrate a process or procedure. Moving images are useful for technical communicators in many situations, and video and animation are often integrated into the mainstream workflow in many technical workplaces.

Uses of Video and Animation

Technical fields and industries use moving images for computer-based training (CBT) to illustrate mechanical processes, document safety procedures, provide technical information, and offer software training to users. Animation is sometimes used in conjunction with video in these applications. CBT programs may use animations to illustrate a concept or process, add an interactive illustration, or provide additional technical information to users between sections of live video. Video clips are short sections that can be included in a web page, presentations, and other digital documents. Video clips are also quite common across the Internet for news and entertainment, appearing on sites such as cnn.com, sports pages, and YouTube.

Technical communicators might be asked to create storyboards for a training video, to design and test the usability of educational software, or to help design videos and animations for software training purposes. As the hardware and software become less expensive and easier to use, full motion and interactive documents have become easier to produce in house and distribute to users outside the organization. The specialized types of video often created by technical communicators include videos used for education and training.

Technical Training Videos

This still frame comes from a computer-based training video created for the offshore drilling and refineries industries. When users select a link, the file downloads and a small window opens the video in an appropriate media-playing software plugin, such as Apple Quicktime. Conventional tools at the bottom of the screen let the user control the video playback. An instructional voice-over begins while the video clip plays for less than one minute.

■ Training video that demonstrates the safe handling of hydrogen sulfide (H_2S)

Features of Training Video

1 The video opens and plays in a small window.

2 Play-pause toggle control lets users stop and start the video.

3 Volume level control allows users to set the voiceover volume.

4 Rewind and fast-forward arrows help control the video playback.

5 Progress bar advances the video manually by dragging the bottom bar, called "scrubbing."

Video clips have a wide range in the window size and video quality. Most video clips use file compression to make file sizes smaller; these changes affect the size and the quality of the resulting clip. Video clips for the Web must be short and the files highly compressed for users to download and view using a media player. Even "streaming" video formats where videos begin to play while downloading, restrict size, speed, and quality. Video compression methods, file size, download time, hardware, network connection speeds, and media playback software all must be considered when planning for the final video production. In addition, most videos depend completely on good quality recorded audio—here, of an instructor's explanations—in order to make sense to a user (for more information, see Audio).

Computer-based Software Training

Software training programs, also called tutorials, combine spoken instruction and video to produce a useful training tool. Computer-based training is often distributed on DVD or over the Web. Most training programs provide a video capture of the software program in action with an instructional voice-over narration that explains what happens on the screen while the mouse moves and the selections are made. Nearly all software companies and training businesses offer software tutorials online to their users.

Animation for Educational Applications

Animation can sometimes illustrate what video or still images cannot. For example, animation can illustrate certain mathematical concepts such as three-dimensional rotation processes so that students can learn the rules of rotation by seeing them in action. The example on page 367, created with Macromedia Director and 3D modeling software, is a module from the interactive CD-ROM, *Introduction to 3D Spatial Visualization: An Active Approach* (Wysocki et. al. 2002).

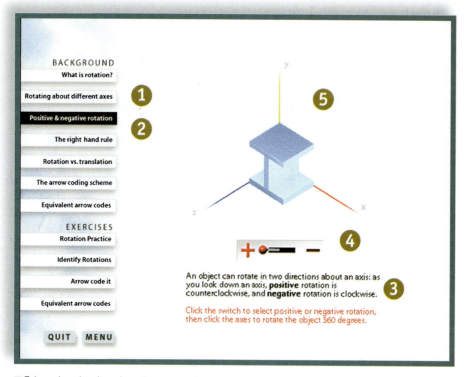

An object can rotate in two directions about an axis: as you look down an axis, **positive** rotation is counterclockwise, and **negative** rotation is clockwise.

Click the switch to select positive or negative rotation, then click the axes to rotate the object 360 degrees.

■ Educational animation about rotation

Features of Educational Animation

❶ Content sections are listed down the left side of the screen.

❷ The current section is highlighted.

❸ Written instructions (red text) tell the user what to click and begin the animation.

❹ Switch graphic, using + for positive, changes the rotation between positive and negative.

❺ Clicking one of the axes (y in yellow, x in red, z in blue) starts the rotation.

Software for Video

Technical communicators who create videos in house may find themselves functioning as producers and directors as well as writers. Video is inexpensive to make, videotape and disk storage space can be reused, and the cost of both digital recording equipment and computer storage continues to decrease. Video editing software makes it fairly easy to work with video files once they have been recorded and saved.

Programs included with personal computer operating systems and entertainment accessories make DV editing fairly simple to learn. These programs include Final Cut Express, Movie Maker, iMovie, and others. All of these programs allow users to make movies, which can combine video clips, still images, sounds, text, and animation. The following example shows the Windows Movie Maker editing screen where all the pieces of a movie are edited together on a timeline:

■ The Movie Maker video editing screen

Features of Video Editing Software

1 The large window displays the movie as it plays various media in sequence.

2 The timeline at the bottom shows each image in order of appearance.

3 The audio appears in its own channel below the video.

4 Drag pictures from your collection onto the storyboard to create a sequence in any order.

5 The photos appear from left to right for the amount of time listed on each item in this sequence.

Media items like these photos of Mexico can be added, moved, edited, and enhanced using the timeline. Various visual effects can be added to each media item and for transitions (like a dissolve) between each kind of media, resulting in a complete movie. The final project, once compiled, edited, and saved, can be exported as a single file in various video formats and then linked to a web site or saved onto a DVD. Saving the video onto DVD has the advantage of holding large amounts of information and longer sections of video, eliminating the need to create small, highly compressed clips for the Web.

Professional video programs like Final Cut Pro use these same basic techniques to create full-screen, broadcast quality videos. These high-end projects usually require expensive and powerful hardware, immense storage space, and sometimes, studio production facilities. Such projects may well require specialists—a camera operator, a film editor, a sound engineer, and so on. Educational software and training programs that allow extensive interaction between users and objects might require a professional animator or an instructional designer as well.

Software for Animation and User Interaction

Simple automated animations are offered in many desktop programs like Power-Point and Word to create text and graphics movement on a screen. DVD authoring programs that come with computers provide the basic tools for creating user interaction and automation. These built-in animation features are limited, however, and don't always translate well into other formats such as the Web. User interactions with text and other objects still need to be created with web programming tools such as Java or PERL scripts, or with more specialized software.

Professional animation programs like Macromedia Director (used to create the rotation animation screen, above) allow users to create extensive user interaction and add many animated processes to training videos and educational software. Applications made in Director can be incorporated into a web site. Similarly, Macromedia's Flash allows users to design user interaction features and other animation for web sites. These programs are expensive, however, and require significant time and effort to learn. Some technical communicators will have the production and design skills for creating highly usable interactive software applications. They rarely work in isolation, however, and a professional quality project may also require other specialists in animation or graphic design.

Guidelines

FOR CREATING VIDEO AND ANIMATION

- Determine your audience, purpose, and context for using an animation or video.

- Decide whether your final application will be used as a stand-alone module, distributed via a web site, or saved onto a DVD.

- Determine whether you want to budget for in-house production or include specialists who can create high-end products.

- Outline or storyboard the video or animation's content, sequence, location, and appearance.

- Involve your target users early and, if possible, involve them in the production process.

- Decide what video and animation software programs you will use and how you will store large files.

- Produce and edit your video, movie, or animation, hiring specialists if necessary, and work together to develop the interactions your users need.

- Save your final movie files and associated media files to a large capacity hard drive, MP3 player, or DVD.

- Save only short video clips and animations on your web site. Provide links and explanations on your site for these examples.

Visual Communication

This section introduces the basic concepts of visual communication commonly used in technical and professional documents. These concepts include the following:

Integrating visuals and text

Visual persuasion

Uses of color

Integrating Visuals and Text

A combination of visual elements with text is often the most effective way to communicate technical information. A picture alone can be worth a thousand words, but usually pictures, diagrams, and other figures communicate best when they are introduced and followed by concise textual explanations. The integration of visual and textual elements in technical documents is most often accomplished through the use of labeled figures. Other visuals such as tables, graphs, and charts that flow with the narrative of the text can also be effective. Another important element in combining visual and textual elements is the layout of the entire document, whether paper or screen based. (For a detailed treatment of these topics, see Document Design and Web Page Design.)

Figures and Tables

Figures refer to all illustrations in a technical document and can include photographs, technical illustrations, or any other visual information that your readers might need (Part 3 includes separate sections on all these types of figures). Always use captions to label your figures so that the content or function of a visual will be clear to your readers. If your document includes more than one or two figures, consider using a numbering system to organize them: Figure 1, Figure 2, and so on. Note that tables often use a separate numbering system (Table 1, Table 2) to distinguish them from the other figures.

A common mistake in the use of figures and tables is not introducing them or explaining them. This error means that the figure or visual lacks context and integration with the rest of the document. At best, a lack of integration between visual and textual elements makes a document sound choppy and disjointed; at worst, it makes a document unclear and unusable. Readers may come across the visual but have no idea how this visual relates to an explanation in the text. Therefore, a clear integration of visual and textual elements is one of the most important features of effective technical communication.

Introduce each visual with a sufficient textual explanation. At minimum, this might be a heading or a phrase. Sometimes a sentence or paragraph will be needed. Use headings, labels, and callouts to clarify the content of a diagram, illustration, chart, graph, or table. The relationship between visual items and textual labels should be clear from their placement, callout lines, or a separate key. Following all visuals include a textual explanation of what they illustrate and their meaning. The amount of detail in your explanation should be dictated by what the reader needs to know to make sense of the visual and its purpose in the document.

The visual and textual information offered in the following example shows how color coding, layout, and careful use of text can present the results of a complex dataset. This NOAA report compares ice cover in the great lakes across several years. Data about duration of ice cover are compared in many maps like this one. Taken as a whole, the figure illustrates some of the best ways to integrate text and visuals together in a technical document.

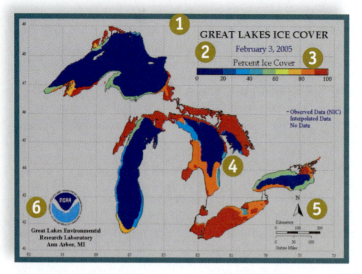

GREAT LAKES ICE COVER

February 3, 2005

Percent Ice Cover

0 20 40 60 80 100

Observed Data (NIC)
Interpolated Data
No Data

N

Kilometers
0   100   200

0   50   100
Statute Miles

NOAA

**Great Lakes Environmental
Research Laboratory
Ann Arbor, MI**

■ Figure that integrates visuals and text effectively

Features of Integrating Visuals and Text

1 A title, in bold and all caps, introduces the overall meaning and relevance of the figure.

2 Text is used to name and clearly identify the date of this data map.

3 An explanation of the color values follows the map heading text with its own clear heading, which helps the reader quickly grasp that colors are associated with percentages.

4 Each map uses contrasting colors to show the amount of ice cover from 0% (dark blue) to 100% (bright red).

5 A legend for the scale and orientation for North appear at the bottom right as important tools.

6 The NOAA logo and name of the research agency at the bottom left both give this study an appearance of technical accuracy and credibility.

The integration of text and visuals in the figure provide a good example of how text and layout work together to present credible, factual, and interpretive information. These maps are compared for ice duration through each year, across the three years and other trends. Based on the data above, which parts of the lakes have the most ice cover? The details would need to be explained in the accompanying text of the report.

Designing Visuals and Text to Work Together

When designing and combining visual and textual elements in technical documents, you should always consider the purpose of the document you are composing and how the elements of the document will support your purpose. As the writer, you are probably quite familiar with the information that you are trying to communicate; in your head, you have the background information as well as your own goals and assumptions.

But readers may not share your knowledge, background, and expertise. During the design process, it is very important for you to envision the perspective of your target audiences and think about what pictures and textual explanations readers will need. Do they need the big picture first? Do they need an overview? How much detail is necessary? These questions can guide you in your selection of visual elements.

Ask yourself what value each visual and textual element provides to the reader. Photographs have an immediate impact and often generate interest and provide emotional appeal. Graphs and charts are excellent for providing big picture trends and comparisons and are often used in executive summaries. Tables and technical diagrams are good for conveying detail (for more information, see Data Display, Part 3).

Finally, decide how figures and tables will be presented in the text. For example, you may decide:

Every visual is a numbered figure or table.

Photographs, graphs, and complex tables will be numbered figures while simple tables will be integrated into the narrative flow without using figure numbers.

Figure numbers are not necessary because figures are integrated clearly into the textual explanation.

Once you choose a method, keep it consistent throughout your document.

Guidelines

FOR INTEGRATING VISUALS

- Aim for a combination of visuals and text that is readable and understandable for the medium and format, whether screen or print.

- Decide which types of visuals will be presented as numbered figures and which will flow as part of the text.

- Introduce each visual with a sufficient textual explanation—a heading, phrase, sentence, or paragraph.

- Use headings, labels, or callouts to clarify the content of figures and tables, if needed.

- Use size, texture, and color coding (if available) to make various parts of the visuals distinct.

- Provide a meaningful caption for every figure.

- Follow all visuals with a textual explanation of what they illustrate and their meaning.

Visual Persuasion

Visual persuasion describes the deliberate use of visual strategies in a document to influence readers or to support the document's main arguments or point. When you illustrate a point of view and try to convince using visuals, you are engaged in visual persuasion. Like any form of persuasion, you will want to use appeals and visual communication techniques together in a convincing and professional manner. Visuals that appeal to logic, emotion, ethics, and shared values will influence your audience and may even move them to do something you want (agree with your reasoning, take a recommended action, make a change).

Forms of Visual Persuasion

Persuasive visuals include many familiar forms (many are discussed in separate entries):

Advertisements

Comics and editorial cartoons

Document (page) design

Graphs, tables, and charts

Logical and mathematical equations

Juxtaposition (aligning and comparing two different items in a visual arrangement)

Symbols and icons

Video and animation

Web page design

As you can see, almost anything that is arranged deliberately and designed convincingly within a documents becomes part of the document's visual persuasion.

Visual Arguments

The phrase *visual rhetoric* is often used in place of visual persuasion because rhetoric describes the field of study associated with argumentation. A visual argument can be made without any words and still offer a particular point of view or position. In a screen-based culture, visual rhetoric is a powerful force in making arguments and creating communication.

This illustration from the Center for Disease Control's good nutrition campaign makes just such an argument without using any words, though the accompanying text does explain its meaning: "Do you ever feel like you can't keep up with the changes in technology? Sometimes it seems that way with dietary advice, as if things are always changing." The baffled young man illustrates this issue:

■ A visual argument

Features of a Visual Argument

1 The picture as a whole shows the concept of good food choices. The spinning circle of items illustrates confusion.

2 Food items include many juxta-posed images to show the wide range of choices—healthy fresh produce, the apparently unhealthy shake and donut with sprinkles, and other unknowns like olive oil and cheese.

3 The young man's facial expression provides an emotional argument by expressing uncertainty and confusion while also looking toward the milkshake.

4 The idea of "good choices" and health in general appeal to the values of the audience.

Cultural Elements of Visual Persuasion

People see visual information and persuasion differently because most visuals carry specific cultural connotations and associations. For example, after O.J. Simpson's arrest for murder in 1994, some readers noticed that a magazine cover deliberately darkened Simpson's skin to make him look more menacing. Sometimes cultural connotations are not just visual, as in a 1980s era Taco Bell commercial that featured a cartoon Chihuahua with an accent suggesting a Mexi-can stereotype to some viewers. When you are not sure about your audience's cultural background, knowledge, or beliefs, try to choose neutral or universally positive symbols and images.

Visual Distortion and Ethics

Readers expect illustrations in technical and public documents to demonstrate the truth as accurately as possible. The altering of illustrations has always taken place, but digital software makes it easy to change elements in a photograph or illustration. There is a fine line, however, between enhancement and visual distortion. If you alter an image extensively in order to change its context or meaning deliberately, you have engaged in unethical behavior. Photojournalists, for example, are required to verify accurate photographic images just as they fact-check other sources. In one famous example, an *L.A. Times* photographer published a photograph of an American soldier pointing a gun at a group of Iraqi civilians. The gun appeared to be aimed at a desperate man holding a child. The photo was actually a composite of two images in which the soldier's placement made him look much more menacing than in the original photo (Irby 2003).

This alteration distorted the meaning of the visual for readers and was deemed an unethical illustration. This false understanding could potentially cause many serious problems, legal difficulties, and misunderstandings. Your credibility as the communicator forms a crucial part in persuasion, because unethical visual persuasion can result in distortions of the truth, stereotypes, or biased propaganda. If that happens, you will lose the trust of your audience.

Guidelines

FOR VISUAL PERSUASION IN TECHNICAL DOCUMENTS

- Choose a form of visual persuasion that makes sense for your purpose and audience.

- Use deliberately arranged visuals to create the desired effect for your audience—whether it is an emotional response or accurate conclusions from a data display.

- Determine whether the cultural associations with the visuals used have symbolic, positive, or neutral connotations for your audience.

- Check to see whether the intended audience for your document understands and responds to your visuals.

- Check your final document to see that all your visuals and arrangements display accurately on paper, on the Web, and on different-sized computer screens.

Uses of Color

Color offers valuable options for highlighting information, designing layouts, creating visuals, and evoking emotions and other responses from users. Designing documents that use color can be complex, though, because a color's appearance varies according to the media and technologies used. You may find that colors in your document do not look the same when printed as they did on a computer monitor. When preparing color documents for page or screen, you need to decide when and how to use color, what colors to use, and also how they will be displayed. In general, follow these three steps:

1. Consider the audience and context for a document when deciding to use color.
2. Apply effective color design techniques.
3. Conduct usability research and test the document with your target audience.

If you decide to use color, select one color for selective emphasis, two colors for contrast, or full-color for printing. Consider other design options as well like type style (such as bold) or a simple design element (a line) instead of color.

Types of Color

This is the traditional color wheel model used by artists and designers:

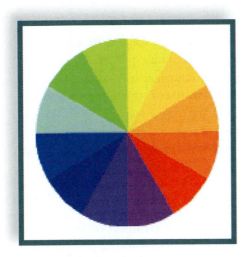

■ Traditional color wheel

Colors are represented on a wheel of 12 colors: three primary colors (red, yellow, blue), three secondary colors (created by mixing primary colors), and six tertiary colors (created by mixing primary and secondary colors). This RYB model (red/yellow/blue) includes the secondary colors of orange, green, and purple, and the tertiary colors in between. Colors directly opposite one another on the wheel are called complementary colors and create the greatest contrast, while colors beside one another, analogous colors, blend harmoniously together. These combinations of colors form a particular kind of color scheme that emphasizes either contrast or harmony.

The color wheel is based on "pure" colors; for every color there are also darker and lighter versions, or *hues* of a color. Darker versions are produced by adding black or removing light, and are called shades. Lighter versions, produced by adding white or additional light, are called tints. The group of colors available to you in a specific application is called a "palette," like an artist's palette of available paints.

Connotations of Color

Human responses to color vary but in general, you can think in terms of warm, cool, and neutral colors when working on a project. Warm colors have a yellow undertone, such as bright red, oranges, yellows, and green-yellows, and are typically thought to express warmth, comfort, and energy. These colors also tend to make things stand out and advance toward you from the page or screen. Cool colors have a blue undertone, and include violets, blues, aqua, and greens. When they are used together, cool colors seem to move away from the viewer, and express coolness, detachment, stability, and calmness. Black, grays, and whites are neutral; browns, beiges, and tans are sometimes considered to be neutral as well. Neutral colors are the most flexible because they often work harmoniously with other colors.

Cultural Associations with Color

Associations with color can differ based on culture. In some cultures, for example, red may indicate warning or caution, while in others, red may indicate celebration or luck. Cultural associations vary widely, and it often is not possible to generalize. Many large, international companies have groups that specialize in translation and localization (see International Communication, Part 1), and these groups are expert in knowing what approach to take when using visuals, color, and other features in documents that have a cross-cultural audience. You can also find useful information about cultural connotations of color by doing research about color theory.

Using Color in Software Programs

Whether working alone or with a graphic designer, technical communicators may need to use the computer-based color palettes in software programs and color printing specifications for multicolor print documents. When designing documents, you may need to decide when to use color and even choose what colors to use for highlighting information. Full-color documents can become quite expensive when printed, but color adds no cost when you're designing exclusively for on-screen use. Color can affect file size, however, and will contribute to downloading time for users. Photographs and illustrations can also be changed to gray scale ("black and white") images, but they may lose some of their impact.

Colors for Print and Screen

Colors don't look the same on a computer screen as when they are printed. Colors may even appear differently from screen to screen. Screens display colors as combinations of red, green, and blue (or RGB). For printing purposes, however, the color model used is the mixture of cyan, magenta, and yellow, plus pure black (or CMYK). This mixture gives the exact specifications for four-color printing.

Adobe Photoshop, a commonly used graphics program for preparing visuals, lets you choose foreground and background colors for a document and also see how these color models work.

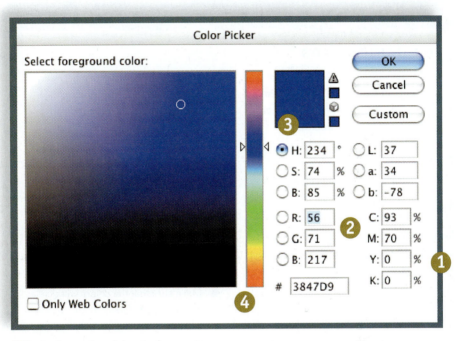

■ Photoshop color picker tool

Features of Color

Note that each color can be described in several ways:

① The mixing of the four components for printing (bottom right corner lists the percent of C, M, Y, and K)

② The "screen" mix (the R, G, B just left of C, M, Y, K)

③ Other details such as the color's hue

④ The specific number (the # box at the bottom) that helps printers and designers identify exactly the right color.

These details can be confusing, but they help ensure that an exact color remains consistent between computers and printings. When possible, write down the exact identifying number of a color you want to use.

The Web can display only a limited number of colors (note the option on the bottom left of the Photoshop color picker shows only web colors when checked). Even if you use thousands of colors in your software program or on your computer screen, the web-based image file formats like JPEG and GIF may reduce the number of colors. Visuals may look quite different on the Web or even become distorted because of the missing colors. If you are designing a document for display on the Web, be sure to choose the web palette or web-only option wherever it appears in your software program.

Guidelines

FOR USING COLOR

- Design with the medium in mind (print, screen, or Web) and remember that for print documents, color can add significantly to your budget.

- Determine the amount of color needed and possible given your budget and design considerations.

- Ensure that the cultural associations with the colors are appropriately symbolic, positive, or universal.

- Use color strategically—for headings and subheadings, for document design elements (such as labels or section changes) or to get the user's attention (as with a stop sign).

- Use color consistently throughout the document.

- Create sufficient contrast between colors (like dark and light) to make elements in the foreground and background stand out.

- Combine harmonious colors, as in the same family, to achieve a blended or subtle effect.

- Use text color that clearly contrasts with the background. Avoid light or hard-to-read colors.

- Determine how users respond to your color choices.

- Make sure that the colors will display accurately on paper, on the Web, or on different computer screens.

See also

Web Page Design

The entry on Web Sites (Part 2) gives an overview of the web and the more common types of sites; this section focuses specifically on page design issues for your site. Strong visual design is especially important for web pages, because of the sheer volume of web-based information that the average Internet user will encounter on a daily basis. As with any type of document, web pages should employ a clear, professional design. Well-designed web pages make information easy to locate, fun to use, and more readily comprehensible. The design of a web page also adds to the credibility of the organization, company, or person who runs the site. Poorly designed web pages not only look bad, they can take away from the credibility of the product or organization. In fact, the Stanford Web Credibility Research Project found that "people quickly evaluate a site by visual design alone" (Fogg 2002).

Web pages can be created by just about anyone who has the right tools and a little computer expertise. Most Internet service providers include server space for posting web pages. Web-authoring software allows even novice web designers to create web pages that include text, color, menus, links, and multimedia elements such as photographs, video clips, animations, and audio files. These programs provide point-and-click options for building web pages while behind the scenes the programs generate codes known as hypertext markup language, or HTML. The HTML code tells the computer how web pages should appear and behave. Other software code, such as Flash or Java, is also employed to create certain interactive effects. A common way to create a uniform appearance among pages is by using cascading style sheets, or CSS. While your software tools may not require you to know anything about coding, your design options and editing capabilities will be limited to what those tools allow. Since building complex web sites involves writing, visual communication, creating various media and coding, most web pages are created through teamwork and collaboration with a design team.

Web pages should rely on many of the layout features and document design principles as the printed page, such as effective use of headings, text, visuals, white space, and margins. But there are also important differences when designing for the Web due to the difference in proportion between a printed page and a computer screen. Also, people read differently when information is on a web page, often scanning the page quickly and skipping between short chunks of information.

Getting Started

The most successful web design projects start with good planning. When designing web pages, it is critical to perform an audience and needs analysis. You need to understand the audience for the web site, the purpose of the site, and the kinds of tasks that users will want to perform on this site. The process described in this book under Audience and Purpose Assessment (Part 1) is a good way to start. The *Web Style Guide* also recommends that you create something more focused, called a "site specification document," which not only addresses audience and purpose but also issues such as goals, budget, and technical aspects of the site (Lynch 2002).

Other special audience considerations for web page design include the following:

- **Demographics:** With an aging population, there is a need for web pages where users can change the font size or otherwise make the information larger and easier to read. Other visual and audio features may need to be designed such that they can be accessed by special software that make the material accessible to people with certain physical limitations. See Accessibility in Part 1.
- **Global reach:** Web pages are often translated into numerous world languages. Page design can influence how the automatic translation software works. See International Communication (Part 1) for more information.
- **Speed of reading:** People do not read Web pages as slowly or in the same linear format as they read print-based materials. Instead, they move quickly between chunks of text, visuals, and links to other information.
- **Writing style:** Writing for the web involves constant attention to the audience and purpose as well as the design of the site. It makes no sense to write large sections of beautifully written prose when, in the end, the text needs to be created in short chunks, linked across the entire web site. Especially on the front page, write in short, crisp sentences, using active voice, and limit the size of text to one short paragraph on one topic. Save longer text for secondary web pages.

Web Page Layout

Certain universal design guidelines should be followed to ensure a web page layout that is professional, easy to follow, suited to the needs of the audience, and accessible to most people. Printed books tend to follow a similar format, usually regardless of audience: table of contents; chapters; pages that consist of text that usually fills up the entire page size; index; and so forth. These features evolved over time, but now, readers have come to expect a book to be printed and designed in a certain way. Similarly, web page design has evolved over the past several years to encompass key features that make the puzzle of figuring out the web site far less complex for readers. Web pages should therefore conform to these basic design and layout principles.

Overall page layout—Web page layouts usually follow a grid-like pattern for aligning the elements. Remember that computer screens are in landscape (horizontal) mode, while printed pages are in usually in portrait (vertical) mode. So don't convert printed material directly into a web page without doing a page redesign. Plan a prototype page first and make your first decisions about the page's overall purpose/functions and aesthetics.

Aesthetically pleasing and functional look and feel—Well-designed web pages have a clean, spare, attractive look that creates credibility and encourages users to stay on the page and explore. Key concepts for this principle include these:

- White space—Use white space to draw the reader's eye to key areas on the page.
- Typography—Use clean, professional fonts that are easy to read. Traditionally, fonts such Times Roman or Helvetica are fine, leaving more decorative fonts for limited uses such as headings or logos.
- Titles and headings—Use these items to break up text and guide readers through information.
- Balance of text and visual material—The page should not overwhelm with either too much text or too many visuals or video clips.
- Attractive and consistent color palette—As part of the planning process, choose a color palette that will be consistent across the entire web site. Use colors to draw readers to key parts of the page.
- Visual elements—Visuals like photographs or charts should fit with the color palette and should be placed with enough white space around the border to provide balance with the text and other page elements.

Linking that is simplified, meaningful, and standardized—The primary difference between web page design and print document design is the means by which information is connected, displayed, and located by readers. Links allow a digital text to become a *hypertext*, whereby users can move freely through the elements of a web site and navigate to additional information, to other web sites, to illustrations and other graphics, to interactive elements such as search boxes, and to multimedia elements such as video or audio. Key guidelines for effective use of links include the following:

- Navigation bars should follow a standard web page format—Navigation bars should be consistent.
- Links should be meaningful for this web page—Do not link to other pages simply because you can. Only use links to direct readers to relevant material.

- Give a clue about what the link will do—An important linking guideline offered by experts Price and Price (2002) is to "tell people about a media object before they download it" (170). Use text or a mouseover feature to explain what people can expect. There is nothing worse than clicking on a link and being directed to a site you didn't want to visit, or having a large music or video file start to download unexpectedly.
- Don't over-link—The page should function on its own and not just be full of links to other sites.
- Use consistent colors for user feedback, such as visited links and unvisited links—Once a link has been clicked, the color should change.

Consistent use of navigation features—Users have come to expect web pages to have key navigation bars along the left margin and sometimes across the top of the page, not scattered throughout the document.

- Use marginal navigation—Use the left margin for links that navigate to other key pages on the web site. Since readers do not access the bottom of the page as frequently as other spots, save that for items that are not used as frequently, such as "privacy policy."
- Provide a search bar—Users expect to be able to search a site. Put the search bar at the top margin or in a prominent place on the front page.
- Use headings to group links into logical categories—Headings can organize links in ways that make the web page more accessible to readers.

An appropriate mix of text, visuals, audio, and other media—Web pages are obviously more flexible than print documents because they support a wider range of visuals and other media. Yet web pages that are truly outstanding manage to find the right mix of media features. Key guidelines for balancing the media on a web site include these:

- Keep textual material short—Especially on the main page, limit the use of text to short chunks, using links to direct readers to other information.
- Test large media files—Large media files can affect the performance of a web site. Response time and performance are critical factors for web users; they do not want to wait for a web site to respond to their requests.
- Limit video clips (such as Flash files) to short segments that can be streamed or downloaded quickly. Keeping video window size small will help limit file size.
- Use photographs that are suited to the site's purpose—Photographs and other images should enhance the message of the web page.
- Provide a mute button and preferably a volume adjustment for any sounds on the page.
- Use clip art or drawings sparingly—The organization's logo, for example, is an appropriate use of art.

nps.gov

National Park Service
U.S. Department of the Interior

Search

NPS HOME
FIND A PARK
HISTORY & CULTURE
NATURE & SCIENCE
FOR KIDS & TEACHERS
GETTING INVOLVED
ABOUT US
NEWS

Site Index
Frequently Asked Questions
Bookstore
Director's Office
Contact Us

Support the
National Park Service

America The Beautiful - National Parks And Federal Recreational Lands Pass

The National Park Service is an important participant in the new Interagency Pass Program which was created by the Federal Lands Recreation Enhancement Act and authorized by Congress in December 2004. Participating agencies include the National Park Service, U.S. Department of Agriculture - Forest Service, Fish and Wildlife Service, Bureau of Land Management and Bureau of Reclamation. The pass series, collectively known as the **America the Beautiful – National Parks and Federal Recreational Lands Pass,** is shown below with a brief explanation of each.

America the Beautiful – National Parks and Federal Recreational Lands Pass – Annual Pass - Cost $80.

This pass is available to the general public and provides access to, and use of, Federal recreation sites that charge an Entrance or Standard Amenity Fee for a year, beginning from the date of sale. The pass admits the pass holder/s and passengers in a non-commercial vehicle at per vehicle fee areas and pass holder + 3 adults, not to exceed 4 adults, at per person fee areas. (children under 16 are admitted free) The pass can be **obtained in person at the park,** by calling 1-888-ASK USGS, Ext. 1, or via the Internet at **http://store.usgs.gov/pass.**

America the Beautiful – National Parks and Federal Recreational Lands Pass – Annual Pass

America the Beautiful – National Parks and Federal Recreational Lands Pass – Senior Pass. - Cost $10.

This is a lifetime pass for U.S. citizens or permanent residents age 62 or over. The pass provides access to, and use of, Federal recreation sites that charge an Entrance or Standard Amenity. The pass admits the pass holder and passengers in a non-commercial vehicle at per vehicle fee areas and pass holder + 3 adults, not to exceed 4 adults, at per person fee areas (children under 16 are admitted free). **The pass can only be obtained in person at the park.** The Senior Pass provides a 50 percent discount on some Expanded Amenity Fees charged for facilities and services such as camping, swimming, boat launch, and specialized interpretive

America the Beautiful – National Parks and Federal Recreational Lands Pass – Senior Pass

■ Example of a well-designed, aesthetically pleasing web page

Features of Web Page Design

1 White space and consistent horizontal alignment allows visual items to stand out.

2 Clean, sans serif font in black on white make body text easy to read.

3 Headings and rules on white background break up the page into logical sections.

4 Attractive and consistent color palette uses greens and browns, with reds and oranges for contrast.

5 Photographs enhance the site's message and work well with the color palette.

6 Navigation bar on left margin follows standard web page layout.

7 Flag across top of page includes organization's name and site address in eye-catching reversed text.

8 Organization's logo is not too large and is well placed in the flag at top right of page.

9 Page layout is for the shape of a screen, not printed book or brochure.

10 Red color in center of page draws the eye to this area first.

11 Text is brief and crisply written. Longer text appears on a secondary page, accessed via a link.

Guidelines

FOR CREATING WEB PAGES

- Make an inventory of assets, including existing media, logos, and organizational documents that you can use.

- Target your primary audience by addressing their needs and interests on the front page.

- Create a prototype page on paper or in a web page authoring program and make preliminary decisions about the page's aesthetics and functions (graphics, color schemes, fonts, links, etc.)

- Use standard linking and consistent navigational features that users expect from your type of web site.

- Balance text and media features carefully on each page.

- Storyboard all your web site pages using pencil and paper (one sheet per screen) or a software tool.

- Build accessibility into your web page design.

- Warn users about large files and file formats available for downloading.

- Test the first version of the web site with members of the organization and of the general public audience before publishing the site on your server.

- Create a plan for storing the files and for regular updates to all pages in the site.

See also

4

Research Strategies and Tools

Abstracts

A bstracts are either a summary of a paper (see Scientific and Technical Journal Articles, Part 4, for an example) or a comprehensive collection of sources that let you search for information in a particular field. Almost every field of study has a collection of abstracts devoted to it. For instance, if you were working on a project about wind turbines, you might try the *Applied Science and Technology Abstracts*, which you can access via a research library.

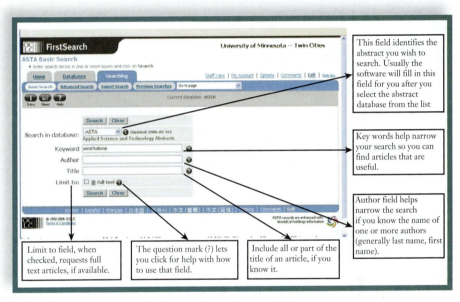

This field identifies the abstract you wish to search. Usually the software will fill in this field for you after you select the abstract database from the list

Key words help narrow your search so you can find articles that are useful.

Author field helps narrow the search if you know the name of one or more authors (generally last name, first name).

Limit to field, when checked, requests full text articles, if available.

The question mark (?) lets you click for help with how to use that field.

Include all or part of the title of an article, if you know it.

■ An abstracts database

Your search would yield many articles, and you would be able to view either just a short summary of the article (also called an "abstract") or, if your library subscribes to the digital version, the full paper. To locate an abstract, you will need to look at all available options, usually by searching the databases and indexes that are provided by your library. See Databases and Indexes (Part 4).

You can follow these general guidelines when working with abstracts.

■ **Finding abstracts:** Your college or university library is the place to start. Ask a librarian to help you use the web site to search through the list of published abstracts. Or, if your library offers it, use an online tutorial to learn more.

- **Narrowing your search:** Work from general to specific. Start with a general engineering abstract such as *Applied Science and Technology Abstracts*. Review the articles you find, and then narrow your use to abstracts such as *Environmental Sciences and Pollution Management Database*.
- **Searching smart:** Experiment with the search fields so that you focus on what you really want to find. In the example above, the search phrase "wind turbine" will give you numerous hits; look over the titles and see how you might tighten your search. For example, try "wind turbine energy output."
- **Keeping track of key words:** As you locate useful articles, keep track of the key words, often listed right at the top of the article. These words will help you narrow your search even further.

Bibliographies

Bibliographies are alphabetical listings of sources, usually restricted to a field of study, a topic, or a research area. For any given area, numerous bibliographies are available on the Internet. (As with all digital research, check to ensure the material is from a credible source.) Bibliographies differ from abstracts in that not all bibliographies are published by academic publishers. In addition, bibliographies are often very specific and do not provide a comprehensive, broad look across hundreds of publications. For instance, the Danish Wind Industry Association publishes a "Wind Resource Bibliography" located at http://www.windpower.org/en/stat/biblio.htm. This bibliography is not a comprehensive abstract, and it is focused specifically on wind resources. From this bibliography, you can select articles that look relevant to your research project and locate these articles at the library.

Journal and book publishers update and publish bibliographies, but even in these cases they are still specific and narrow in focus. These reference sources are useful because you can be sure they have been reviewed by experts in the field. For instance, the book *Business and Technical Communication: An Annotated Guide to Sources, Skills, and Samples*, offers a bibliographic listing related to technical communication research (Belanger 2005). In addition, this book, like many bibliographies, is annotated—that is, the listings come complete with a short abstract summarizing the article or the research findings (see the Annotated Bibliography on Migratory Birds figure, below).

You can create your own bibliography tracking the research you do for a particular project. For instance, you may compile a very long list of interesting material for a project on the viability of wind energy for your hometown; you may not use all of this material, but if you keep a bibliography, you can refer to it later. You can even make it available to others via a web site or blog. A number of software programs allow you to keep your own bibliographies, such as RefWorks, ProCite, or EndNote. RefWorks is used by many college and university libraries and lets you import your citations directly from the library database into your own personal account stored on the university server. EndNote, ProCite, and other programs can be downloaded and installed to store the citations on your personal computer.

Using these programs, you only need to enter the information one time. The software then lets you choose from dozens of citations styles so that your citations and references page is automatically formatted in whatever documentation style (MLA, APA, other) you need to use.

Here is an example bibliography using RefWorks:

Creating Your RefWorks Database [Continued]

Manually Entering References

1. Select *References* from the pull-down menu tool bar and then choose *Add New Reference*.
2. Select a potential bibliographic output style (e.g., APA, Chicago, MLA) under *View fields used by* to enable the AccuCite feature. **C**
3. Designate the type of reference you are entering (e.g. journal, book or dissertation) under *Ref Type*. Field names marked with a green checkmark (✓) indicate recommended information needed to produce an accurate bibliography for the selected output style and reference type. These fields are not required to save the actual reference in RefWorks.
4. Enter information in the boxes provided and click *Save* when finished.

Importing from RSS [Really Simple Syndication]

RefWorks has integrated an RSS Feed Reader so you can easily add your favorite RSS Feeds from publishers and websites, view the information and import data into your RefWorks database.

1. Locate the RSS Feed you wish to include.
2. Right mouse click on the *RSS Feed icon or link* and select *Copy Shortcut*.
3. From within RefWorks, select *RSS Feed* from the *Search* Menu.
4. Paste the shortcut into the text bar and click the *Add RSS Feed* button. **D**
5. Launch the RSS Feed by clicking on the name link. Your feed results will be displayed in a separate window for selection and importing.

Importing from Online Catalogs or Databases

You can use RefWorks as a search interface for a number of online resources. **E** RefWorks provides access to a number of publicly available services such as NLM's PubMed as well as many universities' Online Catalogs. Additionally, institutional subscribers may also provide access to subscription-based online services (e.g., Ovid or ProQuest) through RefWorks.

1. From the *Search* pull-down menu, select *Online Catalog or Database*.
2. Under *Online Database to Search*, select a database from the drop-down menu.
3. Select the *Max. Number of References to Download* from the drop-down menu.
4. Enter terms in either the *Quick Search for:* or in the *Advanced Search for:* box.
5. Click on *Search* to begin your search. A new window will open displaying your search results. **F**
6. Select the references you wish to import into your RefWorks account and click *Import*.

■ Sample bibliography software (RefWorks)

RefWorks and similar tools let you keep track of the bibliographical references for a paper, report, thesis, or other project. You can import citations from an online database, or you can enter citations manually. You can then create a final bibliography in any of dozens of citation styles, from APA and MLA to styles for specific scientific and technical journals.

The following is a sample annotated bibliography using APA style documentation:

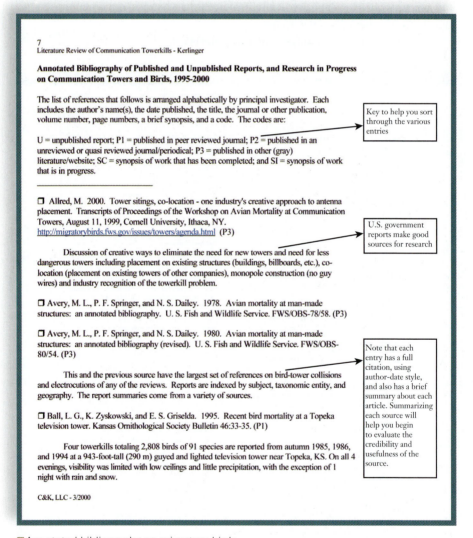

7
Literature Review of Communication Towerkills - Kerlinger

Annotated Bibliography of Published and Unpublished Reports, and Research in Progress on Communication Towers and Birds, 1995-2000

The list of references that follows is arranged alphabetically by principal investigator. Each includes the author's name(s), the date published, the title, the journal or other publication, volume number, page numbers, a brief synopsis, and a code. The codes are:

[Key to help you sort through the various entries]

U = unpublished report; P1 = published in peer reviewed journal; P2 = published in an unreviewed or quasi reviewed journal/periodical; P3 = published in other (gray) literature/website; SC = synopsis of work that has been completed; and SI = synopsis of work that is in progress.

❏ Allred, M. 2000. Tower sitings, co-location - one industry's creative approach to antenna placement. Transcripts of Proceedings of the Workshop on Avian Mortality at Communication Towers, August 11, 1999, Cornell University, Ithaca, NY.
http://migratorybirds.fws.gov/issues/towers/agenda.html (P3)

[U.S. government reports make good sources for research]

   Discussion of creative ways to eliminate the need for new towers and need for less dangerous towers including placement on existing structures (buildings, billboards, etc.), co-location (placement on existing towers of other companies), monopole construction (no guy wires) and industry recognition of the towerkill problem.

❏ Avery, M. L., P. F. Springer, and N. S. Dailey. 1978. Avian mortality at man-made structures: an annotated bibliography. U. S. Fish and Wildlife Service. FWS/OBS-78/58. (P3)

❏ Avery, M. L., P. F. Springer, and N. S. Dailey. 1980. Avian mortality at man-made structures: an annotated bibliography (revised). U. S. Fish and Wildlife Service. FWS/OBS-80/54. (P3)

[Note that each entry has a full citation, using author-date style, and also has a brief summary about each article. Summarizing each source will help you begin to evaluate the credibility and usefulness of the source.]

   This and the previous source have the largest set of references on bird-tower collisions and electrocutions of any of the reviews. Reports are indexed by subject, taxonomic entity, and geography. The report summaries come from a variety of sources.

❏ Ball, L. G., K. Zyskowski, and E. S. Griselda. 1995. Recent bird mortality at a Topeka television tower. Kansas Ornithological Society Bulletin 46:33-35. (P1)

   Four towerkills totaling 2,808 birds of 91 species are reported from autumn 1985, 1986, and 1994 at a 943-foot-tall (290 m) guyed and lighted television tower near Topeka, KS. On all 4 evenings, visibility was limited with low ceilings and little precipitation, with the exception of 1 night with rain and snow.

C&K, LLC - 3/2000

■ Annotated bibliography on migratory birds

☐ Anderson, R., M. Morrison, K. Sinclair, D. Strickland, H. Davis, and W. Kendall. 1999. Studying wind energy/bird interactions: a guidance document. Metrics and methods for determining or monitoring potential impacts on birds at existing and proposed wind energy sites. Prepared for National Wind Coordinating Committee Avian Subcommittee. 87 pp.

> This item sounds interesting. You might need to learn more about the "National Wind Coordinating Commitee" and if they are a lobbying group, an industry research group, or something else. Try searching on the Web and on your library's web site to find out.

  Complex and detailed volume about monitoring birds at wind plants (pre and post construction). Sections on fatality searches are relevant to communication towerkills, although whether other parts of volume are applicable is arguable.

☐ Colson & Associates. 1995. Avian interactions with wind energy facilities: a summary. Prepared for the American Wind Energy Association.

  Summary of early studies of what was known about wind turbine-bird interactions. The review focused primarily on the first commercial wind plants in the United States and, to a lesser extent, Europe. It was conducted before a major wave of studies that commenced in about 1995, so it does not examine research conducted at the new and mostly smaller facilities that use modern wind technology.

> This paper is from a conference proceedings, useful because these often represent new information that may still take time to come out as a journal article.

☐ Cooper, B. A., C. B. Johnson, and R. J. Ritchie. 1995a. Bird migration near existing and proposed wind turbine sites in the eastern Lake Ontario region. Report to Niagara Mohawk Power Corporation, Syracuse, NY. 71 p.

☐ Cooper, B. A., C. B. Johnson, and E. F. Neuhauser. 1995b. The impact of wind turbines on birds in upstate New York. Windpower '95 Proceedings. American Wind Energy Association, Washington, DC.

  Radar and visual observations of migrants at 2, unlighted wind turbines 30 miles east of Lake Ontario, NY. Two migration seasons (spring and fall) of searches under the towers revealed no dead birds.

☐ Leddy, K., K. F. Higgins, and D. E. Naugle. 1999. Effects of wind turbines on upland nesting birds in conservation reserve program grasslands. Wilson Bulletin 111:100-104.

  Grassland nesting birds were studied using an impact gradient design along a transect running away from wind turbines. Reduction in nesting results close to turbines in the first years following turbine installation. Not applicable to communication tower issue.

☐ Proceedings of National Avian - Wind Power Planning Meeting, Denver, CO, 20-21 July 1994. 1995. Prepared for the Avian Subcommittee of the National Wind Coordinating Committee. Resolve, Inc. and LGL Ltd., King City, Ont. 145 p.

■ Annotated bibliography on migratory birds

Compendium of papers and other formatted contributions on wind power and bird issues. Paper by Winkelman from Netherlands is one of few that summarizes fatalities at coastal turbines. Summary of collisions at many European sites showed only small numbers of birds were involved.

❏ Proceedings of National Avian - Wind Power Planning Meeting II, Palm Springs, CA, 20-22 September 1995. Prepared for the Avian Subcommittee of the National Wind Coordinating Committee by Resolve Inc., Washington, D.C. and LGL Ltd., King City, Ont. 152 p.

Compendium of papers and other formatted contributions on wind power and bird issue. Several papers on techniques for monitoring bird movements.

❏ Proceedings of National Avian - Wind Power Planning Meeting III, San Diego, CA, May 27-29 1998. Prepared for the Avian Subcommittee of the National Wind Coordinating Committee by Resolve Inc., Washington, D.C. and LGL Ltd., King City, Ont. In press.

This item is useful because it is from a reviewed publication.

Compendium of papers summarizing the results of studies of fatalities, behavior, and other impacts at wind turbine plants in the United States and Europe.

❏ Osborn, R. G., C. D. Dieter, K. F. Higgins, and R. E. Usgaard. 1998. Bird flight characteristics near wind turbines in Minnesota. American Midland Naturalist 139:29-39.

Study of flight patterns of birds around wind turbines and behavior of birds in close proximity to turbines, similar to earlier European studies. Not applicable to night migrants.

■ Annotated bibliography on migratory birds

Citing Sources

As a researcher, you must cite ideas or direct quotations that are not your own. Citing a direct quote is not a hard idea to grasp; if you use the exact words found in another essay, web site, report, brochure, or anywhere else, you must cite it. But citing another's idea is a more complex concept. You may come up with a great idea, based on all of your readings and research, only to discover later that your idea is very close to something that someone else has published. This is a common problem, and the best way to avoid it is to keep careful notes of the material you read, so that you can go back and look for places where your ideas overlap with others. The University of Minnesota offers a thorough research guide on citing sources and avoiding plagiarism:

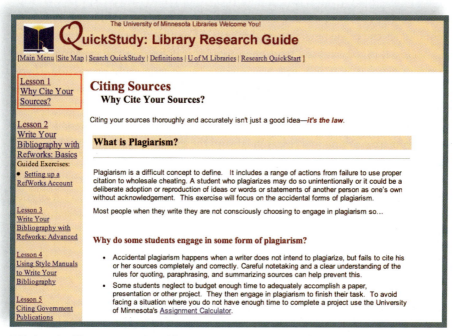

University library research guide

Your college, university, or public library probably offers similar guides to assist you. See Documentation (Part 6) for information on citing sources and on using software on your own computer (such as EndNote or ProCite) to keep track of your sources and create a references page that conforms to the style you are using.

Copyright and Intellectual Property

Copyright law is part of a three-part system of legal protections in the United States (and many other countries) that include trademark, patent, and copyright. The first two cover icons, symbols, and slogans (trademarks) and mechanical inventions and processes (patents). Technical communicators may need to be familiar with all three, depending on where they work. Most organizations have legal counsel to help guide all employees on intellectual property issues.

Copyright provides protection for the creators of original works by providing the copyright holder with exclusive rights to reproduce and distribute the material, prepare or authorize any derivative works, and perform or display the works publicly. Originally written with printed books in mind, copyright covers "the expression of an idea, fixed in a tangible medium," which includes sound recordings, photographs, other images, text, and more. No copyright case is clear-cut, as Napster and other recent recording and downloading cases have demonstrated. New technologies can cause new ways of thinking about copyright.

Material is copyrighted from the moment it is fixed in the medium. But to ensure protection, you or your company may file for copyright protection. If you work as an employee, it is most likely that the work you create is owned by the company, not you, under what is called the "work for hire" doctrine. As a student, normally you, not the university, own your work. Yet this policy differs if, for instance, you are working in a research lab and the inventions are patented or copyrighted, or both, by the university.

In most cases involving commercial use of copyrighted material, you will need to seek permission and perhaps pay a fee. But fair use provides a set of four items that you can use to help determine whether your use is considered fair use and you can waive the permissions requirement. Courts tend to rule in favor of fair use claims if a case meets the following criteria:

The use is educational, not commercial.

The material in question has already been published elsewhere.

Only part of the original material is being used (usually 10 percent or less).

Your use of the material will not affect the market value of the original.

Sometimes, people confuse copyright with plagiarism. You may be able, under fair use doctrine, to use an item without seeking permission (for instance, for a paper for class, which is considered educational). Also, some material is part of what is called "the public domain": no copyright applies because it has either expired or doesn't fall under copyright laws. But to avoid plagiarism, you would still need to provide a reference or citation to the original.

Examples of When Permission and Citation are Needed

Item	Example	Do you need to request permission (so you comply with copyright laws)?	Do you need to cite it (so you don't plagiarize)?
Direct quotation for a report for an engineering class	According to Margaret Emery, "The only reason we will deplete oils supplies in this century is our lack of will" (2000, p. 6).	No	Yes
Paraphrase of the same quotation for class report	According to Margaret Emery, if we can find the will, we can change our energy policy (2000, p. 6).	No	Yes
Paraphrase of a generally accepted idea or knowledge in a discipline for class report	All fossil fuel energy sources, such as oil and natural gas, are based on carbon.	No	No
New research, attributable to a researcher or set of researchers, not yet considered generally accepted knowledge (direct quotation or paraphrased) for class report	All fossil fuel energy sources, such as oil and natural gas, are depleted at an average rate of 10 percent every century (Granger 2007).	No	Yes
Photograph or other visual, found on the Internet, for a class report	Exploded view of the Hubble telescope optics	No (fair use, because you are using it for educational purposes)	Yes (provide a caption plus location where you found image)
Photograph of a work that is in the public domain	The Mona Lisa (original image copyright expired)	No (in the public domain)	Yes
Photograph or other visual, found on the Internet, for a company's annual report	Exploded view of the Hubble telescope optics	Yes (because you are using this for a commercial, not educational purpose)	Yes
Table of data from a published book for a company's annual report	Data collected from the Hubble telescope	Yes (because you are using this for a commercial, not educational purpose)	Yes (provide location where you found table)
Large blocks of text, copied from a web site	See first item in this table. This is the same as taking a direct quotation from a book.	No	Yes

Databases and Indexes

Subject matter databases and indexes are research tools similar to abstracts. Like abstracts, databases can be general or specific. General topic databases include the *Reader's Guide to Periodical Literature*, which covers magazines, newspapers, and other general audience publications (see Abstracts and Bibliographies, Part 4). Other general databases include the Online Computer Library Center (OCLC), a cooperative of libraries worldwide. Specific databases include:

Westlaw (for legal issues)

Medline (from the U.S. National Library of Medicine and the National Institutes of Health)

CE Database (from the American Society of Civil Engineers)

While some of these databases are freely available online, some are only available with a specific user ID and password (or provided if you are faculty or student at a university because your library pays the subscription fee). For instance, WestLaw tends to be accessible only to lawyers at firms that subscribe and pay the fees.

As a student or as a researcher, a great investment of your time would be to spend an hour with a research librarian at a college or university library (or at your local public library). Librarians can help you narrow down the vast array of databases into a few manageable ones that fit your needs.

■ University library web site listing research databases and abstracts that can be accessed and searched

Dictionaries and Thesauruses

Numerous dictionaries can be useful to technical communicators, from the more general "Webster's" types to those that focus on specific areas in science and engineering. Most dictionaries are available in a variety of media: print, CD-ROM, and on the Internet. Use a dictionary to check spelling, word meaning, and usage. Use a thesaurus to look for words that are closely related or are synonyms to each other.

Dictionaries and Related References for Technical Communication

This table represents just some of the available resources. Your library will have specialized dictionaries for medical, scientific, and engineering fields. If you decide to look for a specialized dictionary using a Google search, check the source to be sure it is credible. For instance, a professor of engineering might maintain a very useful online dictionary. But a web site that calls itself a dictionary and is filled with advertisements might be suspect.

Dictionary	Purpose	Web site (if available)
dictionary.com	General purpose dictionary, available online only; provides access to other online dictionaries and resources	http://dictionary.reference.com
Engineering dictionaries online	Look up engineering terms and phrases	Available online via subscription; check with your research librarian or at http://www.oxfordreference.com
Keller and Erb's *Dictionary of Engineering Materials*	Look up engineering terms and phrases	Not available on the Web; available as a printed book (check your library)
Grove Art Online	Dictionary and reference for visual arts	http://www.groveart.com
Merriam-Webster	General purpose dictionary, available in print and online	http://www.m-w.com
Medical dictionary online: Medline Plus dictionary (NIH medical dictionary)	Look up medical terms and phrases	http://www.nlm.nih.gov/medlineplus/mplusdictionary.html

(continued)

Dictionary	Purpose	Web site (if available)
Medical dictionaries in print: Stedman's online medical dictionary	Look up medical terms and phrases, illustrated; available in print and online	http://www.stedmans.com/product.cfm/481
Roget's online thesaurus	General purpose thesaurus, available in print and online	http://www.bartleby.com/62
techdictionary.com	Dictionary for computer terms, online only	http://www.techdictionary.com
Wikipedia + Wiktionary	Fastest growing and most wide-reaching wiki. See Part 2, Blogs and Wikis for more information	http://www.wikipedia.org http://www.wikitionary.org
WordSpy	Devoted to *"lexpionage,"* the sleuthing of new words and phrases	http://www.wordspy.com

Digital Research Strategies

The vast majority of research today starts off as digital research. According to a Pew Internet & American Life survey in 2001, 71 percent of students surveyed said that the Internet was their primary research tool. People use Google to research medical information, financial trends, scientific information, and more. But once you have those 2 million hits, what next? You need to quickly refine your search and hone in on the key ideas and information that will lead you to finding the sources and information you need. Using the "advanced search" feature in Google gives you filters to narrow your search by certain dates, domain types (e.g., .gov) or Boolean operators (e.g., OR, AND).

Yet not all digital information is created equally. With a Google search, you will get more information than you can possibly process, and you'll need to weed out the credible from the ridiculous. With a library search, you will also need to do some weeding out, but most library publications have already been reviewed by experts. If you are a student, you should turn to your college or university library for assistance. Most libraries offer short courses or tutorials on how to use the vast amounts of digital information available to you. Usually these are free if you are a student. Many public libraries also have searchable web sites for their card catalogs, popular research indexes, and so on.

Assume that you are a technical communicator working for an energy resource company. You've been given the assignment of researching and writing a report on the feasibility of using wind turbines in your local community. Your research will need to be very comprehensive. You'll need primary source information—interviews with subject matter experts, site visits to other wind turbine facilities, or surveys of citizens of the community. You'll also need secondary source information: journal articles, newspaper articles, or reports from the government and from private companies. You will work in collaboration with colleagues, and your final product may be printed on paper and also distributed on a web site or as a PDF document.

The following steps will help you with digital research.

1. Starting your search	Use the most specific term you can. "Wind turbine" may be too general, but start there and see what you find; this will help you narrow your search.
2. Narrowing your search	Based on your initial search, try adding other key words. "Wind turbine energy output" will yield more specifics. You can also go into the Google "advanced search" link and indicate that you only want sources from .edu sites.
3. Evaluating the value of online sources	See Evaluating Sources (Part 4).
4. Finding preliminary sources	Look for sources that lead you to other sources. If the other sources are journal articles, keep a list so you can find them in your library or on the library's database site.
5. Determining key sources	Determine which sources are key by noticing how many times they are cited and whether they are scientific, peer-reviewed papers, technical reports, or other credible documents. Look for what are called "meta analyses" (papers where the authors do a thorough job analyzing a vast amount of the scientific work to date). These types of papers can provide you with a wealth of new sources.
6. Using what you find	From the dozens of sources you locate, select the best items. Unless a particular paper appears to be a "classic" (cited widely), you might want to narrow your research down to items that have been published within the past three to five years.

Ethical Issues in Research

Technical communicators can face ethical issues when making decisions about content, visuals, web sites, technologies, products, and services. Taking an ethical stance requires you to make personal decisions about how to balance your ethical and moral beliefs with the realities of your job. You need to consider the impact of your decisions on the users of your product, on your company, on society at large, and on your own job. Sometimes, taking a stand on an ethical issue means that you may lose your job or suffer retaliation from coworkers.

You may have heard the term "whistleblower." It usually refers to a person who, despite risks to the security of his or her own job, will report serious ethical situations, such as safety concerns. The U.S. Department of Labor's Occupational Safety and Health Administration, referred to as OSHA, describes on its web site the Occupational Safety and Health Act, which prohibits discrimination or firing of an employee who reports a serious violation of workplace safety. Yet even with such legal provisions in place, it still takes courage and a strong ethical sense to make these decisions. (For more information on this act, see the U.S. Department of Labor's web site at http://www.osha.gov/dep/oia/whistleblower/ index.html.)

The ethics of conducting research include only one set of decisions you must make throughout the process (see Ethics, Part 1, for more information). As a technical communicator, you might be asked to work on a report on the risks that large wind turbines present to local wildlife, such as migrating birds. If you are working for a company that designs and sells turbines, you may feel pressure to downplay the data about wildlife. Yet is this an ethical decision? Would it be fair to the citizens of the local community if they did not have all of this information?

Many professional associations provide guidelines for ethical behavior in their profession. For instance, the Society for Technical Communication suggests six principles for ethical behavior: legality, honesty, confidentiality, quality, fairness, and professionalism.

STC Ethical Principles for Technical Communicators

As technical communicators, we observe the following ethical principles in our professional activities.

Legality

We observe the laws and regulations governing our profession. We meet the terms of contracts that we undertake. We ensure that all terms are consistent with laws and regulations locally and globally, as applicable, and with STC ethical principles.

Honesty

We seek to promote the public good in our activities. To the best of our ability, we provide truthful and accurate communications. We also dedicate ourselves to conciseness, clarity, coherence, and creativity, striving to meet the needs of those who use our products and services. We alert our clients and employers when we believe that material is ambiguous. Before using another person's work, we obtain permission. We attribute authorship of material and ideas only to those who make an original and substantive contribution. We do not perform work outside our job scope during hours compensated by clients or employers, except with their permission; nor do we use their facilities, equipment, or supplies without their approval. When we advertise our services, we do so truthfully.

Confidentiality

We respect the confidentiality of our clients, employers, and professional organizations. We disclose business-sensitive information only with their consent or when legally required to do so. We obtain releases from clients and employers before including any business-sensitive materials in our portfolios or commercial demonstrations or before using such materials for another client or employer.

Quality

We endeavor to produce excellence in our communication products. We negotiate realistic agreements with clients and employers on schedules, budgets, and deliverables during project planning. Then we strive to fulfill our obligations in a timely, responsible manner.

Fairness

We respect cultural variety and other aspects of diversity in our clients, employers, development teams, and audiences. We serve the business interests of our clients and employers as long as they are consistent with the public good. Whenever possible, we avoid conflicts of interest in fulfilling our professional responsibilities and activities. If we discern a conflict of interest, we disclose it to those concerned and obtain their approval before proceeding.

Professionalism

We evaluate communication products and services constructively and tactfully, and seek definitive assessments of our own professional performance. We advance technical communication through our integrity and excellence in performing each task we undertake. Additionally, we assist other persons in our profession through mentoring, networking, and instruction. We also pursue professional self-improvement, especially through courses and conferences.

Adopted by the STC board of directors
September 1998

■ Society for Technical Communication code of ethics

Evaluating Sources

Given the vast amount of information available these days via the Internet, it is essential that you evaluate sources carefully to ensure accuracy. Some sources are easy to distinguish as obviously credible or not. For instance, the homework assignment from a ninth grader on the subject of wind energy posted on a web site is not a serious source of information, but a peer-reviewed journal article published in a well-established scientific journal is credible (see Scientific and Technical Journal Articles, in Part 4). Many other sources are more difficult to determine as credible or not. What about newspaper articles? A blog written by a well-known engineer or scientist? How can you evaluate the accuracy and reliability of these types of sources?

Questions to Ask	Criteria for Answering
Purpose: Why was the resource written: to inform, to present opinions, to report research, or to sell a product? For what audience is it intended?	Advertisements are a good clue that a site is more commercial than it is educational. Many publications will state at the outset the intended audience and the purpose.
Authority: What are the author's credentials? Are qualifications, experience, or institutional affiliation given?	The author's credentials should be listed on the page (professor, MD, PhD, director). The affiliation (college, university, government agency) should be easy to identify.
Accuracy: Is the information correct and free from errors?	You can tell whether information is accurate by checking against other sources that you know are credible.
Timeliness: Is the information current or does it provide the proper historical context for your research needs?	Many web sites fail to list a date when the document or information was published. Especially with technical information, you need to know the date to assess whether the information is useful.
Coverage: Does the source cover the topic in depth, partially, or is it a broad overview?	If an online document or web site makes claims but without any supporting data, it may not provide enough depth to be useful.
Objectivity: Does the information show bias or does it present multiple viewpoints?	If the web site or document clearly aligns the material with people or organizations that are biased or subjective, you may need to question this information or balance it with other sources.

The University of Minnesota's library (2000) suggests the six evaluation criteria listed in the first column above. The second column provides you with criteria you can use to assess each area.

Along with these questions, you should also consider whether your material is from a *primary source* or a *secondary source*. A *primary source* is a firsthand account or material in original form, whereas a secondary source has usually been subject to editing, interpretation, and other commentary. If you were interested in how citizens in your town felt about wind turbines, you could interview them; the transcripts of your interviews (tape recordings, typed transcripts) would be primary source material. Information obtained by a survey or focus group would also be considered primary material. Or, you could read all of the newspaper articles and other material written about citizens and their thoughts on this topic; this approach would mean that you were using *secondary sources*.

If you were interested in an historic perspective on the environment and energy, you could visit the Library of Congress (in person or online) and read some of the original scientific writings by Benjamin Franklin. These materials would be considered primary sources.

Fluid passing, being the same, and the Quantity of Matter acted upon.

7 Thus the Links of a Brass Chain, &c
with a certain Quantity of Electricity
passing thro' them, have been melted in the
small Parts that form their Contact, while the
rest have been affected.

8 Thus a Piece of Tin Foil cut in this
Form, indeed between two Ends,

and having the Charge of a large Battle sent
thro' it, has been found unchanged between a
and b. melted in Spots between b & c, wholly
melted between c and d, and the Part between
d and e reduced to Smoke by Explosion.

9 The Tin foil melted in the Spots between
b and c; and that whole Space not being melted,
seems to indicate that the Foil in the melted
Parts had been thinner than the rest. on which
thin Parts the passing Fluid had therefore
a greater Effect.

10 Some Metals melt more easily than
others. Tin more easily than Copper, Copper
than Iron. It is supposed (perhaps not
yet proved) that those which melt with the
least of the separating Power, whether that
be common Fire or the electric Fluid, do
also explode more easily or with less of that
Power.

11 The Explosions of Metal like those of
Gunpowder act in all Directions. Thus

Or, you could find a book on this subject, such as a biography of Franklin, and read how that author describes the primary material; that biography would be a secondary source.

Types of Sources

Primary sources:

> Interviews with subject matter experts
>
> Interviews with end users, citizens, or students
>
> Surveys and questionnaires
>
> Observations (of meetings, discussions)
>
> Focus groups
>
> Manuscripts and other archival material

Secondary sources (available in print or electronically):

> Abstracts, bibliographies, databases, and indexes
>
> Journal articles
>
> Newspaper articles
>
> Published reports
>
> Blogs and wikis
>
> Films, television programs, and other published media
>
> Lectures and presentations
>
> Books and pamphlets
>
> Credible web sites

Focus Groups

Focus groups are small group sessions that provide you with the opportunity to learn more about your customers and end users. Normally, a focus group consists of six to ten people with a group leader or interviewer. Focus groups can be run informally or formally. An informal focus group include might include an open-ended discussion about a concept for a new product, with the moderator recording the session or taking notes on the overall impression, interest, and sense of the customers. A more formal focus group would be run with the moderator working from a script of specific questions, designed to get input on key issues, design choices, or branding.

Focus groups are sometimes conducted by the marketing department, but also are part of usability and user testing (Part 4). These group sessions are not the same as having actual users test a document, but they can provide you with a lot of information, which you can then narrow down and use as questions in interviews or surveys.

Image Searches on the Web

Users of the Web often want to search for specific images online. Images are easy to find, copy, and download, but like other visual media, images that others own or created are usually subject to copyright laws. They sometimes require a fee to purchase or license them. Creating your own original visuals, especially if they are simple forms or illustrations, is the best alternative to paying fees. When you need to find images or illustrations for your documents, keep in mind that most sources will cost you money, and that you must always give credit to the owners or authors in your document (see also Copyright and Intellectual Property, Part 4).

Search engines offer easy ways to find visuals on the Web. Here is Google's popular image search site, which provides useful filters and can constrain file size, image type, and more.

■ Google's advanced search page for finding images

The following table lists some of the web sites you might use to search for visuals. Your university, college, or company might also have a digital library of images and visuals for students and employees to access. Government web sites, especially those in science, health, and engineering, usually offer copyright-free image libraries that you can use free of charge because these sources are in the public domain (for more information, see Copyright and Intellectual Property, Part 4).

Popular Image Searches

Image Site	Web Address	Features	Using It
Popular image sources and commercial stockhouses			
Google Images	http://images.google.com	Searches the Internet using Google	Check to see where the image came from, date, and copyright status
Getty Images	http://creative.gettyimages.com	Wide range of subject matter and images	Several categories of use: rights managed and royalty free. Be sure to read their license agreements in advance of use
Flickr	http://www.flickr.com	Search for images and store your own photos	Requires you to sign up for the service
New York Public Library Picture Collection Online	http://digital.nypl.org/mmpco/	Over 30,000 images from books, magazines, photographs, postcards, and more	Based on the collections of the New York Public Library
Corbis	http://pro.corbis.com/	Wide range of subject matter and images	Several categories of use: rights managed and royalty free. Be sure to read their license agreements in advance of use
Selected scientific and government sources			
National Oceanic and Atmospheric Agency (NOAA)	http://www.photolib.noaa.gov	Scientific images from NOAA research and educational work	Because they are government materials, most NOAA images are in the public domain. You still need to provide credit

Image Site	Web Address	Features	Using It
NASA Image exchange (NIX)	http://nix.nasa.gov	"A web-based search engine for searching one or more of NASA's online multi-media collections"	Because they are government materials, most NASA images are in the public domain. You still need to provide credit
Science source	http://www.sciencesource.com	A division of Photo Researchers, Inc.	A commercial site with a wide range of photographs
Insect images	http://www.insectimages.org	"A joint project of the University of Georgia and the USDA Forest Service."	Over 13,000 images of insects from around the world

Regardless of whether you use a commercial or a government site, you need to create a photo credit whenever you use an image. Often, the web site will give you specific instructions on how to word the credit and where to locate it in relation to the image. If not, make sure you give credit to the photographer or organization and note where you found the image. Also, make sure you give each photograph a caption that helps readers see how the photo or image is related to the text. If you are not sure about whether a site is commercial, educational, copyright-free, or for other concerns, look for the links that say "about us" or "copyright policies."

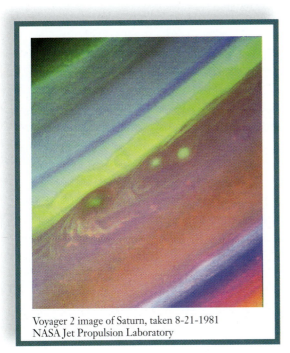

Voyager 2 image of Saturn, taken 8-21-1981
NASA Jet Propulsion Laboratory

■ NASA photo and caption as it might appear in a research paper

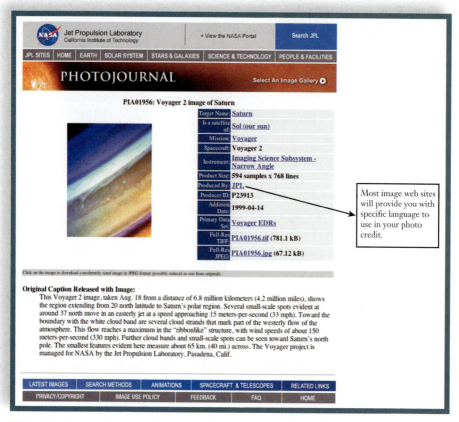

Sample photo and caption from Photojournal, NASA's image access site

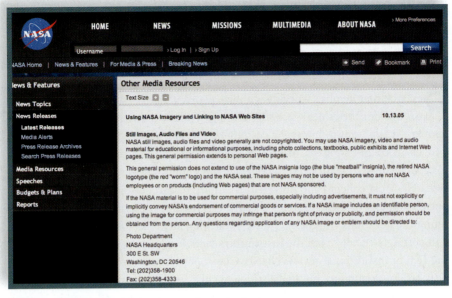

■ NASA web page explaining terms of use for images

Using Images in Technical Communication

Use advanced or specialized search engines whenever possible to get the best image search results.

Try several online sources to find alternative or free images.

Make sure your selected images (clipart, photographs, technical illustrations) are compatible with the purpose and style of your document.

Check with the organization's Web site or offices about whether the images you use are subject to permissions and copyright fees.

Budget for some permissions payments in your writing or research project.

Give proper credit to the source and copyright information for each image in its caption.

Informed Consent

Research that involves working with live people ("human subjects") requires you to consider the ethical issues involved. If you are doing research that involves people—if you want to interview customers, conduct a survey of students, conduct a focus group of end users, or do some usability testing on the intended audience for a web site—you will need to consider the human subject implications of your research. Most colleges and universities require researchers and students to obtain permission to work with people. Usually, this involves describing your study, outlining the questions you intend to ask, indicating any risks that might be involved, and getting a review from the institution's *human subjects review board* (also called the IRB or institutional review board). Class or student projects usually qualify for a classroom exemption, but make sure you check with your instructor. On the job, your company will have a procedure for informing users and getting written permission in advance. If you are a student and want to do research at a private company, you may need permission from both your school and the company. Often, a condition of the research is that you provide people with a clear description of the research, explain the study, give them the option to not participate at any time, and obtain their written signature. Known as *informed consent forms*, you can find examples of these online or at your school or company. The following explanation of informed consent is from the U.S. Office of Health and Human Services:

> Informed consent is a process, not just a form. Information must be presented to enable persons to voluntarily decide whether or not to participate as a research subject. It is a fundamental mechanism to ensure respect for persons through provision of thoughtful consent for a voluntary act. The procedures used in obtaining informed consent should be designed to educate the subject population in terms that they can understand. (1993)

CONSENT TO ACT AS A SUBJECT IN A RESEARCH STUDY

TITLE: **Problem-Solving In College Students**

PRINCIPAL INVESTIGATOR: Jim Jones, Graduate Student
3811 O'Hara Street, Pittsburgh, PA 15213; Phone: 412.624.2413
e-mail: jones@upmc.edu

FACULTY MENTOR: Christopher M. Ryan, Ph.D., Professor of Psychiatry
3811 O'Hara Street, Pittsburgh, PA 15213; Phone: 412.624.2963
e-mail: ryancm@upmc.edu

The purpose of this study is to identify differences in problem-solving strategies in males and females. Approximately 100 college students, at least 18 years of age or older, will be invited to participate in this research study, which is being supported by funding from the Department of Psychology. If you agree to participate, you will complete a brief survey and be asked to solve a series of complex verbal and visual problems presented on a computer screen. After you solve each problem, you will be asked questions about the approach you took. These tasks will take you less than 2 hours to complete.

There is little risk involved in this study. No invasive procedures or medications are included. The major potential risk is a breach of confidentiality, but we will do everything possible to protect your privacy. Another potential risk associated with your participation is the frustration some people experience when they attempt to solve difficult problems. This is not unusual, and if you like, we will discuss your feelings and concerns when you have completed the tasks.

There are no costs to you for participating in this study, and you will receive no direct benefit from participating in this study. If you are a student in a psychology course, and are completing this experiment for course credit, you will receive 2 hours of credit. As your course instructor has pointed out, there are other ways you can satisfy your course requirements, and so you are not obligated to participate in this study. If you are not completing this for course credit, you will receive coupons for 3 Blockbuster video or DVD rentals.

All records pertaining to your involvement in this study are kept strictly confidential and any data that includes your identity will be stored in locked files, and will be retained by us for a minimum of five years. Your identity will not be revealed in any description or publications of this research. Results will not be shared with your instructors or University administrators, and will have no effect on your standing at this University. It is possible that authorized representatives from the University of Pittsburgh Research Conduct and Compliance Office (including the University of Pittsburgh IRB) may review your data for the purpose of monitoring the conduct of this study. In very unusual cases, your research records may be released in response to an order from a court of law. Also, if the investigators learn that you or someone with whom you are involved is in serious danger of potential harm, they will need to inform the appropriate agencies, as required by Pennsylvania law.

Your participation in this study is completely voluntary. You may refuse to take part in it, or you may stop participating at any time, even after signing this form. Your decision will not affect your relationship with the University of Pittsburgh.

Page 1 of 2   Subject's Initials _____
Approval Date: 00/00/00   Consent Form Version: 1.0   IRB # 101203

Annotations (left margin):

Title of project and names of student and faculty investigators are clearly stated at the top of the form.

First paragraph explains the study in clear language.

Second paragraph explains any level of risk involved to the subject.

Costs and record keeping are explained.

Clear statement that participation is voluntary.

■ Informed consent example

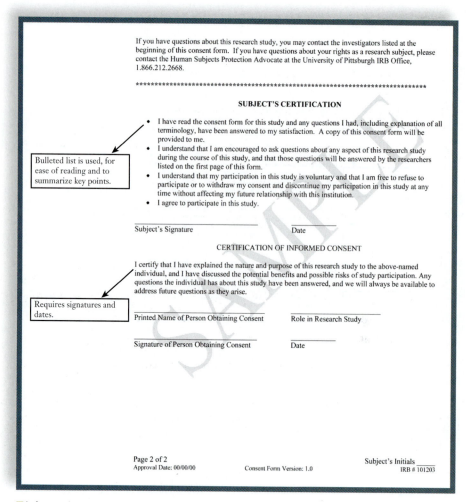

If you have questions about this research study, you may contact the investigators listed at the beginning of this consent form. If you have questions about your rights as a research subject, please contact the Human Subjects Protection Advocate at the University of Pittsburgh IRB Office, 1.866.212.2668.

****

SUBJECT'S CERTIFICATION

- I have read the consent form for this study and any questions I had, including explanation of all terminology, have been answered to my satisfaction. A copy of this consent form will be provided to me.
- I understand that I am encouraged to ask questions about any aspect of this research study during the course of this study, and that those questions will be answered by the researchers listed on the first page of this form.
- I understand that my participation in this study is voluntary and that I am free to refuse to participate or to withdraw my consent and discontinue my participation in this study at any time without affecting my future relationship with this institution.
- I agree to participate in this study.

Bulleted list is used, for ease of reading and to summarize key points.

_____   _____
Subject's Signature   Date

CERTIFICATION OF INFORMED CONSENT

I certify that I have explained the nature and purpose of this research study to the above-named individual, and I have discussed the potential benefits and possible risks of study participation. Any questions the individual has about this study have been answered, and we will always be available to address future questions as they arise.

Requires signatures and dates.

_____   _____
Printed Name of Person Obtaining Consent   Role in Research Study

_____   _____
Signature of Person Obtaining Consent   Date

Page 2 of 2
Approval Date: 00/00/00

Consent Form Version: 1.0

Subject's Initials _____
IRB # 101203

■Informed consent example, *continued*

Integrating Sources

Once you have a good selection of source material for your report or project (see Evaluating Sources, Part 4), you need to integrate the source material into your document in a way that is appropriate as well as fair. There are several approaches to integrating sources. You might use a direct quotation. Or, you might paraphrase the original idea. Or you might do both and offer some analysis about the source material.

Incorporating Source Material

There are many ways to incorporate the source material for a report or other technical document. These ways include taking a direct quotation, paraphrasing, summarizing, and citing a source to support an idea or argument. The following technical report illustrates some of these ways of incorporating sources so that the theme of the paper or report does not disappear by way of too many sources.

A Review of Wind-Resource-Assessment Technology

Shikha Singh[1]; T. S. Bhatti[2]; and D. P. Kothari[3]

Abstract: Generation of electrical energy from wind can be economically achieved only where a significant wind resource exists. Because of the cubic relationship between wind velocity and output energy, sites with small percentage differences in average wind speeds can have substantial differences in available energy. Therefore, accurate and thorough monitoring of wind resource at potential sites is a critical factor in the siting of wind turbines. An accurately measured wind-speed frequency spectrum at a site is another important factor. For assessment of the wind-power potential of a site, most investigators have used simple wind-speed distributions that are parametrized solely by the arithmetic mean of the wind speed. Assessment of power output of a wind turbine will be accurate if the wind speeds measured at the hub height (30–50 m) of a wind turbine-generator are known. However, the existing wind data available at most of the meteorological stations worldwide is measured at a height of 10 or 20 m above the ground. Therefore, wind speeds measured at anemometer heights are extrapolated to the hub height of the wind turbine. Many investigators have proposed simple expressions for height extrapolation of wind speeds. This paper reviews wind-speed prediction and forecasting, and development of techniques for accurate assessment of wind-power potential. Also, the need of wind-resource assessment and the techniques and methods used for it are highlighted.

DOI: 10.1061/(ASCE)0733-9402(2006)132:1(8)

CE Database subject headings: Wind direction; Wind speed; Wind energy; Wind pressure; Wind loads.

Introduction

Since 1947, the year India got freedom, total installed generating capacity has gone up from 1,360 MW to 105,000 MW. However, the share of renewable energy today is only 3% (around 3,000 MW), which is likely to become 10% by 2012. Wind power has entered the new millennium with a lion's roar. In the power-starved developing countries, especially in Bangladesh, India, and China, wind power is a viable source of electricity, and can be installed and transmitted very rapidly even in remote, hilly, and inaccessible areas (Shikha et al. 2000).

The current slump in the solar energy industry worldwide does not seem to affect the windmill industry. In fact, what is on the rise is the number of "wind farms" set up specifically for supplying the energy needs of isolated consumers or for feeding the grid system. Wind energy has come to stay as an important and viable alternate source of energy. The literature abounds with windmill applications in various locations. To cite a few: Chang et al. (2003) assessed the wind characteristics and wind turbine characteristics in Taiwan. In 1995, a software design venture was

established in Taiwan to design and produce data collection and analysis software in English, Chinese, and Spanish. This has included large-scale networked systems using Microsoft NT technology as well as simple single-site systems. This area has grown to include software for air-quality monitoring systems using both analyzers and meteorological equipment. It provides for recording and reporting of air-quality and meteorological data from a variety of sources.

Wind resources in China is enormous, estimated to be exploitable at around 250 GW by the Ministry of Electric Power. It is concentrated in the north and west regions of China and along the coast, and is suited for both remote-village power development and for large-scale, grid-connected electricity production. The wind resource has been surveyed on a national level by the National Meteorological Bureau and by the Meteorology Institute, the data being available on a fee basis. Published isopleth maps of the wind-energy distribution throughout China are available. There is a consistency in the national assessment of territorial regions in China that have the highest wind-energy potential. The criterion for determining wind-energy potential is based on a simple set of parameters, including average wind-energy density (W/m²), annual hours of wind speed above 6 m/s, and annual hours of wind speed above 3 m/s. Much more detailed data is available at specific meteorological stations and in the databases of several major wind demonstration sites, located throughout China (Zhou 1996).

Similar reports on Netherlands (Bruijse 1995), South Africa (Linda 1996), Romania (Gârbacea 1996), and Denmark (Frandsen and Andersen 1996) illustrate the scale of interest in wind energy. Wind program technological developments of the United States were outlined by Ancona et al. (1997). Wind prospects and activities of countries like New Zealand (Dawber and Drinkwater 1996), Bangladesh (Rahman 1996), Turkey (Hanagasioglu 1999), and Japan (Ushiyama 1999) have been of great interest. During

[1]Research Scholar, Centre for Energy Studies, Indian Institute of Technology, Hauz Khas, New Delhi - 110 016, India (corresponding author). E-mail: aceenu@rediffmail.com

[2]Professor, Centre for Energy Studies, Indian Institute of Technology, Hauz Khas, New Delhi - 110 016, India. E-mail: tsbhatti@ces.iitd.ernet.in

[3]Professor, Centre for Energy Studies, Indian Institute of Technology, Hauz Khas, New Delhi - 110 016, India. E-mail: dkothari@ces.iitd.ernet.in

Note. Discussion open until September 1, 2006. Separate discussions must be submitted for individual papers. To extend the closing date by one month, a written request must be filed with the ASCE Managing Editor. The manuscript for this paper was submitted for review and possible publication on May 7, 2003; approved on April 27, 2004. This paper is part of the *Journal of Energy Engineering*, Vol. 132, No. 1, April 1, 2006. ©ASCE, ISSN 0733-9402/2006/1-8~14/$25.00.

Annotation (top left): This citation indicates that the idea in the previous sentence comes from the cited authors (Sikha et al.) and is placed at the end of the sentence to indicate this.

Annotation (middle left): This citation to Change et al. is placed at the beginning of the clause, because the authors wish to make clear that this study is an example of what is referenced in the previous sentence ("The literature abounds with windmill applications in various location. To cite a few...").

Annotation (bottom right): This "clustered list" of citations, in one sentence or paragraph, indicates that for every country listed, the authors are aware of at least one representative study.

■ A technical report on wind energy

These citations play a similar role to those shown in #3. They reinforce the theme of this paragraph: "The potential of wind energy in countries...."

Wind-Resource Assessment

Wind-Speed Prediction and Forecasting

This citation references both a direct quotation ("regional roughness length") as well as an idea (the "which" clause that followed this quote).

■ Technical report, continued

For more information on integrating sources, see Citing Sources (Part 4) and Documentation (Part 6).

Interviews

Interviews can be a rich source of information for any research project. Although they can be time consuming, interviews provide a primary source of information, direct from the people you interview. There are many methods of conducting interviews, each unique to the particular situation and what you need to know to write your report. For instance, if you want to get an idea of how citizens in the community feel about having a wind turbine, you might attend a public forum and then interview some of the attendees afterward. If you wanted to learn firsthand from engineers about the technical challenges of wind turbine energy, you might decide to interview members of the engineering design team.

Interviews can be conducted as open ended, where you start with a few general questions and let people talk about whatever comes to mind, or more close ended, where you have a fixed set of interview questions that you ask each person. Open-ended interviews can provide you with useful information to narrow down and use as survey questions. Similarly, surveys and questionnaires are sometimes followed up with a series of in-depth interviews with the respondents. If you are going to be conducting research that needs approval from your university or from a company, you may be required to come up with questions in advance and submit them for approval.

Your college or university, or the company you are working for, will have procedures and guidelines about doing research with *human subjects*. A review from the institution's *human subjects review board* (also called the IRB or institutional review board) requires significant time and paperwork, even if you are exempt from IRB oversight. Make sure you follow the guidelines and submit paperwork for approval before beginning your interviews.

Interviews

What do I want to learn?	Make a list of the kind of information that you can usefully obtain by interviews and do background research in advance. Keep in mind that people are busy and that an interview should be brief.
Who should I interview?	What kind of people can best help you answer your questions? Subject matter experts? Citizens? Students?
Should I keep the interview open ended?	If you don't know what questions to ask, but you want to get a feel for the "big picture" about what might be on someone's mind, generate a few brief interview questions and then let people speak freely.
Should I generate a list of specific questions in advance?	If you know what to ask, make a list of questions in advance. You may need to have these questions approved by the company, university or other organization.
Should I do interviews in person, on the phone, or over the Internet?	Determine what will give you the best outcome. If you need to interview a lot of people quickly, and across a wide geographic area, phone or the Internet may be best. If using the Internet, you can send interview questions in advance, then use IM or another chat tool to do the interview. If you want a conversational interview, consider interviewing in person with a recording device.

Libraries

Libraries provide a gateway to numerous research tools, information, and services. From books to magazines, newspapers to electronic databases, search tools to electronic card catalogs, libraries are at the heart of any research project. (Libraries and librarians are always on the cutting edge of digital and search technology, so much so that degrees in library science are now almost all called "library and information science.") There are several ways to categorize libraries.

Public Libraries

Public libraries serve communities and their citizens. Most public libraries offer reading material for a general audience, but they usually also have a web site and section of the library devoted to research, with some of the more general indexes such as the *Reader's Guide to Periodical Literature*. Public libraries might also house special collections of materials related to the region, such as historic papers or books by local authors.

College and University Libraries

College and university libraries provide a wealth of resources. Depending on the size of the institution, a university library might be a resource not only for its faculty and students but also for the state, the region, and for other nearby colleges and institutions of higher learning. If you have access to a college or university library, you have access to a world of information and information services. If you have already graduated, you may be able to retain some library access through your alumni society.

Specialized Libraries and Collections, Including Archives

Within colleges and universities, there are often specialized libraries, such as medical law, or architecture libraries. You may not have access to these resources unless you work directly with that unit. But you may be able to view and borrow materials from these libraries if you are working on a special project. Check with a librarian to find out. Many libraries also house special collections of materials, such as antique maps, rare books, manuscript archives, digital archives, or computer information. For instance, the University of Minnesota's Charles Babbage Institute (http://www.cbi.umn.edu) collects "primary source materials and rare publications documenting the history and development of information technology."

Private Libraries

Many large organizations have libraries with materials available only to employees. For instance, most large computer and software companies have technical libraries. If you are doing research as part of your job, be sure to find out if your company has a technical library and, if so, how you can access it.

Library of Congress

The Library of Congress (http://www.loc.gov) offers an amazing range of materials, most available for viewing on the Internet. From maps and original papers to books and sound recordings, this site is a vast resource, especially if your research requires any historic information.

Literature Reviews

Literature reviews are documents written to demonstrate the overall scope and major themes of a body of scholarship. Literature reviews serve two purposes. First, they demonstrate that the authors of the report are familiar with research in this area and have examined the major lines of scholarship (significant experiments, published reports, and other work) relative to the problem or issue they are addressing. Second, literature reviews provide a quick way for researchers to make a case about what, in their view, is happening in a field and why the issues they are about to raise, or attempt to solve, are important in the context of the work to date.

A good literature review has a theme, and it tells a story. Literature reviews should not be paragraph upon paragraph about how author x said one thing, author y said another, and so on. A literature review should put these ideas into a context for readers and make some connections between themes.

For example, in doing research for a report on wind turbines, you might find three major themes in the literature:

1. Wind turbines can only contribute a small percent of energy to the overall community needs at this time.
2. Wind turbines can create excess energy on days when energy use is down, but there is no way to store this energy.
3. Wind turbines often are of concern to environmentalists, related both to the look of a wind turbine on the landscape and to the potential they have to interfere with bird migrations.

Your literature review should use these three key themes to guide how you present the literature you find. You should present the findings in a style that demonstrates that you have done a thorough job; you should also describe the materials in a neutral manner.

Summary and Conclusions from Recent Literature and Current Research

The results of the literature and current research review are divided into sections, each pertaining to a different aspect of the problem or how the problem has or is being examined.

Literature review is chunked by major areas: fatality studies, lighting studies, and so on. This approach (organizing the literature into key themes) is far more useful in terns of what you learn and for your readers than simply listing every article you find.

Fatality Studies

There have been few systematic or quantitative towerkill studies in the past 5 years that have focused solely on determining the numbers of fatalities at given towers. There are, however, promising areas where there is strong interest among qualified researchers who wish to pursue the towerkill issue. These researchers (reference Appendix III) are now collecting information on bird kills at towers. This information can be used to test hypotheses that have or are being proposed. These researchers are currently managing projects in West Virginia, New York, and Kansas in which several towers are searched for fatalities. Search schedules vary greatly among the studies, with some towers being searched only when weather conditions suggested a mortality event (low ceiling and poor visibility due to rain or fog). Other studies in this group used more frequent sampling with a relatively constant interval between sampling.

The author uses headings to show these themes: electromagnetic and other radiation studies, methods for studying, and results of interviews.

Though they have not been published, studies now being conducted in 3 states (Appendix III) suggest that towers less than 400 to 500 feet in height are not as dangerous to migrating songbirds, especially neotropical species, as towers greater than 500 feet in height. The basis for this statement is a small database from West Virginia (Canterbury, personal communication), New York (Evans, personal communication and data on the <www.towerkill.com> website), Kansas (Young, personal communication), Florida (Engstrom personal communication), and Minnesota (Cuthbert, personal communication). See Appendix III for details of these studies in progress. In these situations, towers less than 500 feet have generally experienced very few kills, while under taller towers larger numbers of dead birds were found. There is 1 notable exception. On Jan. 22, 1998, a kill of between 5-10,000 Lapland Longspurs and a few other birds occurred at a series of 3 communication towers and a natural gas pumping facility tower near Rochester, KS. The tallest of these towers is 420 ft. AGL. In most of the studies there generally has not been what many call a mortality event or large kill involving more than a several dozen or one hundred birds in a single night.

The author summarizes and explains conclusions that can be drawn from the literature instead of just listing author names.

The fact that between an estimated 5,000 - 10,000 Lapland Longspurs and others were killed at a series of 3 communication towers and a natural gas pumping facility tower – the tallest tower 420 feet AGL – in mid-winter is problematic because this species has rarely been reported from towerkills. This event may be an anomaly in some ways and should be treated differently from the mortality events involving Neotropical and North American migrants that are normally found in the literature, although the mechanisms or circumstances may be the similar.

C&K, LLC - 3/2000

Sample literature review for a research report on wind turbines

Another seemingly important result from some researchers is the fact that the number of fatalities seems to be declining. Arthur Clark (Appendix III) reports that in recent years, the numbers of birds under the towers he searches has dropped precipitously. Mr. Clark has been studying towerkills at several communication towers in the Buffalo, NY, area for well over 33 years. There is speculation among several other researchers that towerkills in general decline a few years after a new tower is erected. Explanations of this phenomenon range from the fact that Neotropical migrants have declined in number over the past 40 years, to the fact that there are more towers – numbers currently increasing at an exponential rate the past 3 years or so – and that the kills may be more dispersed. All explanations are speculative, although many years ago researchers noticed that fatalities decreased at towers, particularly several years after initially large kills.

Lighting Studies

In the past 5 years there have been no definitive or suggestive studies regarding how or if lights disorient or attract songbirds to towers. At least one study was published (Bruderer et al. 1999) in which a spotlight trained on migrating birds disoriented them, but this may not be comparable to towerkill issue. Bruderer was attempting to find ways to haze birds away from aircraft, not attract them to towers. Information that is forthcoming from the few studies now being conducted may help us understand the role of lights of different color in attracting birds, but it is more likely that specific research is needed to address this problem.

Despite a lack of empirical evidence or studies, there seems to be a degree of consensus among experts, based on past data collection or experience that white strobes are less hazardous to migrating songbirds than are white or red blinking lights. The fact that several researchers believe strongly enough to suggest or recommend strobe over other tower lighting types, suggests that research efforts focus on the difference. This promises to be fruitful research that could have direct impact on numbers of birds killed at towers. To date, however, there very few or no published papers or recent databases that substantiate the fact that white strobes are less dangerous than other color or type of lights, other than what was presented at the August 11, 1999, workshop at Cornell University on "Avian Mortality at Communication Towers," transcripts of which are currently available at <www.towerkill.com> and at <http://migratorybirds.fws.gov/issues/towers/agenda.html>. No data were presented in that paper and the results should be considered speculative.

There is a body of information of recent literature from Europe in which migrants of several species and Homing Pigeons were studied in controlled situations in which various color lights were used in an effort to override or disorient birds' magnetic compasses. This literature strongly suggests that birds exposed to red lights in laboratory or controlled conditions may not be able to use magnetic cues as well as birds exposed to green or white lights. The applicability of these studies, at least in the immediate future, is worthy of consideration, especially in light of

■ Sample literature review for a research report on wind turbines (*continued*)

speculation that red lights are more dangerous than white strobe lights. However, the underlying mechanisms behind the disorientation of songbirds at lighted communication towers during times of poor visibility (precipitation, fog, low ceiling) may be related to the findings of these studies. If birds are attempting to use magnetic cues in times of poor visibility, red lights may disorient them.

Electromagnetic and other Radiation Studies

The literature on applicable electromagnetic radiation and radio frequency influences on migrating birds is nearly nonexistent. By "applicable," I refer to studies in which radio frequencies or geomagnetic fields that are similar to those created by communication towers were investigated. Only one study really addressed the influence of short-wave radio waves on Homing Pigeons (Bruderer and Boldt 1994), not migrating birds or species of birds that migrate at night.

Interviews with researchers who are knowledgeable about migration or study migration provided little insight into the question. However, most seemed to doubt whether the strength and type of radio frequencies or electromagnetic fields around communication towers could disrupt the orientation/navigation of migrants. They stated that the earth's magnetic field was likely to be much stronger than that of communication towers. Several of the researchers referred to the Project Seafarer study in which electromagnetic pulses, similar to those experienced in nuclear explosions, were investigated by avian researchers. These pulses are so strong as to not be applicable to communication tower situations, according to those interviewed. The behavioral data from migrants tested in cages were so variable as to be inconclusive (F. Moore, personal communication).

Methods for Studying Fatalities and Behavior at Communications Towers

There are currently no standard metrics or metrics being used to evaluate towerkills. Researchers use different methods to search for birds and report their results in different fashions. These differences may lead to different results. For example, some researchers look on mornings after nights in which the visibility was poor. Others search once per week or once per month. Thus, there is no way of comparing results. Similarly, the area searched under towers is not standardized by researchers and in some cases the search area is not even reported.

Regarding the study of behavior of night migrants that is applicable to studying how birds behave around communication towers, several studies appeared between 1995 and 1999. These studies included several methods: radar (tracking, marine surveillance, and NEXRAD), infrared (LORIS) devices, acoustical devices, ceilometers, and moon watching. Very briefly, radar uses microwave radiation that is reflected by the water in the bodies of birds to determine their location over the ground. Tracking radar locks on to individual migrants and provides very

■Sample literature review for a research report on wind turbines (*continued*)

detailed behavioral information. Marine surveillance radar (portable via truck or boat) provides a two dimensional picture (X,Y coordinates) of migrants as they move through a several square mile area. NEXRAD is a weather radar technology that provides a macro image of bird migration over several hundred square miles. Infrared devices track birds by imaging the heat differences between birds and the air they are in. Infrared devices provide a detailed picture of birds migrating by a location at night. Acoustical devices sense birds via their flight calls, enabling researchers to determine the species composition of birds as they fly within the first thousand or so feet overhead. Ceilometers are pencil beam lights that are pointed vertically in the night sky. As birds pass through the beam they are counted and their direction is determined. Moon watching is a means of determining the number migrants and their directions as they pass through the disk of the moon, as observed with a spotting scope.

Each of these methods has advantages and disadvantages. It is likely that the most useful methods, especially for studying the reactions of birds close to towers, and therefore testing hypotheses about how birds react to lighting or other stimuli at towers, will be tracking and marine surveillance radars and infrared devices. Acoustical devices may be useful in some situations and for answering some questions. Moon watching is not likely to prove useful, although ceilometer may be useful if volume of migration at low altitude is the desired measurement. The references provided above will be of use to researchers who wish to investigate or consider various methodologies for studying behavior of migrants in close proximity to towers.

The study of behavior of captive birds, as with the magnetism and light studies (see Wiltschko and other references above), may be useful in situations where the perceptual ability of birds is investigated.

Results of Interviews

Responses of those interviewed were almost uniform. None of those interviewed were currently working on communication tower projects or projects related to communication towerkills. They also knew of very few (almost no) studies that were currently being conducted (or researchers doing studies). Of those questioned, Gauthreaux stated that he planned to publish a paper on lighting and towerkills and Engstrom stated that he would be publishing a paper shortly on the 30+- year towerkill project he is conducting with R. Crawford in northern Florida (Also see Appendix III).

C&K, LLC - 3/2000

| The tone of the text remains neutral: "may be useful in situations where the perceptual ability of birds is investigated." |

■ Sample literature review for a research report on wind turbines (*continued*)

The following guidelines, illustrated by sections from the literature review on pages 422-425, will help you think about how to organize and write a literature review.

Start with an introductory paragraph or two stating the purpose of the review and describing the categories or areas you plan to cover.	This introduction is nice and short, but it would be more useful if it provided an overview of the categories: "This review groups the literature into five areas: fatality studies, lighting studies, electromagnetic and other radiation studies, methods and metrics, and interviews."
Use headings to separate major areas in the literature.	These headings are underlined and separated by white space. Traditional academic literature reviews are often done without headings but with a clear transition between topics: "After fatality studies, the next major body of research is on lightning."
Let the research literature do the talking.	A good literature review is written around topics or themes, not authors. Notice that in this review, the paragraphs begin with ideas, not author names.
If appropriate, write a short, succinct conclusion.	This review does not have a conclusion, but for some purposes, such as an academic literature review, you might want to conclude with a short paragraph, such as "This review illustrates that major work on migratory birds and wind turbines to date is focused primarily on three areas. . . ."

Outlines

Outlines provide an overview of the main sections of a document. Most technical communicators use outlines to organize their research materials and develop a structure for a document before they begin drafting. In some cases, you may need to provide an outline as part of a research proposal before you get approval for a project. Developing an outline helps you to see the shape of your document, identify topic areas where you may need to do more research, and decide what organizational strategy to use for a particular document. Beginning technical writers sometimes prefer to plunge in and start drafting at the outset; experienced writers know that it is easier and more efficient in the long run to develop an outline first. Drafting is easier once you have an outline, because the outline allows you to focus on one section or topic at a time. Your detailed outline often becomes the basis for your table of contents that will appear in your finished document.

A basic outline is often called a *topic outline*, because it uses brief keywords and short phrases to suggest main topics and supporting detail in an organized hierarchy. Here is an example:

I.   Introduction: financial and environmental costs of gasoline autos

II. Possible solutions

   A. hybrid autos

   B. electric autos

   C. public transit

III. Hybrid autos

   A. Hybrids currently on the market

      1. Toyota Prius

      2. Honda Civic

      3. Biodiesel and hydrogen prototypes

   B. Electrics currently on the market ...

Example of a topic outline

A *sentence outline* uses complete sentences to represent each topic or section. If you find that you often have problems with logical transitions, paragraphs, coherence, or find yourself restating your main points over and over, you may want to try developing a sentence outline before drafting. Many writers find that a sentence outline helps to see the direction and steps in an argument. Each point in a well-developed sentence outline can become a topic sentence for a paragraph in your final report.

I. The financial and environmental costs of gasoline-powered autos have risen to a point where alternatives must come more available across all makes and models.

II. At least three possible solutions are currently being explored on a limited basis: hybrid autos, electric autos, and new or expanded public transit networks.

A. Hybrid autos have been introduced and marketed successfully by several major manufacturers.

B. Electric autos have also been re-introduced to the commercial market.

C. Portland, San Diego, Minneapolis and other cities have begun major construction projects to build new public transit infrastructure.

III. Hybrid autos appear to be a viable solution in many communities.

A. Several hybrid autos now on the market have been both successful and reliable.

1. The Toyota Prius, the most popular hybrid auto in 2005, has been an overwhelming market success and demand continues to exceed supply in most areas.

2. The Honda Civic hybrid has also begun to capture the public's attention.

3. Other models include even luxury manufacturers like Lexus and Infiniti.

B. Electric cars have been introduced but so far battery life and recharging difficulties have made them less popular than hybrids.

■ Example of a sentence outline

Ideas for Creating and Using Outlines

Use a topic outline for short documents and informative reports when the structure of your document is relatively simple.	Group your main ideas and notes by topic, and then decide how to arrange them in a sequence and hierarchy with main points and supporting detail.
Use a detailed sentence outline to organize a longer report or for any persuasive document when you need to clarify the steps of your argument.	Review your outline to evaluate your organization for continuity and completeness before you begin drafting.

Plagiarism

Plagiarism is a topic of increasing complexity and importance in our digital age. Especially for research-based writing in college classrooms and for scholarly work, plagiarism is something we all wish to avoid. Yet doing so may be harder than we think. The basic rule of thumb is to give credit when the words, or the ideas, belong to someone else. But this guideline can be more challenging than it appears, especially in an environment where researchers move quickly between digital documents, cutting and pasting material and not paying careful enough attention to the source of the material.

Consider plagiarism in the context of these general categories:

1. **Exact wording**: When you cut and paste, or photocopy, or retype, or in any other way use the exact wording from another source, you must cite this material. For example, for research on wind energy, you locate the following newsletter from the American Wind Energy Association:

JUNE 2005

Senate Passes Energy Policy Bill 85-12;
Bill Contains 3-Year PTC and 10% RPS
Energy Legislation Now Set to Enter Final Phase

By Kathy Belyeu
AWEA Staff

On June 28, the U.S. Senate overwhelmingly approved a wide-ranging energy policy bill containing a three-year extension of the wind energy Production Tax Credit (PTC) and a national Renewables Portfolio Standard (RPS) that would require 10% of the nation's electricity to be generated by renewables by the year 2020.

Passage of the bill, H.R. 6, signifies the third time in five years that the Senate has passed broad energy policy legislation. The previous two bills failed to become law when they became stuck in unresolved negotiations with vastly different energy legislation approved by the House of Representatives. The House passed an energy policy bill earlier this year. Both bills now advance to a House-Senate conference committee, to be reconciled, and ultimately presented to President Bush to sign into law. The President set the entire process in motion over five years ago with the release of the administration's national energy plan. While there is no firm schedule for the conference committee, observers expect its members to be announced soon and its first meeting to take place shortly after the 4th of July holiday. Congress is only in session for three weeks between Independence Day and the start of its break for all of August.

Despite some remaining hurdles, prospects for a compromise version of the bill becoming law this year have increased. Last year the Senate bill had little to no support from Senate Democrats. This year, the bill claims support from both Democrats on the Senate Energy Committee as well as Senate Democratic Leader Harry Reid of Nevada. In addition, over the last few weeks, President Bush has repeatedly called on Congress to deliver an energy bill to his desk for signature by early August. One of the biggest remaining roadblocks is trying to break the impasse over whether to exempt manufacturers of a formerly used gasoline additive (MTBE) from lawsuits over groundwater contamination.

3-Year Production Tax Credit
The 3-year PTC contained in the Senate energy bill would extend the placed-in-service date for gaining access to the credit from December 31, 2005, to December 31, 2008. The credit would be maintained at its current value of 1.9 cents per kilowatt-hour and it would continue to be adjusted annually for inflation. In addition, once qualifying for the credit, individual wind turbines would continue to generate credits for a term of 10 years. The House bill does not address the PTC extension or the 10% national renewables requirement.

■A newsletter research source

You decide that this first paragraph is very useful in your research paper, and you think the exact wording will have the most impact in your paper. Here is how you would use and cite this material properly using the APA citation style:

There are many types of energy policy legislation at both the state and federal levels, and these are often well supported by constituents and may have tax benefits. For instance, according to the American Wind Energy Association

(2005), "[o]n June 28, the U.S. Senate overwhelmingly approved a wide-ranging energy policy bill containing a three year extension of the wind energy Production Tax Credit" (p. 1).

Note the square brackets ([ ]), used to indicate when you have made a change to the original material. For more information on these types of details, consult Part 6 of this handbook or the style manual used for your class or organization. Also, see Citing Sources in Part 4 of this handbook.

2. **Paraphrasing**: When you paraphrase material, but don't use the exact wording, you still need to cite this material. Here is an example.

   There are many types of energy policy legislation at both the state and federal levels, and these are often well supported by constituents and may have tax benefits. For instance, in June 2005, the Senate approved an energy policy bill that provides for a variety of tax credits and other incentives (American Wind Energy Association, 2005).

   Paraphrasing becomes complicated when you are working with a variety of materials, some online, some from hard copies, and so on. You may have cut and pasted material from a variety of reports, abstracts, technical papers, newspapers, and other web-based sources, but you need to carefully track the material's source. One easy way to do this is to put the exact citation right next to the material that you are keeping in your notes. For instance, the exact citation of this source in APA style would be as follows:

   American Wind Energy Association. (2005, June). Senate passes energy policy bill 85-12. *AWEA Windletter*, p. 1. Retrieved from http://www.awea.org/windletter/wl_05jun.html.

3. **Material that is common knowledge**: You do not need to cite material that is common knowledge. For instance, if you wrote that the United States is in North America, you would not need to cite a source because that general information can be found in many source locations. However, what is common knowledge and what is information you garnered when doing research can become blurred unless you keep careful documentation as you do your research. For example, it is common knowledge that many U.S. states are looking into producing alternative energy, including wind power. But specific statistics, or information about legal policies or state guidelines, would not be common knowledge. These data may feel like common knowledge to you, because you have been doing so much research on this topic. Check your source information to find the original source of the information, and cite sources as needed.

4. **Digital material**: It is especially easy to inadvertently plagiarize when using material from a digital source, such as an electronic journal article, a digital

newspaper, a web page, or any other source available in electronic format. The Internet makes cutting and pasting very quick and efficient, but in that quickness, you may forget to cite the source or place quotation marks around the direct quotation. Or, the original material may flow nicely with the material you have already written, and it is a temptation to just paste in a sentence here and there. But keep in mind that you are using someone else's work, and their ideas, as your own. Not only is this unethical, it is also disrespectful to the other author. In some cases it is also illegal. Just because the technology makes it simple does not mean that you should fall prey to cutting and pasting without properly citing sources.

5. **Collective authorship and workplace writing**: In the workplace, it is quite common to reuse previously written material in new documents. For instance, sentences from the executive summary of a company's annual report might be very good sentences to include or modify for a marketing brochure or customer handbook. Most materials written in a full-time workplace setting are owned by the organization, not the individual. In addition, technical content must often be carefully reviewed and approved by lawyers and technical experts, and so any rewriting could put the company at risk. For all these reasons, workplace materials are frequently reused in new documents.

For more information about plagiarism, see **Citing Sources** (Part 4) and **Ethical Issues in Research** (Part 4).

Way to Avoid Plagiarism	
Carefully track of the source of all research materials.	Whether using an exact quotation or a paraphrase, you must cite the source when the idea is not your own original.
If you cut and paste material from the Web, make sure you include a complete citation.	

Scientific and Technical Journal Articles

In the section on Abstracts (Part 4), the search example yielded a number of articles, including the following:

> Hermann, T.M., Mamarthupatti, D., and Locke, J.E. (2005). Postbuckling analysis of a wind turbine blade substructure. *Journal of Solar Energy Engineering*, 127–44, 544–52.

This citation is a typical scientific journal article. The *Journal of Solar Energy Engineering* is a peer-reviewed journal and the paper presents the results of a scientific study. *Peer reviewed* means that an appointed group of subject experts, or editorial board, use a rigorous process to evaluate and accept the papers to this journal. The first part of the paper is a short summary, also called an abstract:

> Postbuckling analysis of composite laminates representative of wind turbine blade substructures, utilizing the commercial finite element software ANSYS, is presented in this paper. The procedure was validated against an existing postbuckling analysis. Three shell element formulations, SHELL91, SHELL99, and SHELL181, were examined. It was found that the SHELL181 element with reduced integration should be used to avoid shear locking. The validated procedure was used to examine the variation of the buckling behavior, including postbuckling, with lamination schedule of a laminate representative of a wind turbine blade shear web. This analysis was correlated with data from a static test. A 100% postbuckling reserve in a composite structure representative of a shear web was quantified through test and analysis. The buckling behavior of the shear web was improved by modifying the lamination schedule to increase the web bending stiffness. Modifications that improved the buckling load of the structure did not always equate to improvements in the postbuckling reserve.

Reading an abstract will help you to determine whether the material presented in the paper will be useful for your research. Scientific papers may be written in dense prose, using passive voice and other features of writing style that can make the content hard to grasp. In the above abstract, a careful reading tells you that the engineers tested three types of composite material to determine which type would best prevent something called "shear locking." You would have to read the entire paper to learn more. But if your research project required you to find out about the various strengths of materials used to build wind turbines, this paper would probably be a good source.

Surveys and Questionnaires

Surveys and questionnaires provide a structured and efficient way to learn about what people think about products, concepts, or ideas. Unlike interviews or focus groups, where the data is qualitative, surveys and questionnaires can yield both qualitative and quantitative data. Interviews and focus groups can be a great way to let people talk freely and get ideas for the kinds of questions you might ask on a survey or questionnaire.

Surveys are formal and well planned, with an intended purpose and serious consideration given to the number of questions, wording of the questions, and so on. Questionnaires, on the other hand, are informal, short, and not intended to be analyzed with statistics. On the Internet, some sort of "survey" is always going on (news sites ask you to rate a story or sports pages that ask you to take a guess at who will win the NCAA basketball tournament), but usually, these are just polls, not surveys.

A vast amount of information is available on how to design and administer a survey (see Surveys, Part 2, for how to design one). You may want to talk with a survey researcher or someone with ample survey experience before you get started. How you state your questions can have a major impact on the value of the information that you get. For instance, if your questions are obviously biased or if they are written using language that is unclear or ambiguous, you will not get useful information. You can create a short but useful survey if you follow some basic guidelines. You can use a web site such as surveymonkey.com to set up and distribute the survey.

Surveymonkey is an online tool for creating and administering surveys

Survey Guidelines	Questions to Ask Yourself
Determine exactly what you want to learn from this survey—keep your survey focused on one topic	Do I want to know how citizens feel about an issue or event?
	Do I want to know how much they know about an issue or event?
	(It's best to focus the survey on just one theme.)
Identify the group of people you want to survey	Citizens of the community?
	Politicians?
	Students?
	Customers or end users?
	(It's easier to compile results if you have one target group in mind.)
Choose the best way to reach these people	Email to reach them?
	A web site for the survey?
	(It's easy to reach people via the Internet.)

continued

Survey Guidelines	Questions to Ask Yourself
Design a form that explains the purpose of the survey and, if required, asks for participants' informed consent	How can I let people know the purpose of the survey, where and how I will use the results, if I intend to keep their identity private, and so on? (It is important to keep people informed, and it may be required by your school or company.)
Create survey questions that will get you the information you need	Open-ended questions: lots of information about audience ideas but hard to categorize and count. Likert scales ("on a scale of 1–5..."): easy to count up and find the average but no room for explanations. Yes/no questions: easy to count but not a lot of room for any middle ground. Multiple choice: easy to count and room for some variation in responses.
Keep it short and simple	People get asked to take surveys dozens of times a day via email. Keep your survey short. Users should be able to complete it within five to ten minutes.
Make questions and all wording clear and unambiguous	An unclear question or phrasing will yield results that you can't use.
Test the survey on a small group of people first	Testing the survey will help you clean up any bad questions, unclear wording, and so on.

Types of Research

Clear and accurate technical communication should be based on solid, credible research. Different forms of communication require different kinds of research. For example, a manual or set of instructions to accompany a new home computer would need to be researched to ensure that it's written with information that users need and that such information is accurate and doesn't give the user any surprises. As the writer, you might need to run a usability test on the manual in advance to make sure it's accurate.

Wind turbines can be researched in a variety of ways.

A report designed to analyze the capability of wind turbines to produce electricity for a community, in contrast, would require research in wind power, conversion of wind to electricity, the engineering specifications of a wind turbine, and, perhaps, policy and environmental issues. It might need to be reviewed by a team of technical and public policy experts. You might need to interview local citizens to get their input.

Whether you are doing research for a class project or for a project on the job, it is important to do the kind of research that will provide you with what you need to produce a clear, accurate, fact-based communication product. Research spans a range of approaches, from database searches to usability testing to interviews. In most scientific and technical settings, you will need to talk with subject matter experts (SMEs), such as the product development engineers and scientists who know the technical specifications and details about the product and can give you the information you need to make the communication (web site, user manuals, instructions) accurate.

Usability and User Testing

U sability describes the type of research that you can use to study your end users or customers and then learn how to make information easy for them to perform tasks or meet their own goals. All technical documents—reports, feasibility studies, online help, user manuals, and instructions—need to be *usable* to the reader; otherwise, the information will not be of much value. Usability encompasses a wide range of methods and tools that can be employed to test whether products and information are understandable and help the end user perform a task or learn new information. For example, you can test the layout of a web site, its technical content, or specific navigation tasks that users need to perform.

Usability is most often associated with the computer and software industries, where usability specialists (often called human factors engineers or usability engineers) are involved, usually right from the start, with the design and specifications of documentation, screen design, and other aspects of the user interface. Different kinds of research work best at different stages in the writing process. User testing can also be conducted in a usability lab, using software and video equipment designed to record user behavior for direct observation and later analysis. Such tests can include *think-aloud protocols*, which ask people to talk through their thoughts and feelings as they work with the product. Usability research may also include focus groups, interviews, and surveys and questionnaires as methods for collecting user interactions with a document.

Methods for Usability Research on a Wind Energy Reference Manual

In the following example, imagine you are going to test the usability of the Danish Wind Industry Association (DWIA) Wind Energy Guided Tour and Reference Manual, located at http://www.windpower.org/en/tour. You have several points you need to consider before deciding on the best methods for your usability research:

1. What research **question** are you trying to answer? What **features** of the document do you want to test?
2. What is the appropriate research **method** that would provide the best kind of feedback about usability of the site? What purpose does each method serve?
3. What is your research **goal**? How will your goal help you select the best methods for performing your usability research?

The following examples illustrate common methods and goals for how usability research would be conducted on an online user manual. For each example page, these three issues—research question, method, and goal—are explained.

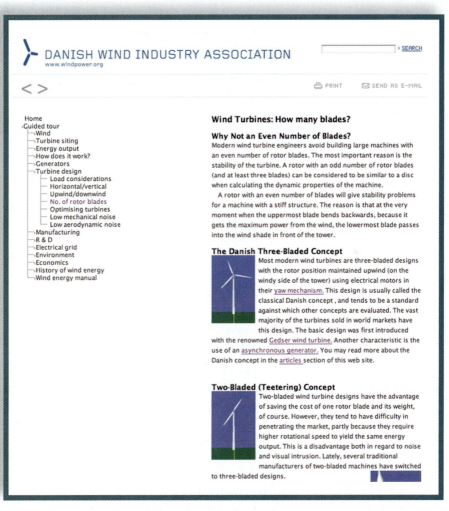

DANISH WIND INDUSTRY ASSOCIATION
www.windpower.org

〈 〉   🖶 PRINT   ✉ SEND AS E-MAIL

Home
›Guided tour
   ⊢→Wind
     →Turbine siting
     →Energy output
     →How does it work?
     →Generators
     →Turbine design
       ├─ Load considerations
       ├─ Horizontal/vertical
       ├─ Upwind/downwind
       ├─ No. of rotor blades
       ├─ Optimising turbines
       ├─ Low mechanical noise
       └─ Low aerodynamic noise
     →Manufacturing
     →R & D
     →Electrical grid
     →Environment
     →Economics
     →History of wind energy
     └─Wind energy manual

Wind Turbines: How many blades?

Why Not an Even Number of Blades?
Modern wind turbine engineers avoid building large machines with an even number of rotor blades. The most important reason is the stability of the turbine. A rotor with an odd number of rotor blades (and at least three blades) can be considered to be similar to a disc when calculating the dynamic properties of the machine.

    A rotor with an even number of blades will give stability problems for a machine with a stiff structure. The reason is that at the very moment when the uppermost blade bends backwards, because it gets the maximum power from the wind, the lowermost blade passes into the wind shade in front of the tower.

The Danish Three-Bladed Concept
Most modern wind turbines are three-bladed designs with the rotor position maintained upwind (on the windy side of the tower) using electrical motors in their yaw mechanism. This design is usually called the classical Danish concept , and tends to be a standard against which other concepts are evaluated. The vast majority of the turbines sold in world markets have this design. The basic design was first introduced with the renowned Gedser wind turbine. Another characteristic is the use of an asynchronous generator. You may read more about the Danish concept in the articles section of this web site.

Two-Bladed (Teetering) Concept
Two-bladed wind turbine designs have the advantage of saving the cost of one rotor blade and its weight, of course. However, they tend to have difficulty in penetrating the market, partly because they require higher rotational speed to yield the same energy output. This is a disadvantage both in regard to noise and visual intrusion. Lately, several traditional manufacturers of two-bladed machines have switched to three-bladed designs.

■ DWIA Wind Energy Guided Tour: Number of rotor blades

1. **Question:** For this page about rotor blades, you might want to research the effectiveness of the layout, graphics, and writing style.
2. **Best Method:** Questionnaire. A written test with both close-ended and open-ended questions would best help you evaluate layout elements such as the headings and blue animated graphics. See Surveys and Questionnaires (Part 4) for example questions and more information.
3. **Goal of the research:** Evaluate participant thoughts and perceptions about the document; get data about users.

DANISH WIND INDUSTRY ASSOCIATION
www.windpower.org

< >

🖨 PRINT   ✉ SEND AS E-MAIL

Selecting a Wind Turbine Site

Photograph
Soren Krohn
© 1997 DWIA

Wind Conditions

Looking at nature itself is usually an excellent guide to finding a suitable wind turbine site.

If there are trees and shrubs in the area, you may get a good clue about the prevailing wind direction, as you do in the picture to the left.

If you move along a rugged coastline, you may also notice that centuries of erosion have worked in one particular direction.

Meteorology data, ideally in terms of a wind rose calculated over 30 years is probably your best guide, but these data are rarely collected directly at your site, and here are many reasons to be careful about the use of meteorology data, as we explain in the next section.

If there are already wind turbines in the area, their production results are an excellent guide to local wind conditions. In countries like Denmark and Germany where you often find a large number of turbines scattered around the countryside, manufacturers can offer guaranteed production results on the basis of wind calculations made on the site.

Look for a view

As you have learned from the previous pages, we would like to have as wide and open a view as possible in the prevailing wind direction, and we would like to have as few obstacles and as low a roughness as possible in that same direction. If you can find a rounded hill to place the turbines, you may even get a speed up effect in the bargain.

Grid Connection

Obviously, large wind turbines have to be connected to the electrical grid.

For smaller projects, it is therefore essential to be reasonably close to a 10-30 kilovolt power line if the costs of extending the electrical grid are not to be prohibitively high. (It matters a lot who has to pay for the power line extension, of course.)

■DWIA Wind Energy Guided Tour: Turbine siting

1. **Question**: An appropriate research question about this page would be one that seeks to determine whether readers comprehend the information; for instance, factors that affect how wind turbines are sited.

2. **Best Method:** Tests and Quizzes. Written tests about material presented in the document would best help you evaluate user concept retention and clarity of the ideas. These tests usually include some close-ended questions (such as multiple choice) and open-ended, short-answer questions.

3. **Goal of the Research**: Your goal would be to measure what readers have learned from using the document. How well do they grasp a certain concept?

Standard Wind Class Definitions (Used in the U.S.)

Class	30 m height		50 m height	
	Wind speed m/s	Wind power W/m^2	Wind speed m/s	Wind power W/m^2
1	0-5.1	0-160	0-5.6	0-200
2	5.1-5.9	160-240	5.6-6.4	200-300
3	5.9-6.5	240-320	6.4-7.0	300-400
4	6.5-7.0	320-400	7.0-7.5	400-500
5	7.0-7.4	400-480	7.5-8.0	500-600
6	7.4-8.2	480-640	8.0-8.8	600-800
7	8.2-11.0	640-1600	8.8-11.9	800-2000

■ DWIA Wind Energy Reference Manual: Wind energy concepts

1. **Question:** For this table, you may want to learn whether readers can describe when to use the Standard Wind Class Definitions table.
2. **Best Method:** Focus Groups. To answer this question, you could lead a small group discussion to discover user needs and behaviors in situations in which the document would be used. See Focus Groups (Part 4) for more information.
3. **Goal of the Research:** The goal here is to evaluate participant feelings, priorities, and perceptions.

DANISH WIND INDUSTRY ASSOCIATION
www.windpower.org

> SEARCH

< >

🖶 PRINT   ✉ SEND AS E-MAIL

Home
Guided tour
→Wind
→Turbine siting
→Energy output
→How does it work?
→Generators
→Turbine design
→Manufacturing
→R & D
→Electrical grid
→Environment
→Economics
→History of wind energy
→Wind energy manual
— Index
— Wind energy concepts
— Energy and power
— Proof of Betz' law
— Wind turbine acoustics
— Electricity
— 3 phased electricity
— 3 phased connection
— Electromagnetism 1
— Electromagnetism 2
— Induction 1
— Induction 2
— Environment and fuels
— Bibliography
— Glossary

Wind Energy Reference Manual

1. Wind Energy Concepts
   1. Unit Abbreviations
   2. Wind Speeds
   3. Wind Speed Scale
   4. Roughness Classes and Roughness Lengths
   5. Roughness Class Calculator
   6. Roughness Classes and Roughness Length Table
   7. Density of Air at Standard Atmospheric Pressure
   8. Viscosity of air
   9. Power of the Wind
   10. Standard Wind Class Definitions (Used in the U.S.)
2. Energy and Power Definitions
   1. Energy
   2. Energy Units
   3. Power
   4. Power Units
3. Proof of Betz' Law
4. Wind Energy Acoustics
   1. dB(A) Sound Levels in decibels and Sound Power in W/m^2
   2. Sound Level by Distance from Source
   3. Adding Sound Levels from Two Sources
   4. How to add sound levels in general
5. Wind Energy and Electricity
   1. Three Phase Alternating Current
   2. Connecting to Three Phase Alternating Current
   3. Electromagnetism Part 1
   4. Electromagnetism Part 2
   5. Induction Part 1
   6. Induction Part 2
6. Wind Energy, Environment, and Fuels
   1. Energy Content of Fuels
   2. CO_2-Emissions from Fuels
7. Bibliography
8. Wind Energy Glossary

■ DWIA Wind Energy Reference Manual: Navigational methods

1. **Question:** This site would be a great place to test the web pages' two main navigational methods: the clickable outline and textual links on the right of the screen, and the collapsing navigation tree on the left of the screen.

2. **Best Method:** Structured Walkthrough. In a structured walkthrough, a small group uses the manual (or a prototype of the site) to test navigation and then has a structured discussion. This method usually includes props, such as storyboards or prototypes, and includes a moderator to keep the discussion focused.

3. **Goal of the Research:** The goal of a structured walk through is to gather early feedback before major drafting begins and while design and format can be modified.

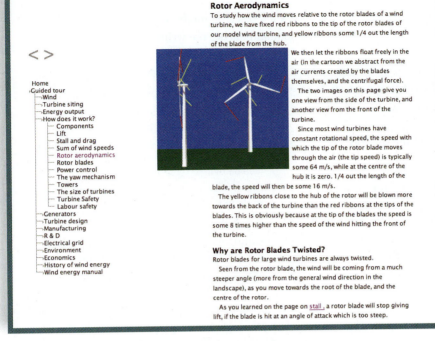

Rotor Aerodynamics

To study how the wind moves relative to the rotor blades of a wind turbine, we have fixed red ribbons to the tip of the rotor blades of our model wind turbine, and yellow ribbons some 1/4 out the length of the blade from the hub.

We then let the ribbons float freely in the air (in the cartoon we abstract from the air currents created by the blades themselves, and the centrifugal force).

The two images on this page give you one view from the side of the turbine, and another view from the front of the turbine.

Since most wind turbines have constant rotational speed, the speed with which the tip of the rotor blade moves through the air (the tip speed) is typically some 64 m/s, while at the centre of the hub it is zero. 1/4 out the length of the blade, the speed will then be some 16 m/s.

The yellow ribbons close to the hub of the rotor will be blown more towards the back of the turbine than the red ribbons at the tips of the blades. This is obviously because at the tip of the blades the speed is some 8 times higher than the speed of the wind hitting the front of the turbine.

Why are Rotor Blades Twisted?

Rotor blades for large wind turbines are always twisted.

Seen from the rotor blade, the wind will be coming from a much steeper angle (more from the general wind direction in the landscape), as you move towards the root of the blade, and the centre of the rotor.

As you learned on the page on stall , a rotor blade will stop giving lift, if the blade is hit at an angle of attack which is too steep.

■ DWIA Wind Energy Guided Tour: Rotor aerodynamics

1. **Question**: Here, you may wish to see if users can find the section on rotor aerodynamics.

2. **Best Method:** User Testing. Empirical tests of a small sample of users will allow you to collect and analyze data in several ways.

   - For instance, you can watch users find this section, under "How does it work," and observe as they try navigating their way through the manual. You may give them a written test or use talk-aloud protocols to provide additional data.

   - If you have a usability lab, you can observe readers interacting with the document and record their actions using a video camera, a screen-capturing program, or a keystroke-recording program.

3. **Goal of the Research**: The goal in this area is obtain feedback and quantifiable measures of a document's probable success. You can also discover levels of success or failure as people attempt to use the site's features.

Timing of Usability Research

Depending on your purpose, each research method has an optimal timing during the development of a document and during the writing process.

Usability Research Method	Stages of the Writing/ Development Process
Questionnaires	Any time during the design phase, during drafting, or even at project completion
Tests and Quizzes	Later in the design phase, early in the drafting or prototyping phase, and usually during user testing (see below)
Focus Groups	During planning or after final delivery (to get feedback for the next version)
Structured Walkthroughs	During the planning and design phase
User Testing	Early in the drafting or prototyping phase, or when an early version (called *alpha* or *beta*) is released

Early feedback on a product, document, or prototype is the most useful type of usability testing because it identifies common problems early in the development of a technical document. User-centered design, for example, involves end users in every stage of the design process and has become increasingly popular across technical fields. Ideally, audience feedback should be gathered at multiple points in a technical writing project.

Wikipedia and Online Encyclopedias

Wikis are web pages that allow multiple users, from any location, to add, edit, archive, and update information (see Part 2, Blogs and Wikis). The most popular wiki is Wikipedia, an online encyclopedia where entries are written, updated, and revised by anyone around the globe who has access. The idea behind Wikipedia is that the information is created collaboratively, and that the power of having so many people involved provides for instant fact-checking and updating.

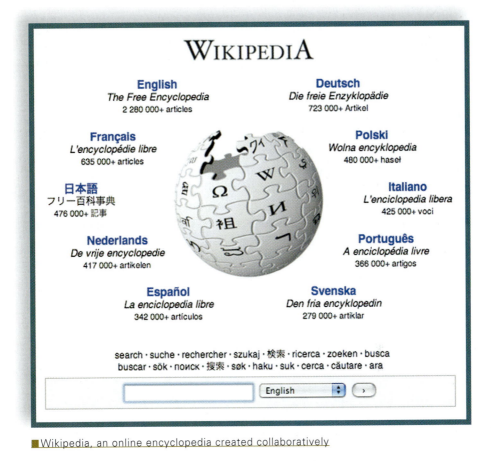

■ Wikipedia, an online encyclopedia created collaboratively

Many people, from students to journalists to researchers, use Wikipedia, often as a starting place on a particular subject. The positive aspects of this approach are that Wikipedia is quick and easy to use and can be accessed from any computer with an Internet connection. In addition, the entries are typically well written, formatted for each of use, and designed in a similar fashion so that one page resembles another. Also, entries are updated constantly, so especially on entries where the subject matter is changing (an evolving scientific theory, for example), Wikipedia offers advantages over print-based encyclopedias. Negative aspects of Wikipedia as a research tool includes the fact that since Wikipedia is open to anyone to revise and edit, there is no central editorial system to check content, as there would be with a peer-reviewed journal or a newspaper. As Wikipedia itself states, "it is important to use Wikipedia carefully if it is intended to be used as a research source, since individual articles will, by their nature, vary in standard and maturity" (Wikipedia: Research with Wikipedia, 2007). You should read this entry before using Wikipedia for your research.

Any encyclopedia, digital or otherwise, should be considered as a starting point, not an end point for your research. The footnotes and links in a Wikipedia entry can lead you to other articles, books, and journals that you can access through a library. Encyclopedias and other reference sources do not usually count as credible primary and secondary sources, but rather as a means to educate yourself and find those sources.

For instance, if you search on "wind energy" in Wikipedia, you will find a very long entry with a list of references. One such reference is an article from the journal *Nature*, a peer-reviewed scientific journal (see http://www. nature.com/ nature/journal/v447/n7141/full/ 447126a.html). You may be able to access this article online, or you may need to access it through your library. This highly credible source will in turn have references, which should lead you to more information from a variety of sources. For more information on finding and using credible sources, see Evaluating Sources (Part 4).

5

Guidelines for Grammar, Style, and Punctuation

Abbreviations and Acronyms

Abbreviations and acronyms are shortened forms of longer names or terms. Use these in ways your audience will understand. For instance, the common abbreviation "Dr." is well understood, but the abbreviation for a technical part or a company name may not be understood by those outside the area.

Abbreviations in Names

Use an abbreviation for the following designations preceding names:

Mr.	Messrs.	St. (Saint)
Mrs.	Mmes.	Mt. (Mount)
Ms.	Dr.	Rev. (unless preceded by *the*)

Is *Ms.* Jones our new chemistry instructor?

Avoid using an abbreviation for any other designation preceding a name.

NOT: Are you campaigning for *Sen.* Jones?

BUT: Are you campaigning for *Senator* Jones?

NOT: In his new book, *Prof.* Rosenthal discusses the use of laser surgery to correct the most common forms of astigmatism.

BUT: In his new book, *Professor* Rosenthal discusses the use of laser surgery to correct the most common forms of astigmatism.

Use an abbreviation preceded by a comma for a designation or an academic degree following a name.

Jr.	Ph.D. or PHD (Doctor of Philosophy)
Sr.	
B.A. or BA (Bachelor of Arts)	Ed.D. or EdD (Doctor of Education)
M.A. MA (Master of Arts)	
M.S. or MS (Master of Science)	M.B.A. or MBA (Master of Business Administration)
D.D.S or DDS (Doctor of Dental Science)	
D.D. or DD (Doctor of Divinity)	J.D. or JD (Doctor of Jurisprudence)
M.D. or MD (Doctor of Medicine)	D.V.M. or DVM (Doctor of Veterinary Medicine)

> The speaker will be Thomas Dean, *Jr.*
>
> The academy announced the appointment of Marion Unger, *Ph.D.,* as chair. (*or PhD without periods*)

Avoid using the abbreviation *Dr.* before a name that is followed by an abbreviation denoting a doctoral degree. Using both is redundant.

Acronyms and Organizational Abbreviations

Use abbreviations without periods for many well-known agencies, organizations, and businesses and for other familiar abbreviations using capital letters.

> The newspaper accused the *CIA* of covert activities in that country.
>
> The *YMCA* is presenting a revival of Arthur Miller's *All My Sons*.
>
> The candidate sought the support of the *AFL-CIO*.

An acronym is a pronounceable word made from initials or parts of words. Consult your dictionary about its capitalization and be sure to explain, in parentheses, an acronym that might be unfamiliar to your readers.

> The next meeting of *OPEC* (Organization of the Petroleum Exporting Countries) will be an important one.
>
> Bring your *scuba* (*s*elf-*c*ontained *u*nderwater *b*reathing *a*pparatus) equipment with you.
>
> The lecture explained *quasars* (quasi-stellar objects).

Abbreviations in Dates and Numbers

The abbreviation *B.C.* or *BC* should be put after the date, and the abbreviation *A.D.* or *AD* is usually put before the date. However, the practice of putting *AD* after the date is now also considered acceptable. *BC* and *AD* are sometimes replaced with the abbreviations *B.C.E.* (Before the Common Era) and *C.E.* (Common Era), respectively.

A.M. or *AM* and *P.M.* or *PM* may be written with either capital or lowercase letters, but be consistent within a single piece of writing. Lowercase letters require periods—*a.m.* or *p.m.*

Use the following abbreviations for common Latin words and expressions when appropriate:

> *c.* or *ca.* (about)   *etc.* (and others)
> *cf.* (compare)   *i.e.* (that is)
> *e.g.* (for example)   *viz.* (namely)

> Ada Lovelace (*c.*1815–1852) is considered the first female computer programmer.
>
> Scientific disciplines (*e.g.,* biology and chemistry) use similar deductive methods.

Do not overuse these Latin abbreviations. Where possible, substitute the English equivalent.

Spell out the names of days and months.

> NOT: The first game of the World Series will be played on *Oct.* 11.
>
> BUT: The first game of the World Series will be played on *October* 11.
>
> NOT: It snowed heavily on the first *Sat.* in *Dec.*
>
> BUT: It snowed heavily on the first *Saturday* in *December.*

Spell out the names of cities, states, and countries, except in addresses.

> NOT: He came to *N.Y.C.* to study music.
>
> BUT: He came to *New York City* to study music.

Avoid Unconventional Abbreviations

Spell out first names.

> NOT: The editor of the collection is *Thom.* Webster.
>
> BUT: The editor of the collection is *Thomas* Webster.

In names of businesses, spell out the words *Brothers*, *Corporation*, and *Company*, except in addresses or in bibliographic information in research papers.

In formal, nontechnical writing, spell out units of measure.

| Not: | The report claims that anyone who is more than ten *lbs.* overweight is a candidate for heart attack. |
| But: | The report claims that anyone who is more than ten *pounds* overweight is a candidate for heart attack. |

In technical writing, abbreviations are acceptable and often preferred.

Adjectives

An adjective is a word that modifies, or describes, a noun or pronoun. It limits or makes clearer the meaning of the noun or pronoun.

The *efficient office administrator* organized the schedule. (*modifies a noun*)

He is *efficient*. (*modifies a pronoun*)

They are still *popular* though *their composer* died in 1937. (the first *modifies a pronoun*, the second *modifies a noun*)

Forms of Adjectives

Most adjectives have a comparative form to compare two things and a superlative form to compare three or more.

	Comparative	Superlative
rich	richer	richest
beautiful	more beautiful	most beautiful
bad	worse	worst

Placement of Adjectives

Adjectives can usually be identified by their position in a sentence. For example, an adjective will fit sensibly into any of the following blanks.

The _____ person way very _____.

The _____ object was removed.

It seems _____.

Adverbs

An adverb is a word that modifies, or limits the meaning of, a verb.

During the Harbor Festival the tall ships *sailed gracefully* into the bay.

My brother-in-law *always drives fast* on the highway.

Kinds of Adverbs

An adverb modifies a verb by answering one of the following questions: (1) "When?" (2) "Where?" (3) "How?" (4) "How often?" and (5) "To what extent?"

Adverbs of time answer the question "When?"

We *will discuss* the matter *soon*.

Environmentalists warn that we *must eventually reach* an equilibrium with nature.

Adverbs of place answer the question "Where?"

The ambassador *encountered* many welcoming citizens *everywhere* in the five countries he visited.

Faith healers *look upward* and *inward* for cures for disease.

Note that nouns can function as adverbs of time and place.

> The symposium *was held yesterday*.
> Some students *go home* every *weekend*.

Adverbs of manner answer the question "How?" They tell in what manner or by what means an action was done.

> The movie going public *enthusiastically embraces* disaster films.

Adverbs of frequency answer the question "How often?"

> The teacher *repeatedly praised* his students for their achievements.

Adverbs of degree answer the question "To what extent?"

> My friends and I *thoroughly enjoyed* the new Stephen King novel.

Forms of Adverbs

Many adverbs can be formed by adding the suffix -*ly* to an adjective. An additional spelling change is sometimes required.

> The general's actions seemed *heroic*. (*adjective*)
> The general acted *heroically*. (*adverb*)

Most adverbs have comparative and superlative forms.

	Comparative	Superlative
profoundly	more profoundly	most profoundly
fast	faster	fastest
well	better	best

Ampersand and Special Characters

The ampersand (&) symbol means "and." It is a substitute for *and* in places such as book titles, company names, or in an in-text citation using APA style (see Part 6). Do not use the ampersand in place of the word *and* in regular writing. Do use the ampersand if it is used as part of a company name or other formal title and you are citing that title.

The other special characters (@, #, ˆ, %) should be used carefully and precisely.

The at (@) symbol is used as part of email addresses. It should not be substituted for the word *at* in formal writing, although in playful writing, logo design, and other places, @ is quite common.

The pound (#) symbol is used to indicate numbers or pounds.

The percent (%) symbol is used to indicate percentages. Style guides give differing advice on how percentages should be written. In general, if you use the numeral, use the percent sign (1%). If you use the word, use both words (one percent). (See also **Numbers and Dates.**)

Copyright can be indicated by (c) or by the copyright symbol ©. Trademark is indicated with the symbol ™.

Most word processing programs offer a variety of other specialized symbols and usually tell you what the symbol means.

Apostrophe

The apostrophe has several uses, which are explained here.

Showing Possession

The possessive case forms of all nouns and of many pronouns are spelled with an apostrophe. The following rules explain when to use an apostrophe for possessive forms and where to place the apostrophe.

Note: Do not be misled by the term *possessive*. Rather than trying to see "ownership," consider the possessive case form as an indication that one noun modifies another in some of the ways adjectives do. Consider these examples:

a *week's* vacation   the *flower's* fragrance
Egypt's history   a *car's* mileage
the *river's* source   your *money's* worth

Singular Nouns

To form the possessive of a singular noun, add an apostrophe and an *s*.

Anthony's review of the new science fiction film emphasized the *director's* outstanding use of special effects.

Usage varies for singular nouns ending in *s*. Many writers follow the regular rule for singular nouns and add an apostrophe and *s*.

The *boss's* salary is three times that of her assistant.

We discussed *Fermat's* theorem in class today.

But it is also acceptable to form the possessive of such singular nouns by adding only an apostrophe.

The *witness'* testimony seemed vague and unconvincing.

The engineering team visited *Gates'* home office today.

Plural Nouns

To form the possessive of a plural noun that does not end in *s*, add an apostrophe and *s*.

The *women's* proposal called for a daycare center to be set up at their place of employment.

To form the possessive of a plural noun that ends in *s*, add an apostrophe alone.

> The *doctors'* commitment to their patients was questioned at the forum.
>
> The course highlights two *biologists'* studies—Bayer and Cohen.

Noun Pairs or Nouns in a Series

To show joint possession for noun pairs or nouns in a series, add an apostrophe and *s* to the last noun in a pair or a series.

> Sociologists were concerned that the *royal couple's* announcement would set off a new baby boom in Britain.
>
> *Felts and Smith's* research had a dramatic effect on their contemporaries.

To show individual possession, add an apostrophe and *s* to each noun in a pair or series.

> The *president's* and the *vice president's* duties are clearly defined.

Personal Pronouns

Add an apostrophe and *s* to form the plural of words being named, letters of the alphabet, abbreviations, numerals, and symbols. The apostrophe may be omitted with capitals without periods: TVs, UFOs, PhDs.

> One drawback of this typeface is that the capital *i's* and the lowercase *l's* look exactly alike.
>
> Avoid weakening your argument by including too many *but's* and *however's*.

Replace Missing Letters

Use an apostrophe to indicate that part of a word or number has been omitted. Be careful that your word processing program does not substitute an open single quote (').

> In *'64* the Beatles invaded the United States with a new style of rock *'n'* roll.
>
> The manager told the singer to "go out and knock *'em* dead."

In Contractions

A contraction is a shortened form of a word or words. Contractions are widely used in speech and in informal writing.

Use an apostrophe to indicate a missing letter or letters in a contraction.

> Winning at cards *wasn't* Tony's only claim to success.
>
> Enrique claimed that enough attention *isn't* being paid to the threat of environmental pollution.

Note: *It's* is a contraction meaning "it is" or "it has." *Its* is a possessive pronoun meaning "belonging to it" and does not require an apostrophe. See Commonly Confused Words.

> *It's* your turn to drive today.
>
> The dog covered *its* right eye with *its* paw.

Articles

Articles refer to the words *the*, *a*, and *an*. Each can precede a noun. *A* and *an* are referred to as indefinite articles. A noun may refer to a single indefinite, unspecified person, place, or thing. Such count nouns are indicated as singular in English by the use of one of the **indefinite articles** *a* or *an* (meaning any unspecified *one*): *a glass*, *an owl*. These nondefinite singular count nouns require an indefinite article or some other determiner in front of them. Use *a* before words beginning with a pronounced consonant and *an* before words beginning with a vowel sound: *a* house, *an* hour.

Nondefinite count nouns that are plural can be used without any determiner: *Glasses* break easily; *owls* live in barns.

The is referred to as the definite article. A noun may refer to a definite person, place, or thing. Such nouns—count or noncount—are signaled by the use of the definite article *the*. The definite article can be used with either singular or plural nouns. Use the definite article in the following instances:

FIRST MENTION: We saw two cars in the driveway. Put rice in a bowl.

SUBSEQUENT MENTION: *The* cars belong to Juanita. Stir *the* rice with a spoon.

Brackets

This section describes square brackets, angle brackets, and braces.

Square Brackets [ ]

In writing, you will use square brackets with inserted information and inside parenthetical information. (Square brackets are also used in mathematics and in computer code.)

With Inserted Information

Use brackets to enclose information inserted into direct quotations for clarification.

The comedian quipped, "From the moment I picked your [S. J. Perelman's] book up until I laid it down, I was convulsed with laughter. Someday I intend reading it."

"Government [in a democracy] cannot be stronger or more tough-minded than its people," said Adlai Stevenson.

Use brackets to enclose editorial comments inserted into quoted material.

According to Clarence Darrow, "The first half of our lives is ruined by our parents [how many people under twenty agree with this!] and the second half by our children."

Notice that the exclamation point is placed inside the closing bracket because it is part of the editorial comment.

The word *sic* or *thus* enclosed in brackets is used to indicate that an incorrect or seemingly incorrect or inappropriate word is not a mistake on the part of the present writer but appears in the original quotation.

> Anthony sent a memo requesting a new workstation so that he could "work with more effecency [*sic*]."

With Parentheses

Use brackets to replace parentheses within parentheses.

> Some humpback whales reach a length of over fifty feet. (See p. 89 [chart] for a comparison of the sizes of whales.)

Notice that the period is placed *inside* the closing parenthesis because the entire sentence is enclosed by parentheses.

> The reading list contains several books dealing with the issue of freedom of the press (for example, Fred W. Friendly's *Minnesola Rag: The Dramatic Story of the Landmark Supreme Court Case that Gave New Meaning to Freedom of the Press* [New York: Random House, 1981]).

Notice that the period is placed *outside* the closing parenthesis because only part of the sentence is enclosed by parentheses.

Angle Brackets < >

Angle brackets are used in the computer field, usually as part of a programming language, to enclose a programming command. They are also used in HTML and XML coding, usually to enclose a tag, command, or URL. For instance, in the following sentence, the angle brackets demarcate the section that will appear in bold:

> Before you connect to the Internet, make sure to install anti-virus software.

Sometimes, angle brackets are used when referring to URL as part of a document.

> To learn more about the use of angle brackets in XML, see **<**http://www.xml.com/pub/a/2001/08/29/anglebrackets.html**>**.

Angle brackets are also used in statistics and mathematics to indicate greater than (>) and less than (<).

Braces { }

Also called "curly brackets," braces are used in the computer field, usually as part of a programming language, to enclose a programming command. They can also be used as part of musical notation or to notate sets in mathematics.

Buzzwords and Phrases

Buzzwords are words that become fashionable in business, politics, management, or advertising and may be widely used for a time but do not have a clearly defined meaning. Also, such usage starts to lack precision and become a cliché. For instance, the phrase "Think outside the box" does not have any specific meaning. When possible, avoid such buzzwords and phrases, using instead words and phrases that are as precise and unambiguous as possible.

> Buzz: Think outside the box
>
> Specific: Address this problem with original ideas and new concepts.

Capitalization

In marketing and advertising, capital letters are often used in a whimsical manner. In technical writing, however, capitalization follows these guidelines.

First Words of Sentences

Capitalize the first word of a sentence, the pronoun *I*, and the interjection *O*.

> *M*any game designers have created imaginary worlds.
>
> *D*o *I* think that life exists on other planets?
>
> *T*hese creatures, *O* mighty Gork, come from the other side of the universe.

Capitalize the first word of a direct quotation that is a complete sentence.

> At the start of the seminar, Jamil said, "*T*his software will change your life."

Titles, Names, and Words Based on Names

For titles of books or articles, capitalize the first and the last word and all other important words, including prepositions of five or more letters.

Avoid capitalizing articles, short prepositions, or coordinating conjunctions that do not begin or end a title. A short preposition is one that has fewer than five letters.

> *The Principles of Beautiful Web Design*
>
> "*E*nvironmental *P*lanning and *I*mpact *S*tatements"
>
> "*O*n the *D*ay *A*fter the *S*torm"

Capitalize both parts of a hyphenated word in a title.

> "*H*ome-*T*houghts from *A*broad"   "*G*ood-*B*ye, *M*y *F*ancy!"

Capitalize abbreviations and designations that follow a name. Do not capitalize titles used as appositives.

> Eugene Anderson, *Jr.*
>
> Anne Poletti, *Ph.D.* (PhD)
>
> Louise Tate, *A*ttorney at *L*aw

Capitalize proper nouns.

Sharon	the Middle Ages	the Empire State Building
Andrew Jackson	Snoopy	the Revolutionary War
Portuguese	Lake Michigan	
Greta Garbo	Hawaii	

Avoid capitalizing compass points unless they are part of a proper noun: *northwest of Chicago* but the *Pacific Northwest*.

Capitalize an official title when it precedes a name.

The guest speaker will be Congresswoman Katherine Murphy.

The nation mourned the death of President Lincoln.

The changes were supported by Governor Celeste.

Note: Do not capitalize a title that does not name a specific individual.

Seymour Rosen, a chemistry professor, submitted an article to the magazine.

Willie Mae Kean, a first-year law student, won the award.

Spike Kennedy, an intern at the hospital, was interviewed on television.

Capitalize the title of a relative when it precedes a name or is used in place of a name. Do not capitalize the title if it is used with a possessive pronoun.

Aunt Joan	Cousin Mary
Uncle Carlos	Grandfather Tseng

Is Grandmother coming to visit?

You look well, Grandpa.

My uncle Bill could not come to the performance.

Capitalize proper adjectives.

Barringer meteorite crater	Machiavellian goals
Parisian style	Argentinian wine
Islamic teacher	Tibetan prayer flags

Conventional Capitalization

Capitalize words naming deities, sacred books, and other religious documents and names of religions, religious denominations, and their adherents. Note: Pronouns referring to deities are often capitalized; this usage may differ based on the style guide or other conventions.

the *B*ible	*A*llah	*B*uddha
*C*atholicism	the *K*oran	the *L*ord
*C*hrist	*M*uslim	*G*od

Capitalize names of months, days of the week, and holidays. Note: Do not capitalize the names of the seasons.

*A*pril	*D*ecember	*J*anuary
*T*uesday	*S*aturday	*W*ednesday
*H*alloween	*N*ew *Y*ear's *E*ve	the *F*ourth of *J*uly

Capitalize the abbreviations *A.D.* (*AD*) and *B.C.* (*BC*). Note: The abbreviations *C.E.* (common era) and *B.C.E.* (Before the Common Era) can be substituted for *A.D.* and *B.C.* for notations that are more religiously neutral. Small caps (as shown here) are often used for these abbreviations. This convention is optional.

A.D. 172	500 *B.C.*
AD 356	275 *BC*

Clarity and Accuracy

In technical writing, it is important to be clear and accurate. Readers want information that is easy to understand and factual.

Ambiguous Reference

Do not use a pronoun that could refer to either of two or more antecedents.

Ambiguous:	Malcolm told Henry that *he* had won a trip to France.

The pronoun *he* could refer to either Malcolm or Henry. If it refers to Malcolm, rewrite the sentence to make this reference clear.

Clear:	Malcolm told Henry, "I have won a trip to France."
Or:	Malcolm, who had won a trip to France told Henry the news.

If the pronoun refers to Henry, rewrite the sentence a different way.

Clear:	Malcolm told Henry, "You have won a trip to France."
Or:	Malcolm knew that Henry had won a trip to France and told him so.
Ambiguous:	Marika met Dr. McCluskey when *she* visited the lab last week.

The pronoun *she* could refer to either Marika or Dr. McCluskey.

Clear:	When Marika visited the lab last week, she met Dr. McCluskey.
Clear:	When Dr. McCluskey visited the lab last week, Marika met her.
Ambiguous:	In the saga, Luke Skywalker and Han Solo are at first rivals. Both want to win the affection of the princess. As the saga progresses, however, the two young men gain respect for each other, until finally the rivalry ends when *he* discovers he is Leia's brother.

The pronoun *he* could refer to either Luke or Han.

Clear:	In the saga, Luke Skywalker and Han Solo are at first rivals. Both want to win the affection of the princess. As the saga progresses, however, the two young men gain respect for each other, until finally the rivalry ends when *Luke* discovers he is Leia's brother.
Ambiguous:	Fourteenth-century Europe was scarred by war and plague. It is hard to tell which was worse. The figures given by the chroniclers differ, but according to some accounts, *it* reduced the population by a third.
Clear:	Fourteenth-century Europe was scarred by war and plague. It is hard to tell which was worse. The figures given by the chroniclers differ, but according to some accounts, *plague alone* reduced the population by a third.

Vague Reference

In general, do not use a pronoun to refer to the entire idea in a previous sentence or clause or to an antecedent that has not been clearly stated.

> VAGUE: Ahmed usually taps his feet, rolls his eyes, and fidgets when he is nervous, *which* annoys his girlfriend.

The pronoun *which* refers vaguely to the entire idea of Ahmed's behavior when he is nervous.

> CLEAR: Ahmed's habit of tapping his feet, rolling his eyes, and fidgeting when he is nervous annoys his girlfriend.
>
> VAGUE: Lou is an excellent mechanic, and she uses *this* to earn money for college.

The pronoun *this* refers vaguely to the idea of Lou's skill as a mechanic.

> CLEAR: Lou is an excellent mechanic, and she uses her skill to earn money for college.
>
> VAGUE: The tourists stared in awe as the great Christmas tree in Rockefeller Center was lit. They listened in rapt attention to the speeches and sang along with the carolers. *It* was something they would tell their friends about back home.
>
> CLEAR: The tourists stared in awe as the great Christmas tree in Rockefeller Center was lit. They listened in rapt attention to the speeches and sang along with the carolers. The spectacle was something they would tell their friends about back home.
>
> VAGUE: Now that her children were away at school, she felt free to pursue her own interests for the first time in years. Perhaps she would get a job. Perhaps she would go back to school. Suddenly she felt alive again. Until this moment, she hadn't realized how badly she had needed *this*.
>
> CLEAR: Now that her children were away at school, she felt free to pursue her own interests for the first time in years. Perhaps she would get a job. Perhaps she would go back to school. Suddenly she felt alive again. Until this moment, she hadn't realized how badly she had needed a change in her life.

Gobbledygook

Gobbledygook, or inflated diction, is stuffy, pretentious, inflated language that often contains an abundance of jargon. It is found in much government, legal, and academic writing, as well as in many other places, and it is sometimes called *governementese* or *legalese*. Avoid gobbledygook, because it obscures meaning and may add a pompous quality to your writing.

NOT:	It behooves all involved in the acquisition and culmination of out-of-the-box organizational options to employ high-end, functional flexibility.
BUT:	Everyone who is helping us think about new human resources should be as smart and as flexible as possible.

Colon

The colon introduces elements that explain, illustrate, or expand the preceding part of the sentence. It calls attention to the word, phrase, clause, or quotation that follows it.

Introducing a Statement, Quotation, or Items in a List

Use a colon when formally introducing a statement or a quotation. Capitalize the first word of a formal statement or a quotation.

> One of the guiding principles of our government may be stated as follows: All people are created equal.
>
> Though Murphy's identity is not known, Murphy's Law seems to be a truth: "If anything can go wrong, it will."

Use a colon when formally introducing a series of items.

> From 1933 through 1981, unsuccessful assassination attempts were made on the lives of the following presidents: Franklin Roosevelt, Harry Truman, Gerald Ford, and Ronald Reagan.

Avoid using a colon after a form of the verb *be*, after a preposition, or between a verb and its object.

NOT:	Three devices the ancient Romans used to tell time were: sundials, water clocks, and sand-filled glasses.

Linking Two Sentences

Use a colon between two independent clauses when the second clause explains or expands the first.

> The personal computer was more than a new machine: it was the start of a communications revolution.
>
> After reading the letter, he did something that surprised me: he laughed.

Introducing an Appositive

An appositive is a clause that provides additional information about the preceding part of the sentence. Use a colon before a formal appositive, including one beginning with a phrase such as *namely, that is, specifically, or in other words.*

In many cases, an em dash (—) would also be appropriate in this situation.

> The author wrote mysteries for one reason and one reason only: to make money.
>
> In 1961, Kennedy made one of the toughest decisions of his presidency: *namely,* to back the invasion at the Bay of Pigs.

Notice that the colon in the third example appears *before* the word *namely.*

In Salutations or Bibliographic Entries

Use a colon after a salutation in a formal letter or speech.

> Dear Dr. Jacoby:
>
> Members of the Board:

Use a colon between the city and the publisher in a bibliographical entry.

> Boston: Allyn & Bacon
>
> Chicago: The University of Chicago Press

Use a colon between a title and its subtitle.

> *Tutankhamen:* The Untold Story
>
> *Nooks and Crannies:* An Unusual Walking Tour Guide to New York City

Commas

The comma functions in a number of ways, including joining sentences with a conjunction, listing items in a series, modifying a noun with more than one adjective, providing introductory material, and nonessential clauses. Commas are frequently misused; you can use the guidelines below to help avoid common mistake.

Joining Sentences

Use a comma between two independent clauses (sentences) joined by a coordinating conjunction—*and, but, for, nor, or, so, yet.*

> Many Caribbean people emigrated to other countries, and 700,000 of these emigrants settled in the United States within a ten-year period.

The comma may be omitted between two short independent clauses.

> *She handed him the note and he read it immediately.*
>
> *Take notes in class and study them.*

Listing Items in a Series

Use commas to separate three or more items—words, phrases, or clauses—in a series.

> **Words in a Series**
>
> A zoo veterinarian is called on to treat such diverse animals as elephants, gorillas, and antelopes.
>
> The sporting goods store carries equipment for skiing, track, hockey, and weight lifting.

Phrases in a Series

The subway carried children going to school, adults going to work, and derelicts going nowhere at all.

The children playing hide-and-seek hid behind boulders, under bushes, or in trees.

Running in the halls, smoking in the bathrooms, and shouting in the classrooms are not allowed.

Clauses in a Series

John Roebling designed the Brooklyn Bridge, hundreds of workers constructed it, and thousands of people crossed it in its first few weeks.

Foster stole the ball, he passed it to Kennedy, and Kennedy made a basket.

The supermarket tabloid proclaimed boldly that the man's character was hateful, that he was guilty, and that he should be punished severely.

Note: The comma before the conjunction with items in a series is often omitted in newspapers and magazines. Most handbooks and technical writers recommend using the comma because it prevents misreading, as in the following cases.

Harry, Anita, and Jayne have left already.

Without a comma before the conjunction, it is possible to read such sentences as directly addressing the first person named.

Harry, Anita and Jayne have left already. (*Someone is addressing Harry and giving him information about Anita and Jayne.*)

The menu listed the following sandwiches: bologna, chicken salad, pastrami, ham, and cheese. (*five sandwiches*)

The menu listed the following sandwiches: bologna, chicken salad, pastrami, ham and cheese. (*four sandwiches*)

Modifying a Noun with More than One Adjective

Use commas between coordinate adjectives that are not joined by *and*.

Coordinate adjectives each modify the noun independently.

> The technical writer was praised for her precise, crisp prose.
>
> The traveler paused before walking into the *deep, dark, mysterious* woods.
>
> The advertisement requested a *cheerful, sensitive, intelligent* woman to serve as governess.

Avoid using a comma between cumulative adjectives. Cumulative adjectives each modify the whole group of words that follow them.

> He gave her a *crystal perfume* bottle.
>
> On top of the stove was a set of *large shiny copper* pots.
>
> She carried an *expensive black leather* briefcase.

How can you distinguish coordinate adjectives from cumulative adjectives? In general, coordinate adjectives would sound natural with the word *and* between them, since each modifies the noun independently.

> The technical writer was praised for her precise and crisp prose.
>
> The traveler paused before walking into the deep and dark and mysterious woods.
>
> The advertisement requested a cheerful and sensitive and intelligent person to serve as governess.

In addition, coordinate adjectives would sound natural with their order changed or reversed.

> The technical writer was praised for her crisp, precise prose.
>
> The traveler paused before walking into the mysterious, dark, deep woods.
>
> The advertisement requested an intelligent, sensitive, cheerful person to serve as governess.

The order of cumulative adjectives cannot be changed. For example, the following sentences make no sense:

> He gave her a perfume crystal bottle.
>
> On top of the stove was a set of copper shiny large pots.

Providing Introductory Material in a Sentence

Use a comma after an introductory word or expression that does not modify the subject of the sentence.

> Why, we didn't realize the telegram was merely a hoax.
>
> Yes, Washington did sleep here.
>
> Well, that restaurant is certainly expensive.
>
> On the other hand, its prices are justified.
>
> By the way, what were you doing last night?

Use a comma after an introductory verbal or verbal phrase.

> Smiling, she greeted us at the door.
>
> While sleeping, Matthew conceived the idea for his next experiment.
>
> To sketch a tree accurately, you must first study it closely.

Use a comma after an introductory series of prepositional phrases.

> Under cover of night, the secret agent slipped across the border.
>
> In this CD from the 1990s, you will find several traditional songs.

Use a comma after an introductory phrase if there is a possibility that the sentence will be misread without it.

> The day before, he had written her a letter.
>
> After the tournament, winners received trophies and certificates.
>
> Without backups, most computers would be susceptible to serious down time.

Use a comma after an introductory adverb clause.

> Although the alligator once faced extinction, its numbers are now increasing dramatically.
>
> If the earth were to undergo another ice age, certain animals would flourish.

Contrasting and Interrogative Elements

Use a comma to set off an element that is being contrasted with what precedes it.

> Mark Graves claims that he writes user manuals for profit, not pleasure.
>
> Birds are warm-blooded animals, unlike reptiles.

Use a comma before a short interrogative element (question) at the end of a sentence.

> I don't know anyone who hasn't see at least on of Hitchcock's films, do you?
>
> The new version of Firefox is great, isn't it?

With Nonessential Clauses

Use commas to set off a nonessential clause. Do not use commas to set off an essential clause. A nonessential appositive clause gives additional information about the noun it refers to but is not essential to identify this noun.

> Duke Ellington, *a famous composer and bandleader,* helped gain acceptance for jazz as a serious musical form.

An essential, or restrictive, appositive identifies the noun it refers to. As its name suggests, it is essential to the meaning of the sentence. In the following sentences, the essential appositives appear in *italics:*

> My friend *George* works in a bookstore.
>
> The word *nice* has undergone many changes in meaning.

Use commas to set off a nonessential adjective clause. Do not use commas to set off an essential adjective clause. A nonessential, or nonrestrictive, adjective clause provides extra information about the noun it modifies but is not essential to identify the noun it modifies. In the following sentences, the nonessential adjective clauses appear in *italics:*

> Norman Borlaug, often referred to as the father of the Green revolution, received his PhD from the University of Minnesota.
>
> Cardiac nursing, which involves many areas of expertise, is a profession on the rise.

An essential, or restrictive, adjective clause limits or identifies the noun it refers to. It is essential to the meaning of the sentence. In the following sentences, the essential adjective clauses appear in *italics:*

> The person *who buys the first ticket* will win a trip to Mexico.
>
> The scientist *whose research is judged the most important* will be given a grant.

Note: The pronoun *that* is used only with essential clauses.

> The car that gets the highest mileage will win the award.

The pronoun *which* may be used with either essential or nonessential clauses. Here is an example of *which* in a nonessential clause.

> The car, which was only available for viewing for one day, got outstanding gas mileage.

Common Mistakes with Commas

The comma is probably the most commonly misused punctuation mark in the English language. Here are some tips for avoiding incorrect comma usage.

Avoid using a comma to separate a subject from its predicate.

> Not: The album that he recorded last year, sold over a million copies.
>
> But: The album that he recorded last year sold over a million copies.

Avoid using a comma to separate a verb from its complement.

NOT:	Did you know, that chimpanzees can communicate through sign language?
BUT:	Did you know that chimpanzees can communicate through sign language?

Avoid using a comma between cumulative adjectives.

NOT:	She declared him to be a handsome, young man.
BUT:	She declared him to be a handsome young man.

Avoid using a comma to separate the two parts of a compound subject, a compound verb, or a compound complement.

NOT:	High ceilings, and cathedral windows are two features I look for in a house.
BUT:	High ceilings and cathedral windows are two features I look for in a house.

Avoid using a comma to separate two dependent clauses joined by *and*.

NOT:	They promised to obey the laws of their new country, and to uphold its principles.
BUT:	They promised to obey the laws of their new country and to uphold its principles.

One way to check for dependent clauses is to look at the part of the sentence after the word *and*. If that part cannot stand on its own, then no comma is needed. If that part is a complete sentence and can stand on its own, then you would use a comma before the *and*.

Avoid using a comma to separate the parts of a comparison.

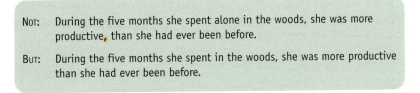

NOT:	During the five months she spent alone in the woods, she was more productive, than she had ever been before.
BUT:	During the five months she spent in the woods, she was more productive than she had ever been before.

Conciseness

Technical writing is most effective when it is concise. Choose words that express ideas precisely, accurately, and crisply. Avoid padding your writing with extra words.

Use Action Verbs

Action verbs give power and precision to your writing, whereas the overuse of the verb *be* weakens your writing. Note how the sentences in the following examples are strengthened when the verb *be* is replaced with an action verb:

> WEAK: Her resume was a long list of outstanding qualifications.
>
> STRONG: Her resume listed many outstanding qualifications.
>
> WEAK: The terms of agreement were no longer in place.
>
> STRONG: The terms of agreement expired.

Use Active Voice

Active voice is clearer read and easier to read and understand, because active voice makes explicit the action and the actor. The general rule of thumb in technical writing is to use active voice whenever possible.

Active voice usually assumes a simple subject, verb, object structure:

> ACTIVE: The field biologist developed a new procedure.

Passive voice involved the verb to be plus a past participle, and the subject is acted upon instead of doing the acting. See also **Use Passive Voice Sparingly.**

> PASSIVE: A new procedure was developed.
>
> PASSIVE: A new procedure was developed by the field biologist.

Avoid Empty Intensifiers

Words such as *incredibly* or *intensely* may add meaning, but often times, they do not. For example, the following sentence is just as effective without the word *incredibly*.

> The project was incredibly interesting.

See also Intensifiers.

Avoid Expletives and Redundancy

Try to avoid the constructions *it is, it was, there is,* and *there was.* Like the passive voice, these constructions are sometimes useful and appropriate, but often they are an unnecessarily wordy way of introducing an idea. Notice how eliminating the unnecessary constructions improves the following sentences:

WORDY:	It is known that there is a need for security on the Internet.
CONCISE:	The Internet needs to become more secure.
WORDY:	It is a fact that the painting is a forgery.
CONCISE:	The painting is a forgery.
WORDY:	There is a need among modern people to gain an understanding of the risks of modern technology.
CONCISE:	People need to understand the risks of modern technology.

Another cause of wordiness is redundant elements, words, or phrases that repeat the idea expressed by the word to which they are attached. For example, the phrase *to the ear* is redundant in the expression *audible to the ear* because *audible* itself means "able to be perceived by the ear." Here is a list of some other common expressions that contain redundant elements:

Redundant	Concise
and etc.	etc.
bibliography of books	bibliography
mandatory requirements	requirements
refer back	refer
tall in height	tall
collaborate together	collaborate

(continued)

Redundant	Concise
visible to the eye	visible
repeat again	repeat
advance forward	advance
negative complaints	complaints
humorous comedy	comedy
close proximity	proximity
expensive in price	expensive
past history	history
continue to remain	remain
component parts	components
free gift	gift

Avoid Noun Strings

Long strings of nouns can confuse the reader, because they lack any action (verb). In general, try not to have more than three nouns in a row.

Not: High-end security system backup procedures

But: Security procedures to back up high-end computers

Eliminating noun strings may result in longer sentences, but the goal is to have a sentence that is easy to understand.

Avoid Nominalizations

Nominalizations are nouns or adjectives that can function equally well as verbs. Often they are easy to spot because of the *-tion* ending. Changing a nominalization to a verb requires you to rewrite the sentence, but the outcome will be a clearer, more direct sentence. In general, technical writers favor verbs over nominalizations, because verbs help readers understand action.

Nominalization	Verb
decision	decide
reaction	react
conclusion	conclude
investigation	investigate
applicability	applicable
difficulty	difficult

Notice the difference in these two sentences.

> Not: We made a decision to initiative an investigation.
>
> But: We decided to investigate.

Use Passive Voice Sparingly

Passive voice involves the verb *to be* plus a past participle, and the subject is acted on instead of doing the acting.

> Passive: A new procedure *was developed*.
>
> Passive: A new procedure *was developed* by the field biologist.

Readers find passive voice difficult to understand. The general rule of thumb in technical writing is to use active voice whenever possible. Yet there are times when you may need to use passive voice:

When the intent is to play down the subject:

> A decision *has been* made.

When the subject is not known:

> A surprise party *was given* on July 18.

When the object, not the subject of the sentence, is more important. This is the approach used in many scientific and technical journals, although even there, styles are changing.

> The data *were examined* and *analyzed*.

Avoid Unnecessary Words and Phrases

One way to achieve conciseness is to eliminate wordy expressions. Notice how each of the following phrases can be changed to a single-word equivalent.

Wordy	Concise
at all times when	whenever
at that point in time	then
at this point in time	now
because of the fact that	because
be of the opinion that	think
bring to a conclusion	conclude
by means of	by
due to the fact that	because
during the time that	while
have a conference	confer
in the event that	if
in spite of the fact that	although
make reference	refer
on a great many occasions	often
prior to this time	before
until such time as	until

You can also make your writing more concise by deleting superfluous words, using exact words, and reducing larger elements to smaller elements. Notice how the following sentence is improved when the author uses these revision strategies:

WORDY: In the month of December in the year 1991, those who were flying in space on board the spaceship that was named *Endeavor* made an attempt to catch hold of and perform a repair job on a satellite that had been disabled.

CONCISE: In December 1991, astronauts aboard the spaceship *Endeavor* attempted to grab and repair a disabled satellite.

The words *the month of* and *in the year* add nothing to the meaning of the sentence; they simply fill up space and can be deleted. The noun phrase *those who were flying in space* can be replaced by one exact noun—*astronauts*. The words *that was named* are also deadwood. The phrase *made an attempt* can be reduced to the more direct *attempted*, *catch hold of* to *grab*, and *perform a repair job on* to *repair*. The clause *that had been disabled* can be reduced to the single word *disabled*.

Avoid the wordiness that comes from overuse of prepositional phrases, the weak verb *be*, and relative pronouns.

Conclusions

Many documents require a formal conclusion. Research reports, for example, conclude with specific results, recommendations, or summaries. The main argument, point, or thesis of the document should be reflected in some way in the conclusion, not by a word-by-word repetition of the thesis, but rather by reaffirming the document's main idea. When writing a conclusion, remember to look back. The final paragraph or final sentence should not contain arguments, perceptions, expansions, or new ideas that do not already exist within the essay. Such ideas belong in the body of the document, where they can receive adequate explanation and support.

Although a conclusion might contain a prediction or suggest a future action, the grounds for such elements should be clearly established in the body of the document. In other words, the conclusion should refer to "what I have written" rather than to "what I can write next."

As with introductions, it is not necessary to come up with the perfect conclusion during the drafting stage. If an idea occurs to you while you are writing, jot it down and save it for later. You can rework it or even replace it with something more effective as your report document develops.

Some introductory strategies—questions, quotations, descriptions—are also useful in bringing an essay to a close. These methods should be combined with a restatement or reinforcement of your main idea. Here are some other strategies that are appropriate for conclusions.

Summary

If your document is long and complex, your readers might appreciate a summary of the major ideas. You want to be careful, however, not to write an ending that sounds forced and simply repeats your introduction. It's a good idea to combine your summary with another strategy, such as a call for action, as is done in this example.

> This report has analyzed current fuel usage patterns across the planet and offered key recommendations for how the largest users of fossil fuels could change within the next 20 years. Yet these recommendations are only the beginning. It will take an incredible effort on the part of all countries to work toward a common vision that will sustain human life as we know it into the coming centuries. In addition to the recommendations called for here, energy analysts should take additional steps to ensure that a global approach is used.

Suggestions

If you are writing an analysis or a persuasive piece, a useful closing strategy involves offering suggestions—possible solutions for problems discussed in the paper.

> This report has analyzed current fuel usage patterns across the planet and offered key recommendations for how the largest users of fossil fuels could change within the next 20 years. Yet these recommendations are only the beginning. It will take an incredible effort on the part of all countries to work toward a common vision that will sustain human life as we know it into the coming centuries. In addition to the recommendations called for here, energy analysts can take these five steps to ensure that the problems discussed in this paper will be addressed worldwide [here, you would list the five steps].

Consequences and Implications

Think about the long-term implications of what you have said in your document. You might want to conclude by suggesting possible benefits or issuing a warning. Most scientific papers conclude by suggesting implications for future research, for example:

> This report has analyzed current fuel usage patterns across the planet and offered key recommendations for how the largest users of fossil fuels could change within the next 20 years. Yet these recommendations are only the beginning. It will take an incredible effort on the part of all countries to work toward a common vision that will sustain human life as we know it into the coming centuries. If we do not take steps now, the situation will be dire. Energy reserves will dry up, and world economies will suffer. Future research on this topic should be conducted in an interdisciplinary manner—by economists, political analysts, chemists, and futurists.

Conjunctions

Conjunctions are words that are used to join other words, phrases, clauses, or sentences. The three types of conjunctions are coordinating conjunctions, correlative conjunctions, and subordinating conjunctions.

Coordinating Conjunctions

A coordinating conjunction joins elements of equal grammatical rank. These elements may be single words, phrases, or independent clauses. The common coordinating conjunctions follow:

and	for	but	nor
or	yet	so	

In the following sentences, the coordinating conjunctions are printed in color, and the elements being joined are printed in italics:

Some enjoy rock music primarily for its lyrics, but *others respect rock more for the beat and sound.*

The flax is then soaked *in tanks, in streams,* or *in pools.*

Conjunctive Adverbs

Words like the following, called conjunctive adverbs, may make clear the connection between independent clauses (clauses that can stand by themselves as sentences), but they cannot—as conjunctions can—join the clauses.

accordingly	hence	otherwise
also	however	still
besides	moreover	therefore
consequently	nevertheless	thus
furthermore		

In the following sentences, the conjunctive adverbs are in color and the independent clauses in italics. Notice that a semicolon precedes a conjunctive adverb that appears between independent clauses.

She wanted to photograph the building in the early morning light; therefore, she got up at dawn on Saturday.

For years the elderly have moved from the north to Florida to retire; however, today many are returning to the north to be near their children.

Correlative Conjunctions

Correlative conjunctions are coordinating conjunctions that are used in pairs. The most common correlative conjunctions follow:

both . . . and   not only . . . but also
either . . . or   whether . . . or
neither . . . nor

In the following sentence, the correlative conjunctions are in color:

Whether you go *or* stay makes no difference to us.

Subordinating Conjunctions

Subordinating conjunctions join clauses that cannot stand by themselves as sentences. They join subordinate, or dependent, clauses to main, or independent, clauses. The following are some common subordinating conjunctions. These words help you make transitions between thoughts.

after	if	than
although, though	in order that	that
as	in that	unless
as if	inasmuch as	until
as long as	now that	when
as much as	once	where
because	provided that	whereas
before	since	wherever
even though	so long as	whether
how	so that	while

Consistency

Readers expect technical prose to be consistent in its use of verbs and sentence structures. This section explains the primary methods of achieving consistency in writing.

Matching Verb with Actor

The verb should match, or agree, with the actor, or subject of the sentence. To agree with a singular or plural verb form, the noun or pronoun in the subject must also show a matching singular or plural form. Sometimes, these situations can be confusing. Here are a few rules to help you choose the appropriate verb form. The agreement of singular or plural subjects with singular or plural verbs requires only one form change in standard dialect verbs. This change occurs in the present tense and with a third-person singular subject. The change is the addition of –s or -es to the basic present tense form.

> The cushion *feels* soft. The tomato *tastes* ripe.
> The goose *flies* south. He *brushes* his hair.

No change in verb form occurs with other singular subjects or with plural subjects.

	Singular	**Plural**
FIRST PERSON:	I *feel*	We *feel*
SECOND PERSON:	You *fly*	You *fly*
THIRD PERSON:	It *tastes*	They *taste*

Except for the verb *do*, the modal auxiliaries (*may, can, will*, and so on) do not add -s or -es in third-person singular present tense. The auxiliary *have* changes form to *has*.

The verb *be* changes to indicate number in both the present tense and the past tense and in both the first person and the third person.

Present Tense		**Past Tense**	
I am	we are	I was	we were
You are	you are	you were	you were
he/she/it is	they are	he/she/it was	they were

Compound Subjects Using *and*

In general, use a plural verb form with a compound subject joined by the word *and*. A compound subject consists of two or more nouns that take the same predicate.

> *Sbatonda and Sal make* films for a living.
>
> *History and biology were* his best subjects.
>
> *Ted and his friends are supporting* Lopez for mayor.

Use a singular verb form with a compound subject joined by *and* if the compound is considered a single unit.

> *Pork and beans is* a popular dish.
>
> The *bow and arrow is* still regarded as a useful weapon.

Use a singular verb form with a compound subject joined by *and* if the parts of the compound refer to the same person or thing.

> My *friend and guest is* the web designer Moose Cramden.
>
> His *pride and joy was* his rebuilt Vespa scooter.

Compound Subject with *or* or *nor*

With a compound subject joined by or or nor or by either . . . or or neither . . . nor, make the verb agree with the subject closer to it.

> The *cat* or *her kittens have* pushed the vase off the table.
>
> Either *the employees* or *their supervisor is* responsible.
>
> Neither the *camera* nor the *lenses were* broken.

Intervening Phrases or Clauses

Make the verb agree with its subject, not with a word in an intervening phrase or clause.

Intervening phrases

> Several *people* in my club *subscribe* to that magazine.
>
> The *books* by that writer *are* very popular.
>
> The *picture* hanging between the windows at the top of the stairs *is* a portrait of the artist's mother.

Phrases introduced by *together with, as well as, in addition to, accompanied by,* and similar expressions do not affect the number of the verb.

> The emerald *bracelet,* as well as her other jewels, *is* in the safe.
>
> The *novel,* together with the plays that she wrote when she was much younger, *establishes* her reputation.
>
> His *wit,* accompanied by his excellent grasp of the facts, *makes* him a sharp interviewer.

Intervening clauses

> The *books* that are in my briefcase *are* about Russian history.
>
> The *people* who came to the concert that was canceled *are receiving* rain checks.
>
> The *medical doctor* who is attending these patients *is* Ellen Okida.

Collective Nouns

A collective noun may take either a singular or a plural verb form.

Usually, a collective noun refers to a group of people or things as a single unit. When this is the case, the collective noun is singular and the verb form should be singular.

> The *army needs* the support of the civilian population.
>
> The *flock is heading* toward the west end of the lake.
>
> The *group is selling* tickets to raise money for charity.

Sometimes a collective noun refers to a group of things or people as individuals. When this is the case, the collective noun is plural and the verb form should be plural.

> The *jury are arguing* among themselves; six believe the defendant is guilty, two think he is innocent, and four are undecided.
>
> The *congregation disagree* about whether to keep the church open during the week.

Some people think that using a plural verb form with a collective noun sounds awkward. Avoid this problem by inserting "the members of" or a similar expression before the collective noun.

> The *members* of the jury *are arguing* among themselves; six believe the defendant is guilty, two think he is innocent, and four are undecided.
>
> The *members* of the town board *disagree* about whether to keep the community center open during the week.

Nouns that Are Plural in Form but Singular in Meaning

Use a singular verb form with nouns plural in form but singular in meaning. The following are some common words that are plural in form but singular in meaning:

checkers	ethics	molasses	pediatrics
civics	mathematics	mumps	physics
economics	measles	news	statistics

> *Checkers is called* draughts in Great Britain.
>
> *Measles is* a contagious childhood disease.
>
> *The news is broadcast* around the clock on some Internet sites.

The words *pants, trousers,* and *scissors* are considered plural and take a plural verb form. However, if they are preceded by the words *pair of,* the verb form is singular, since *pair* is the subject.

> The *scissors need* to be sharpened.
>
> The *pair* of scissors *needs* to be sharpened.
>
> The *pants match* the jacket.
>
> The *pair* of pants *matches* the jacket.

Indefinite Pronoun Subjects

The following indefinite pronouns are considered singular. Use a singular verb form with them.

anybody	either	neither	one
anyone	everybody	nobody	somebody
each	everyone	no one	someone

Neither is willing to go with me.

Everybody is going to vote on Tuesday.

Do not be confused by prepositional phrases that follow the indefinite pronoun. The verb must agree with its subject not with the object of a preposition.

Each of the apartments in the north wing of the building *has* a fireplace.

Either of those methods *is* feasible.

The following indefinite pronouns are considered plural. Use a plural verb form with them.

both	few	many	several

Few are certain enough of their beliefs to take a stand.

Several are riding their bicycles to school.

Both of the paintings *were* sold at the auction.

The following indefinite pronouns may be singular or plural. If the noun to which the pronoun refers is singular, use a singular verb form. If the noun is plural, use a plural verb form.

all	any	enough	more	most	some

> *All* of the *money was* recovered. (singular)
> *All was* recovered. (*singular*)
>
> *All* of these *records are* scratched. (*plural*)
> *All are* scratched. (*plural*)
>
> *Most* of the *cake was* eaten. (*singular*)
> *Most was* eaten. (*singular*)
>
> *Most* of the *guests were* hungry. (*plural*)
> *Most were* hungry. (*plural*)

The indefinite pronoun *none* is considered singular because it means "no one." It takes a singular verb.

> *None* of the books *was* missing.

Relative Pronoun Subjects

A verb whose subject is a relative pronoun should agree with the antecedent of the pronoun.

> The man *who narrates* the film has a raspy voice. (*singular antecedent*)
>
> The radios *that were made* in China are selling well. (*plural antecedent*)
>
> The newspaper, *which was founded* in 1893, is closing. (*singular antecedent*)

The phrase *one of* is worth mentioning. Usually, the relative pronoun that follows this phrase is plural because its antecedent is a plural noun or pronoun. Therefore, the relative pronoun takes a plural verb form.

> The man is one of the hostages *who are* in most danger.
>
> Ralph is one of those *who* never *gain* weight.

However, when the words *the only* come before this phrase, the relative pronoun is singular because its antecedent is *one*. Therefore, it takes a singular verb form.

> This is the only one of Mary's songs *that has been published.*
>
> Mitch is the only one of those men *who is* athletic.

Inverted Sentence Order

Use a verb that agrees with its subject, even when the subject follows the verb.

> Outside the building *were crowds of* spectators.
>
> From the chimneys *rises* thick black *smoke.*
>
> On the wall *are* whiteboards of the schematic drawings.

Do not be confused by sentences beginning with the expletives *there* and *here.* These words are never the subject.

With Titles

Use a singular verb form with a title, even if the title contains plural words.

> *Smoke and Mirrors was* a popular Web site with engineers.
>
> *60 Minutes is* on television tonight.
>
> *Zen and the Art of Motocycle Maintenance tells* the story of a father and son's journey.

With Measurements, Time, or Amounts of Money

Use a singular verb form with a plural noun phrase that names a unit of measurement, a period of time, or an amount of money.

> *Five miles is* too far to walk to school.
>
> *One hundred years is* the usual life span for the crocodile.
>
> *Fifty-three thousand dollars is* a good salary for this job.

Mood

Mood refers to the way in which a verb is used to express a statement, a command, a wish, an assumption, a recommendation, or a condition contrary to fact. Mood contributes to the tone of the sentence.

Indicative mood is used to make a factual statement or to ask a question.

> Bill Gates *lives* in the Seattle area.
>
> *Did* you know that Bill Gates lives in the Seattle area?

Imperative mood is used to express a command or request. In a command, the subject *you* is often not stated.

> *Insert* the DVD into the drive.
>
> Please *close* the door!
>
> *Don't* ever *come* over here.

Subjunctive mood is used to indicate a wish, an assumption, a recommendation, or a condition contrary to fact.

> She wished she *were* in Hawaii right now. (*wish*)
>
> If this *be* true, the validity of the collection is in doubt. (*assumption*)
>
> It is mandatory that he *dress* appropriately. (*recommendation*)

Avoid shifting verbs awkwardly between indicative, imperative, and subjunctive moods, as in the following examples.

INCONSISTENT:	First *brown* the onions in butter. Then you *should add* them to the beef stock.
CONSISTENT:	First *brown* the onions in butter. Then *add* them to the beef stock.
INCONSISTENT:	If I *were* president of this club and he *was* my second in command, things would be very different.
CONSISTENT:	If I *were* president of this club and he *were* my second in command, things would be very different.

Number

Avoid shifting pronouns awkwardly and inconsistently between the singular and the plural. Many shifts of this kind are actually problems with pronoun–antecedent agreement.

INCONSISTENT:	Just before *a person* speaks in public, *they* should do several relaxation exercises.
CONSISTENT:	Just before *a person* speaks in public, *he or she* should do several relaxation exercises.
OR:	Just before speaking in public, a person should do several relaxation exercises.
INCONSISTENT:	*A warthog* may appear ungainly, but *these animals* can run 30 miles an hour.
CONSISTENT:	*A warthog* may appear ungainly, but *this animal* can run 30 miles an hour.
OR:	*Warthogs* may appear ungainly, but *these animals* can run 30 miles an hour.
INCONSISTENT:	*Anyone* who travels to Greece will see many sites about which *they* have read.
CONSISTENT:	*People* who travel to Greece will see many sites about which *they* have read.
OR:	*Travelers* to Greece will see many sites about which *they* have read.

Parallelism

Use the same grammatical form for elements that are part of a series or a compound construction. Sentence elements that have the same grammatical structure are said to be *parallel*.

> The speech was *concise, witty,* and *effective*.
>
> Today's "supermom" is both *a mother* and *an executive*.
>
> He tried to be honest *with himself* as well as *with others*.

When elements that are part of a series or a compound construction do not have the same form, a sentence is said to have **faulty parallelism.**

Repeat articles, prepositions, and the word *to* before the infinitive to make the meaning of a sentence clear.

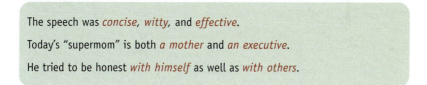

> The audience applauded the guitarist and vocalist.

The preceding sentence is clear if the composer and the lyricist are the same person. It is misleading if they are not the same person. Repeat the article *the* to indicate two people.

> The audience applauded *the* guitarist and *the* vocalist.

UNCLEAR: She was a prominent critic and patron of young artists.

CLEAR: She was *a* prominent critic and *a* patron of young artists.

UNCLEAR: He quickly learned to supervise the maid and cook.

CLEAR: He quickly learned to supervise *the* maid and *the* cook.

CLEAR: He quickly learned *to* supervise the maid and *to* cook.

UNCLEAR: His mother had taught him to shoot and ride a horse.

CLEAR: His mother had taught him *to* shoot and *to* ride a horse.

Place elements joined by a coordinating conjunction in the same grammatical form. Balance a noun with a noun, an adjective with an adjective, a prepositional phrase with a prepositional phrase, and so on.

NOT PARALLEL: The scientific community in general regarded him

        adjective   adjective   noun
         ↓   ↓   ↓

    as *outspoken, eccentric,* and a *rebel.*

PARALLEL: The scientific community in general regarded him

        adjective   adjective   adjective
         ↓   ↓   ↓

    as *outspoken, eccentric,* and *rebellious.*

NOT PARALLEL: In *my department*, the manager is a

    noun   noun   clause
     ↓   ↓   ↓

    wife, a *mother,* and *she plays tennis.*

PARALLEL: In *my department*, the manager is a

    noun   noun   noun
     ↓   ↓   ↓

    wife, a *mother,* and a *tennis player.*

(continued)

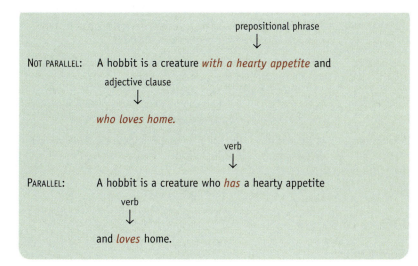

Place elements joined by correlative conjunctions in parallel form.

NOT PARALLEL:	He was not only *her husband* but also *she considered him her friend.*
PARALLEL:	He was not only *her husband* but also *her friend.*
NOT PARALLEL:	Knute Rockne would be either a *science teacher* or *someone who coached football.*
PARALLEL:	Knute Rockne would be either a *science teacher* or *a football coach.*

Point of View

Avoid shifting pronouns awkwardly between the second person and the third person.

All nouns and all indefinite pronouns are in the third person. However, personal pronouns may be in the first person, the second person, or the third person.

	Singular	Plural
FIRST PERSON:	I, me, my, mine	we, us, our, ours
SECOND PERSON:	you, your, yours	you, your, yours
THIRD PERSON:	he, him, his she, her, hers it, its	they, them, their, theirs they, them, their, theirs

INCONSISTENT:	It has been said that unless *you* have a knowledge of history, *a person* is condemned to repeat its mistakes.
CONSISTENT:	It has been said that unless *you* have a knowledge of history, *you* are condemned to repeat its mistakes.
INCONSISTENT:	As *we* read about the emission of greenhouse gases worldwide, *one* becomes appalled by our leaders' lack of vision.
CONSISTENT:	As *we* read about the emission of greenhouse gases worldwide, *we* become appalled by our leaders' lack of vision.

Tone

See **Word Choice**.

Verb Tense

Tense is the time expressed by the form of the verb. The six tenses are simple present, present perfect, simple past, past perfect, simple future, and future perfect. Each of these tenses has a progressive form that indicates continuing action.

	Basic Form	Progressive Form
SIMPLE PRESENT:	compose(s)	is (are) composing
PRESENT PERFECT:	has (have) composed	has (have) been composing
SIMPLE PAST:	composed	was (were) composing
PAST PERFECT:	had composed	had been composing
SIMPLE FUTURE:	will (shall) compose	will (shall) be composing
FUTURE PERFECT:	will (shall) have composed	will (shall) have been composing

Usually, the simple present, the simple past, and the simple future are referred to as the present, the past, and the future tense, respectively.

The time of an action does not always correspond exactly with the name of the tense that is used to write about the action. For example, in special situations, the present tense can be used to write about events that occurred in the past or will occur in the future as well as events that are occurring in the present.

Present Tense

In general, the present tense is used to write about events or conditions that are happening or existing now.

> She *lives* in Austin, Texas.
>
> An accountant *is preparing* our tax returns.
>
> The *are* dissatisfied with their grades.

The present tense is also used to write about natural or scientific laws or timeless truths, events, and habitual action.

> Some bacteria *are* beneficial, but others *cause* disease.
>
> No one *lives* forever.
>
> She *goes* to work every day at eight.

The past tense can also be used to write about events in literature. Whichever tense you choose, be consistent.

The present tense can be used with an adverbial word or phrase to indicate future time. In the following sentences, the adverbs that indicate time are *italicized*:

> This flight *arrives* in Chicago *at 7.30 P.M.*
>
> *Next week* the class *meets* in the conference room.

The verb *do* is used with the present infinitive to create an emphatic form of the present tense.

> You *do know* your facts, but your presentation of them is not always clear.
>
> He certainly *does cover* his topic thoroughly.

Present Perfect Tense

The present perfect tense is used to write about events that occurred at some unspecified time in the past and about events and conditions that began in the past and might still be continuing in the present.

> The usability engineer *has incorporated* theories of psycholinguistics into his testing.
>
> Their new MP3 player *has been selling* well.
>
> The two performers *have donated* the profits from their concert to charity.

Past Tense

The past tense is used to write about events that occurred and conditions that existed at a definite time in the past and do not extend into the present.

> The study *explored* the dolphin's ability to communicate.
>
> The researchers *were studying* the effects of fluoridation on tooth decay.

The word *did* (the past tense of *do*) is used with the present infinitive to create an emphatic form of the past tense.

> In the end he *did vote* against the bill.
>
> Despite opposition, she *did make* her opinions heard.

Past Perfect Tense

The past perfect tense is used to write about a past event or condition that ended before another past event or condition began.

> She voted for passage of the bill because she *had seen* the effects of poverty on the young.
>
> The researches *had tried* several drugs on the microorganism before they found the right one.
>
> He *had been painting* for ten years before he sold his first canvas.

Future Tense

The future tense is used to write about events or conditions that have not yet begun.

> We *shall stay* in London for two weeks.

Future Perfect Tense

The *future perfect tense* is used to write about a future event or condition that will end before another future event or condition begins or before a specified time in the future.

> Before I see him again, the manager *will have read* my feasibility report.
>
> By October, she *will have been singing* with the gospel choir five years.

Some Guidelines for Use of Tense Forms

The following rules will help you to select the appropriate verb form.

Use the past tense form to indicate simple past time.

> NOT: We *seen* him in the library yesterday.
>
> BUT: We *saw* him in the library yesterday.
>
> NOT: His clothing *stunk* from the skunk's spray.
>
> BUT: His clothing *stank* from the skunk's spray.

Use the past participle form with auxiliary verbs *have* and *be*.

> NOT: *Have* you *chose* a major?
>
> BUT: *Have* you *chosen* a major?
>
> NOT: If you don't lock up your bike, it *will be took*.
>
> BUT: If you don't lock up your bike, it *will be taken*.

Use the past participle form with a contraction containing an auxiliary verb.

> NOT: *He's drove* all the way from Miami.
>
> BUT: *He's driven* all the way from Miami.
>
> NOT: *She'd* never *flew* in an airplane before.
>
> BUT: *She'd* never *flown* in an airplane before.

Voice

Voice indicates whether the subject of the clause or sentence performs or receives the action of the verb. In technical writing, pay attention to voice. Technical writers prefer active voice because it gives users a clear sense of who performed the action.

The verb and the clause are in the active voice when the subject performs the action of the verb. (See "Action Verbs" under **Conciseness**.)

> The president *announced* his decision.
>
> The journal *offers* insights into the subject of molecular biology.

When the subject no longer performs the action but is removed or expressed in a phrase, then the verb and the clause are in the passive voice. The passive voice of a verb consists of a form of *be* followed by the past participle of the verb.

> The decision *was announced* by the president.
>
> Insights into the subject of molecular biology *are offered* by the journal.

Many sentences written in the passive voice, like the two preceding examples, contain a phrase beginning with the word *by*. This phrase usually tells who or what actually performed the action, if the phrase has not been omitted, as in

> The decision was announced. [*by the president* is eliminated]

Avoid shifting verbs awkwardly between the active voice and the passive voice.

INCONSISTENT:	Andre-Jacques Garnerin *made* the first parachute jump, and the first aerial photographs *were taken* by Samuel Archer King and William Black.
CONSISTENT:	André-Jacques Garnerin *made* the first parachute jump, and Samual Archer King and William Black *took* the first aerial photographs.
INCONSISTENT:	A group of ants *is called* a colony, but you *refer* to a group of bees as a swarm.
CONSISTENT:	A group of ants *is called* a colony, but a group of bees *is referred* to as a swarm.

See also "Use Passive Voice Sparingly" under **Conciseness.**

Dash

The **dash** is less formal than the colon. It is used to give emphasis or clarity to extra information in a sentence. Dashes should be used sparingly; too much emphasis is not useful to the reader. When typing, produce a dash by two hyphens without a space before, after, or between them (—). Dashes can also be made using shortcut keys in most word processing software.

Dashes with an Introductory Series

Use a dash to separate an introductory series from its summarizing clause.

> His own party, the opposition, and the public—all were astounded by his resignation.
>
> Hawaii, Mexico, and Brazil—these were her favorite travel destinations.

Dashes with Parenthetical Elements

Use dashes to set off a parenthetical element that you wish to emphasize.

> The castle was surrounded by a moat and contained—I found this astounding—an actual dungeon.
>
> On his first day as a volunteer, he fought a fire in—of all places—the firehouse.

Use dashes to clarify a parenthetical element that contains commas.

> Of our first five presidents, four—George Washington, Thomas Jefferson, James Madison, and James Monroe—came from Virginia.
>
> The first recorded Olympic Game—which, you will be surprised to know, this reporter did not see—were held in 776 B.C.

Dashes with Terminal Elements

Use a dash to introduce informally a terminal element that explains or illustrates the information in the main part of the sentence.

> They pledged to prevent what seemed inevitable—war.
>
> He battled his worst enemy—himself.
>
> Willie little appreciated her greatest attribute—her sense of humor.

Use a dash to introduce informally a terminal element that is a break in thought or a shift in tone.

> No one loves a gossip—except another gossip.

Editing Technical Documents

Most technical documents undergo an extensive cycle of editing. Work is edited in different stages, usually starting with content checking by technical experts at an early stage, followed by formal copyediting and proofreading. While some of this work is done with paper documents, in today's networked, global workplace, most editing is done with electronic systems.

Copyediting

Copyediting is the process of reviewing a manuscript before it is set in pages. Copyediting involves reviewing the material for content, formatting, consistency, parallelism, and grammatical errors. Once the copyediting phase is completed, changes are made and the manuscript is formatted.

Copyediting takes place at a higher, less microscopic level than does proofreading. Although copy editors may catch small punctuation and other errors, they must look at the document comprehensively as well.

Items to check for during the copyediting stage, especially with technical documents, include the following:

- Parallelism with headings, titles, sections, and other similar parts
- Consistency in terminology usage
- Conformance of citations and other material to the style guide that is being used (MLA, APA, IEEE; see Part 6 for more on style guides)
- Functionality of web addresses that are used in the document
- Content that is out of place or does not make sense.

See also **Proofreading Technical Documents.**

Electronic Editing

See also **Tracking Changes and Version Control.**

Ellipsis Points

Ellipsis points are equally spaced dots, or periods. They indicate that part of a quotation has been omitted.

Use three ellipsis points within a quotation to indicate that part of the quotation has been left out, or omitted. To distinguish between your ellipsis points and the spaced periods that sometimes appear in written works, put square brackets around the ellipsis dots that you add.

Partial quotations do not need ellipsis points at beginning and end.

> NOT: The engineer declared that most modern bridges have ". . . structural integrity under the most extreme conditions . . ."
>
> BUT: The engineer declared that most modern bridges have "structural integrity under the most extreme conditions."

Use ellipsis points to indicate when you have left material out of the original.

> The engineer declared that most modern bridges have "structural integrity under the most extreme conditions . . . and are generally safe in most climates."

Euphemism

A euphemism is a term that is used in place of one that could be considered offensive or rude. Some euphemisms are used in ways that are thoughtful; for example, it can be more sensitive to use the phrase "he passed away" than it is to say "he died." Other euphemisms, however, sound insincere and make it appear that the speaker or writer is attempting to avoid the issue. For example, if it is clear what the source of a problem is, but this problem is referred to as an "unexplainable technical difficulty," users will become suspicious.

Exclamation Point

Use an exclamation point (!) at the end of a sentence, word, or phrase that you wish to be read with emphasis, surprise, or strong emotion.

> Don't give up!
>
> My new iPod is fantastic!
>
> Impossible!
>
> What a terrible time!

Avoid using the exclamation point to express sarcasm, and try not to overuse the exclamation point; too many can be distracting and ineffective in formal writing. Even in email, too many exclamation points make all the sentences take on equal weight. The more you use, the less effective each one will be.

Hyphen

Hyphens are used to join words. Hyphens are also referred to as dashes but dashes are made up of two hyphen characters (see also **Dashes**).

Joining Compound Nouns

Use a dictionary to determine whether to spell a compound noun with a hyphen. Relatively few compound nouns are hyphenated; most are written either solid (as one word) or open (as two or more separate words). The only kinds of compound nouns that are usually hyphenated are those made up of two equally important nouns and those made up of three or more words.

> philosopher-king   city-state   man-hour
>
> mother-in-law   free-for-all   jack-of-all-trades

Joining Compound Adjectives

Hyphenate two or more words that serve as a single adjective preceding a noun.

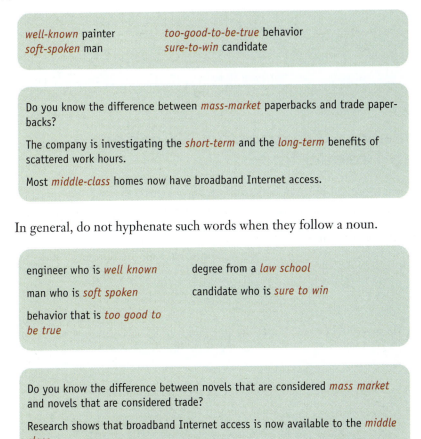

well-known painter   *too-good-to-be-true* behavior
soft-spoken man   *sure-to-win* candidate

Do you know the difference between *mass-market* paperbacks and trade paperbacks?

The company is investigating the *short-term* and the *long-term* benefits of scattered work hours.

Most *middle-class* homes now have broadband Internet access.

In general, do not hyphenate such words when they follow a noun.

engineer who is *well known*   degree from a *law school*

man who is *soft spoken*   candidate who is *sure to win*

behavior that is *too good to be true*

Do you know the difference between novels that are considered *mass market* and novels that are considered trade?

Research shows that broadband Internet access is now available to the *middle class.*

The benefits of scattered work hours are both *short term* and *long term.*

Do not hyphenate two or more words that precede a noun when the first of these words is an adverb ending in *-ly*.

Critics attacked the company's *rapidly expanding* budget.

When the famous rock group first came to this city, the police were called in to restrain the crowd of *wildly screaming* fans.

The government's *widely criticized* policies are the subject of the debate on television today.

Use a hanging hyphen after the first part of a hyphenated compound adjective used in a series, where the second part of the compound adjective is implied but omitted.

> both *paid-* and complimentary-ticket holders
>
> both *short-* and long-term disability
>
> all *first-*, *second-*, and third-year students

Creating Compound Numbers and Fractions

Hyphenate spelled-out numbers from *twenty-one* through *ninety-nine* and spelled-out fractions used as adjectives. In spelled-out numbers larger than *ninety-nine*, do not use a hyphen before or after *hundred*, *thousand*, and so forth.

> *Two hundred fifty-seven* people were killed in the fire.
>
> The installation of computers has effected a *one-third* increase in productivity and an expected *three-quarter* growth in profits.
>
> The novel had *thirty-two* chapters in its first version, but the revised version has *forty-one*.

Do not hyphenate spelled-out fractions used as nouns.

> About *one half* of Serbia is covered with mountains.
>
> Only *three eighths* of the adults in this community voted in the last election.

Applying Some Prefixes and Suffixes

In general, do not use a hyphen between a prefix and its root or a suffix and its root. However, there are a few exceptions. Use a hyphen between a prefix and its root to avoid ambiguity.

> the *re-creation* of the world *but* tennis as *recreation*
>
> to *re-count* money *but* to *recount* an event
>
> a *co-op* apartment *but* a chicken *coop*

Dividing a Word at the End of a Line

Writers no longer need to worry about dividing words at the ends of lines. If a word will not fit on a line, just leave the line short and begin the word on the next line. The word-wrap feature of most word processing and other text formatting programs will take care of this procedure automatically. If you choose to divide a word, consult a dictionary about where the break should occur.

Idioms

Idioms are phrases that if taken literally would make no sense but have over time come to take on certain meanings. For instance, phrases such as "cry over spilled milk"; "kick the bucket"; "when it rains it pours" are commonly understood in American English. Idioms can be difficult for non-native speakers to understand, because every language has idiomatic expressions that one learns while growing up but that do not translate easily. Be careful about using idioms in precise technical and scientific writing, especially in material that will be accessed by international audiences.

Inclusive Language

It is important to use language that does not exclude people. Such language can easily offend or alienate an audience and leave you open to charges of unfairness and bias. There are many ways you can avoid exclusionary language.

Avoid Jargon Except When Appropriate to the Audience and Situation

Jargon is technical and other shared language used in a particular field to communicate within a particular research field or group. Jargon can be a useful way to communicate to this audience, because you don't need to bother explaining concepts or abbreviations if everyone already knows them. The problem with jargon, however, is that people outside the group have trouble understanding it. Language aimed at people with specific technical or professional knowledge is only appropriate when you are sure your audience will understand the meaning. If you are not sure, define the term or create a glossary. See also **Voice.**

Avoid Sexist Language

Sexist language is the use of words that inappropriately call attention to gender. Such language tends to stereotype individuals based on gender and can be demeaning. Using the word *mankind* for instance, when you are referring to all people, discriminates against women by excluding them. Substituting a gender-neutral term such as all people, humanity, or humankind eliminates the sexist language; it is fair to both sexes.

Technical writers strive to be accurate as possible. Avoiding sexist language enables you to fulfill this goal. But sexist language, or any biased language, poses a bigger concern than simply affecting a writer's accuracy. Words have incredible power; the images they create last a lifetime. Special care should be taken to choose the appropriate words that will enable you to communicate effectively and to show consideration for another's point of view. The following guidelines will help you recognize sexist lanauge and eliminate it from your writing.

When discussing all people (both sexes), avoid using the word *man* or nouns ending in *-man*. Replace any masculine-marked words in your writing with gender-free alternatives such as these:

Masculine-marked word	Alternatives
fellow man	other people, humans
forefathers	ancestors
kinsman	relative
layman	ordinary person, nonspecialist
man	person, individual
(to) man	to care for, to work in, to staff
spokesman	speaker, representative
statesman	leader
workmanship	skilled work, quality job

Converting the suffix *-man* to *-person* often works, as in *chairperson* or *salesperson*. However, this practice can be awkward, as in *cave person* or *committee person.* Using sex-free solutions such as *cave dweller* or *committee member* is more effective. Similarly, avoid using feminine-marked terms such as *schoolgirl* or *mother tongue* when *student* or *native language* would work just as effectively.

Rewrite to avoid stereotyping jobs and social roles by gender. Replace the words *salesman, chairman, mailman,* and *businessman* with the gender-free terms *sales representative, chair, mail carrier,* and *business executive.* Similarly, use such terms as *housekeeper* instead of *cleaning woman, homemaker* instead of *housewife,* and *flight attendant* instead of *stewardess* or *steward.*

Watch for the use of language that expresses traditional and outdated assumptions about male and female occupations and roles.

Consider these sentences:

> NOT: A doctor uses his skill and physical strength in the operating room while a nurse uses her compassion at the bedside.
>
> BUT: Doctors and nurses use their skill, physical strength, and compassion in the operating room and at the bedside.
>
> NOT: When cleaning their homes, women should be thorough.
>
> BUT: It's important to be thorough when cleaning house.

Be consistent when naming individuals and identifying their occupations. Failing to do so implies a real or imagined value judgment about the person, the occupation, or both.

> NOT: The corporate president, John Martinez, and his lovely wife Linda, a lawyer and mother of two, attended the gala.
>
> BUT: The corporate president, John Martinez, and his wife, the lawyer Linda Ostacco, attended the gala.

Also, it is appropriate in global workplaces to identify a married woman by her own name instead of her husband's name. When writing the names of male or female authors, artists, composers, and so on, write the full name the first time it appears. Subsequent references should be made by last name only.

Avoid Other Biased Forms of Language

Biased language directed at a group, race, religion, or nationality should be avoided in writing. Also take care to avoid assigning stereotypical physical or behavioral characteristics to members of a particular group. Pay special attention to the language you choose. Keep these guidelines in mind:

■ Mention race and cultural heritage only if there is an important reason for doing so. For instance, an article for the National Society of Black Engineers (nsbe.org) might be focused on an important accomplishment of an African American chemical engineer.

- Use terminology preferred by the people you are writing about. For instance, most gay people prefer the term *gay* to the term *homosexual*.
- Consider using precise terms instead of general ones. For instance, people over age 65 don't want to be called elderly, old, or even senior citizens. Just say "people over age 65."
- Use terms that are broad enough to be inclusive. For instance, if you are working on a program that reads text from a computer screen, this program is not only for people who are *legally blind* but also for those who are *visually impaired*.

Also, remember that we live and work in a highly connected, global environment. Most schools and organizations, especially science and technology companies, are international in scope. Do not assume that all of your readers have the same geographic or cultural background. Use language that is reflective of the international reach of today's information.

Interjections

Interjections are words or phrases that add impact or emotion to a sentence. They include *oh, well, hey, eh, um*, and *ah*. These words are generally not used in technical writing, as they add nothing that is specific or technical and only get in the way of information.

Intensifiers

Intensifiers are adverbs that can be used to expand on the meaning of an adverb or adjective. Use intensifiers carefully and sparingly in technical writing and only as needed to help clarify a point.

With adjective

The culture in the petrie dish was *green*.

The culture in the petrie dish was *very green*.

With adverb

The solar car moved *quickly*.

The solar car moved *extremely* quickly.

Introductions

In most documents the introduction is usually restricted to the first section or paragraph. In technical reports, scientific papers, and professional publications, an introduction might require several paragraphs. The introduction is important because it shapes the reader's first impression and often serves as a deciding factor in whether the reader goes beyond it. The introduction sets the tone, establishes the writer's relation to the reader, and indicates the direction the document will take. Avoid attempting to perfect the introduction in your first draft; many writers actually write the introduction last, once the key themes and findings in the document are clear. Refine your introduction and main point as you work through multiple drafts.

Some of the same strategies for a conclusion—questions, quotations, descriptions—are also useful to open up a document. These methods should be combined with a restatement or reinforcement of your main idea. In technical writing, the old rule of thumb for introductions is "Tell them what you're going to tell them." Here are some other strategies that are appropriate for introductions.

Catch the reader's interest:

> What will it be like to live in a world that no longer uses fossil fuels?

Announce the topic:

> This report will examine ways in which the major users of fossil fuels could make such a change.

State clearly the main points that you plan to make:

> Part 1 of this report analyzes current fuel usage patterns across the planet. Part 2 makes recommendations for how the largest users of fossil fuels could change within the next 20 years.

If appropriate, state your findings:

> In general, our analysis found that a plan that includes applying not only political will but also significant scientific and government resources could ensure that a country such as the United States has energy independence.

Italics and Underlining

Italics are used to highlight or set apart certain titles, words, or phrases. Years ago, when formatting with italics was only available to those who worked in the publishing industry, underlining was considered a comparable marking for typewritten papers. (The same is true today; however, very few people use typewriters.) Underlining is now used mainly to indicate a hyperlink on a web site or within a document.

Using Italics for Titles

> *The Light in the Forest* (book)
>
> *The Wiz* (musical)
>
> the *Mayflower* (ship)

Be careful to italicize only the exact title or name. Do not italicize words added to complete the meaning of the title.

> the *Atlantic* magazine (The word *magazine* is not part of the name.)
>
> the London *Times* or the *Times* of London (*London* is not part of the name.)

Be careful to italicize all the words that make up the title

> *The Decline and Fall of the Roman Empire* (*The* is part of the title.)
>
> *Standard & Poor's New Issue Investor* (*Standard & Poor's* is part of the document's title.)

Using Italics for Words

Italicize foreign words or phrases that are not commonly used in English. In general, a word or phrase need not be italicized if it is listed in a standard English dictionary. For example, the Spanish word "siesta," the French phrase "coup de grace," and the Latin phrase "ad infinitum," are now considered part of English and are not underlined.

> Three commonly confused German words are *Kinder, Kirche,* and *Kliche.*
>
> *Chacun à son goût* proved a difficult principle to apply in this case.
>
> The great English public schools attempted to follow the ideal of *mens sana in corpore sano.*

Italicize letters, words, or phrases being named.

> How many i's are in *Mississippi?*
>
> His life demonstrates the meaning of the word *waste.*
>
> What is the derivation of the phrase *on the ball?*

Using Underlining for Hyperlinks

Most word processing, web design, and other software programs automatically underline any text combination they recognize as a hyperlink

> www. google.com
>
> http://www.noaa.gov/

Users have come to expect hyperlinks to appear as underlined, both in printed documents and in electronic ones.

Nouns

A noun is a word that names. Nouns may name persons, places, objects, events, or ideas. In technical writing, nouns must be used carefully and with precision, so that the correct word is used to identify and explain the concept. In addition, use nouns consistently. Don't confuse and anger readers by calling something a "switch" in one part of the instructions and a "lever" in another. Even if you must repeat the same noun many times, consistency is better than creativity in technical writing.

PERSONS:	Darwin	engineers	women
PLACES:	Pittsburgh	prairie	suburbs
OBJECTS:	reindeer	Bunsen burner	television
EVENTS, IDEAS, ACTIVITIES:	meeting	freedom	frustration fire

Forms of Nouns

Nouns can show number. Most nouns can be either singular or plural.

SINGULAR:	computer	success	criterion	women	sheep
PLURAL:	computers	successes	criteria	women	sheep

Nouns also can change form to show possession (and in this form are considered to show a possessive case); the *cat's* feet (the cat has feet); a *singer's* voice (a singer has a voice).

A noun is the only part of speech that can be preceded by an article or determiner—*a, an,* or *the.* A generalized noun naming something countable can be preceded by an indefinite article—*a* or *an* (indicating "one"): *a* cat; *an* animal: *a* woman. A noun that names something quite specific is often preceded by a definite article—the word *the: the* cat; *the* animal; *the* woman. Nouns also identify something very specific when preceded by certain limiting or determining words such as *this, that, these, those* (*this* cat, *that* animal), or when preceded by indefinite determining words like *first, second, last,* and by relative pronouns like *which, that, whose.*

Kinds of Nouns

Nouns can be classified in various ways.

- Proper nouns name unique and particular persons, places, objects, or ideas. Proper nouns are capitalized: *Einstein, Honda, Martin Luther King Jr.*
- Common nouns name people, places, objects, and ideas in general, not in particular. Common nouns are not capitalized: *motorcycle, preacher*.
- Concrete nouns name generalized things that can be seen, touched, heard, smelled, tasted, or felt: *horizon, marble, music*.
- Abstract nouns name concepts, ideas, beliefs, and qualities. Unlike concrete nouns, abstract nouns name things that cannot be perceived by the five senses: *love, justice, creativity*.
- Count nouns name people or things that are considered countable in English: *eggs, rats, grains*.
- Mass nouns (or noncount nouns) name what is not considered countable: *rice, water, air*.
- Collective nouns are singular in form but plural in meaning: *faculty, tribe, family*.
- Compound nouns consist of more than one word. Some compound nouns are written as one word, some are hyphenated, and some are written as two or more separate words. (When in doubt, check your dictionary.)

bedroom	tomato plant	cathode-ray tube
heartland	fire insurance	father-in-law

Forming Noun Plurals

There are several standard rules for making nouns plural.

For most words, simply add *s*.

pot/pots	lamp/lamps
table/tables	magazine/magazines

For nouns ending in *s, ch, sh, x,* or *z,* add *es*.

toss/tosses	church/churches	dish/dishes
box/boxes	waltz/waltzes	

For nouns ending in *y*, change the *y* to *i* and add *es* if the *y* is preceded by a consonant.

> jelly/jellies   quality/qualities
> theory/theories   enemy/enemies

Add only *s* to proper names ending in *y*.

> Mary/Marys   Germany/Germanys   McNulty/McNultys

And add only *s* if the *y* is preceded by a vowel.

> monkey/monkeys   attorney/attorneys
> display/displays   journey/journeys

For nouns ending in *o*, add *s* when the *o* is preceded by a vowel.

> radio/radios   video/videos   zoo/zoos
> patio/patios   cameo/cameos

But add *es* when the *o* is preceded by a consonant.

> hero/heroes   potato/potatoes
> echo/echoes   mosquito/mosquitoes

Exceptions are *piano/pianos, solo/solos, burro/burros, and memo/memos.*
Some nouns have an irregular plural.

> woman/women   man/men   child/children
> mouse/mice   tooth/teeth

And some nouns have the same form for both singular and plural.

> deer, trout, species, fish, series, moose, sheep

Numbers and Dates

See also **abbreviations** for informations about dates.

Spelled-out Numbers versus Numerals in a Sentence

Use numerals for quantities that cannot be written as one or two words.

> During its first year, the book sold only *678* copies.
>
> Last Saturday this shop sold *1,059* doughnuts.

However, avoid beginning a sentence with a numeral. Recast sentence if possible, or spell out the number.

> NOT: *Nine hundred seventy-six* people bought tickets for the concert, but only 341 attended.
>
> BUT: Although 976 people bought tickets for the concert, only 341 attended.

Spell out all numbers that can be written as one or two words and that modify a noun.

> She presented a list of *sixteen* data points to analyze.
>
> The gestation period for a rabbit is about *thirty-one* days.
>
> We need *one hundred* squares to make this quilt.

Use numerals for decimals or fractions.

> We had *2½* inches of rainfall last month.
>
> What do they mean when they claim that the average family has *2.3* children?

Scientific and technical text is likely to use decimals (2.5" of rain).

When one number immediately follows another, spell out the first number and use a numeral for the second number.

> He ran in *two 50*-meter races.
>
> We have *three 6-foot* ladders in the garage.

For conventional identifiers use numerals for addresses.

> *702 West 74th Street*   *1616 South Street*

However, it is acceptable to spell out the name of a numbered street in an address.

> *417 Eleventh Avenue*   *210 East Seventh Street*

Use numerals to identify pages, percentages, degrees, and amounts of money with the dollar ($) symbol or cent (¢) symbol for U.S. currency. Consult a style manual if you are unfamiliar with the international symbols for various currency.

> The survey found that *70.2 percent* of users surveyed preferred the toolbar on the right side of the screen.
>
> An acute angle is an angle under *90°*.
>
> The computer costs *$1,667.99*.
>
> The estimated cost is *$1 million*.

Use numerals for dates and for hours expressed with *a.m.* (AM) or *p.m.* (PM).

> At *6:07* AM the snow began to fall.
>
> The First International Peace Conference, held at The Hague, began on May *18, 1899*.

Use numerals with units of measurement. However, simple numbers may be spelled out: *six feet*.

> The course is *127* kilometers.
>
> The room is *11'7" 3 13'4"*.
>
> The tree is *6* feet from the garage door.

Use numerals with quantities in a series.

> A grizzly bear can run at a speed of *30* miles per hour; an elephant, *25*; a chicken *9*; but a tortoise, only *0.17*.
>
> The commercial traveler logged his sales for his first five days on the job: *7, 18, 23, 4, 19*.

Use numerals recorded for identification purposes.

> His Social Security number is *142-45-1983*.
>
> For service call the following number: *(800) 415-3333*.
>
> Flight *465* has been canceled.

Paragraphs

A paragraph is a group of sentences that develops an idea about a topic. The word *paragraph* comes from an ancient Greek word referring to the short horizontal line that the Greeks placed beneath the start of a line of prose in manuscripts to indicate a break in thought or a change in speaker. This convention of marking the places in a written work where the sense or the speaker changed was followed by medieval monks, who used a red or blue symbol much like the modern paragraph symbol (¶) in their manuscripts. Today, we indicate such a change in thought by indenting the first line of each new paragraph or adding an extra return (line) after each paragraph.

Although a paragraph is usually self-contained, at the same time it is usually part of a larger work, such as a scientific report or user manual, and depends on the paragraphs before and after it. Each paragraph develops its own point (usually referred to as the controlling idea), which is usually expressed in a topic sentence. Yet for an essay, research paper, or report to function as a whole document, paragraphs must flow well and must work together.

Introductory Paragraph

See **Introductions**.

Topic Sentence and Paragraph Unity

Unity in a paragraph results when all the sentences in the paragraph relate to and develop the controlling idea. In other words, no sentences digress, or go off track. Unity evolves from the use of a topic sentence and relevant support.

The Topic Sentence

A paragraph develops a controlling, or main, idea, which is stated in a topic sentence. Functioning in a paragraph as a thesis statement functions in an essay, a topic sentence establishes the direction for the paragraph, with all the other sentences in the paragraph supporting and developing it. A topic sentence usually appears at the beginning of a paragraph, either as the first or second sentence.

> Anyone majoring in bioinformatics knows how challenging it can be to combine intense study in both biology and computer science. Even in the first year of the program, coursework involves numerous labs, mathematics classes, and programming. The amount of homework can be daunting, and for students with poor study habits, the challenge is even greater. Furthermore, the writing courses, focused as they are on scientific and technical writing, add to the rigor of this undergraduate major.

The first sentence establishes the idea that majoring in bioinformatics is challenging, and the rest of the paragraph provides evidence (coursework, homework, writing courses) to support this claim. The paragraph can be said to have unity because the ideas flow from the topic sentence.

Support your controlling idea with specific information appropriate for the particular audience and the purpose of the report or document. In some cases, specific data, trends, facts, and quotations help support your controlling idea in a way that will be most accurate and convincing to readers.

Sometimes the topic sentence appears in the form of a question. This approach can be useful as a way to engage readers and get them to read further.

When this approach is taken, the rest of the paragraph is usually set up to answer the question.

> Why is there a sudden interest in biofuels when the United States has had ample supplies of corn and other bio-based materials for years now? A brief analysis of economic and political trends yields three possible answers: the wars being fought in the Middle East are a major incentive to reduce reliance on foreign oil; the prices of oil are hard for the United States to control; the world will run out of oil soon.

The Ending Sentence

In order to make paragraphs flow, the last sentence of a paragraph can be used to provide a link to what is coming in the new paragraph. This approach provides what is called cohesion: it allows paragraphs to "stick together" and flow as a logical sequence of ideas.

> Anyone majoring in bioinformatics knows how challenging it can be to combine intense study in both biology and computer science. Even in the first year of the program, coursework involves numerous labs, mathematics classes, and programming. The amount of homework can be daunting, and for students with poor study habits, the challenge is even greater. Furthermore, the writing courses, focused as they are on scientific and technical writing, add to the rigor of this undergraduate major. Yet along with these challenges, students majoring in bioinformatics stand to reap many rewards as well.

As a reader, you would naturally expect the next paragraph to discuss these rewards.

Organizational Strategies

The two most common strategies for organizing a paragraph are general to specific or specific to general. In technical writing, the rule of thumb is that general to specific is easier for readers to follow, because the main point is made first, followed by supporting ideas or evidence. This approach follows the old "tell them what you're going to tell them" axiom of technical writing.

General to Specific/Specific to General

Use this pattern when you want readers to get the main point quickly and when the main point is something that is either very factual or something that readers probably already agree with.

For instance, in this paragraph, part of a longer memo to a building supervisor, the person writing the memo assumes that the supervisor won't need much convincing. The writer gets right to the point.

> The steam heat used for our building needs to be adjusted so that every office receives the same amount of heat. Currently, heat supply is uneven and causes employees to complain, turn the thermostat up and down, and go home early. Correcting this problem is quite simple, according to Facilities Management, and should take less than a week. This memo will outline the steps we have taken and the costs that are remaining. We are hopeful that your office will be able to cover these additional costs.

In this next version, the writer takes the specific to general approach, not reaching the main point until the last sentence. In this case, we can assume that the writer knew that the reader would not be initially receptive to the idea and so needed some "warming up."

> Employees at MidRange Companies enjoy their jobs and want to contribute as much as possible. Our management team works hard to provide an open, comfortable work environment, because we know that if the physical environment is bad, employees will have difficulty performing their jobs. We face a serious workplace challenge with our current steam heat system. Currently, heat supply is uneven and causes employees to complain, turn the thermostat up and down, and go home early. We feel that the steam heat used for our building needs to be adjusted so that every office receives the same amount of heat.

Time Order

Arrange your paragraph in time order if you are describing a sequence of events. Transition words such as *first, second, next, then,* help move the reader through the sequence of events in a logical and flowing manner.

> The human eye is a complex and most impressive organ. The way you end up seeing is through a process by which the eye takes in light and sends information to the brain. First, light enters the eye through the cornea, which then bends the light rays and sends them through the pupil. Second, the eye's lens next focuses the light onto the back of the eye, or retina. Third, the retina contains photoreceptor cells, which translate the light into electrical impulses. Finally, these impulses are sent through the optic nerve through the brain.

Description

Many times a descriptive approach is what is called for. This approach can be important in technical writing when the primary objective is to describe an object or device. You can use an opening sentence to set up the structure of what is to follow.

> The wireless mouse consists of three parts: the main body, the track ball, and the battery section. The main body is 4 inches long and shaped in an oblong manner. The track ball is located directly in the middle of the mouse. Finally, the battery section is located underneath the mouse.

In technical writing, descriptive paragraphs often accompany a diagram or other visual.

Definition

Develop a paragraph through definition when you want to clarify how you are using a term, to assign a particular meaning to a word, or to discuss an abstract concept from a special or unusual point of view. If you think your reader will not know or understand how you are using a term, be safe and define it; however, if appropriate, include some statement about why you are taking the time to define the word or concept.

> It is important to stop and define what we mean by *cat* in this discussion. This particular study of wild cat behavior did not include all wild cats. It was strictly limited to the genus *Lynx*, and although we had hoped to include all four lynx species (*canadensis, lynx, pardinus, rufus*), 80 percent of the cats we studied were *Lynx canadensis*.
>
> In this report, I use the phrase *fuel efficiency*. Efficiency in this study was measured using the carbon balance method.

Classification

Develop your paragraph through classification when you wish to group information into types or classes to help explain patterns. In this example, detail is left out so you can see the general pattern of how such a paragraph would develop.

> Computer programming languages can be classified in a variety of ways but are generally grouped as object oriented, applied, or functional. Object oriented languages include . . . . Applied languages include . . . . Functional languages include . . . . Additionally, there are some "stray" languages that do not classify easily. These languages include . . . .

Comparison and Contrast

Develop your paragraph through comparison and contrast when you want to show the similarities or differences between two or more things.

> There are many similarities but also many differences between the domestic house cat (*Felis silvestris catus*) and the many other wild cats. All cats are similar in that they are true carnivores, the most carnivorous of all mammals. Yet there are many differences between *f. silvestris catus* and its many cousins. One of the most interesting among these differences is not at the behavioral level but at a level only seen by veterinary researchers: the genetic makeup of the retro viruses unique to each species.

Analogy

Develop your paragraph through analogy when you wish to use information that readers will already be comfortable with to introduce new information. Analogy is a powerful tool in technical and scientific writing: often, highly technical concepts are not accessible to lay audiences, but a good analogy can help people understand. Analogies are often used in medical writing when the goal is to convey medical information to patients. Analogies can be very effective, but if they are stretched too far, they can create problems, too. For instance, the heart is a pump, but with its four chambers and its functioning through electrical stimulations, the comparison with a car's gas pump will only be so effective. Sometimes an analogy is a good way to get the paragraph started.

> Your heart is a pump, and just like the pump that moves gasoline from your car's gas tank to its engine, your heart moves blood throughout your body. Like any pump, your heart needs to be maintained. What you eat, how you exercise, and how much stress you have in your life can all have an impact on the heart's ability to do its job. Just as you would not put bad gasoline into your car, you should not put fatty foods into your body.

Example

Develop your paragraph by example when you want to make your ideas specific by supporting them with evidence. Examples are used frequently in technical writing.

> Documents that sound perfectly fine to native speakers of English can contain words and phrases that cause major problems for translation. *For example,* idiomatic phrases such as "when it rains, it pours," has no literal meaning and could cause difficulty if it is part of a set of instructions or other technical document that is being translated from English into other languages.

Process

Develop your paragraph through an explanation of a process when you wish to show how something is done or how something works. This type of paragraph is common in technical writing, often in the form of instructions. Process descriptions can be enhanced by the use of a diagram or other visual.

> Full spectrum lighting works by reproducing the sun's natural light spectrum without any of the dangerous ultraviolet rays. Fluorescent lighting, which works by way of an ionized gas and an electrical charge, provides the best full spectrum artificial light. To set up and use your full spectrum light box, place the metal box, fluorescent tube, and ballast on a clean surface. Insert the ballast into the round hole, then connect the fluorescent tube. Plug the box into a grounded outlet and turn on one switch, then the other.

Cause and Effect

Arrange your information through cause and effect when you want to discuss the causes behind certain effects or the reasons for certain results or consequences. Be certain that you have evidence or can otherwise support the causal relationship you are making. In this example, you will see notations of places where you would insert a citation.

> In the United States during the 1950s, many birds died due to the use of the pesticide DDT [citation]. In particular, bald eagle and other raptor populations suffered greatly, because DDT caused the shells of eggs to be too thin [citation]. DDT was also linked to problems in other species of birds [citation].

Period

The period has a range of uses, from ending a sentence to denoting a technical statement (such as a web address).

End Sentences

Use a period to end a sentence that makes a statement.

> In 1979, the United States ceded control of the Panama Canal Zone to the Panamanian government.
>
> The nursing student washed her hands frequently.

Use a period to end a sentence that makes a request, expresses a mild command, or gives directions.

> Please help the needy.
>
> Open the technical report to page 178.
>
> Turn left at the next corner.

If you want the command or request to be given a great deal of emphasis or force, use an exclamation point instead of a period (see also **Exclamation Point**).

> Sign up now!

Use a period at the end of a sentence that asks an indirect question.

> The editorial questions whether NATO is effective.
>
> The reporter asked how the fire had started.
>
> The doctor wondered why the patient's temperature had risen.

Use a period at the end of a request politely expressed as a question.

> Will you please type this letter for me.
>
> Will you kindly keep your voices down.

See also **Using the Period With Parentheses** and **For Quotations Within Quotations** for other important uses of the period).

Conventional Uses of Periods

Use a period after most abbreviations and initials.

If a sentence ends with an abbreviation requiring a period, use only one period.

> The first admiral in the U.S. Navy was David G. Farragut.
>
> The Marine Corps traces its beginnings to Nov. 10, 1775.
>
> Thomas Jefferson's home, Monticello, is near Charlottesville, Va.

The current trend in abbreviations is away from the use of periods. The following two rules are now considered standard. However, if you are in doubt about whether to use a period after an abbreviation, consult your dictionary.

Avoid using a period when abbreviating a unit of measure.

> 86 *m*   275 *kg*   24 *cm*   20 *ft*
>
> 10.5 *yd*   20 *lb*   20 *cc*

Exceptions: For *mile* use *m.* or *mi.* to prevent confusion with *m* for *meter.* For *inch* use *in* only when there is no possibility of confusion with the word *in;* the abbreviation *in.* avoids all confusion.

Avoid using a period with acronyms or other abbreviations of businesses, organizations, and government and international agencies.

> The *NAACP* has not endorsed a presidential candidate.
>
> My mother served in the *WACs* for eight years.
>
> The impartiality of *UNESCO* is being questioned.

Using the Period with Parentheses

When a parenthetical element is part of sentence, and comes at the end of that sentence, place the period outside the parentheses.

> In the original *King Kong,* the the huge creature climbed what was then the tallest building in the world (the Empire State Building).

When a parenthetical element is a complete sentence on its own, place the period inside the parentheses.

> In the original *King Kong*, the huge creature climbed what was then the tallest building in the world, the Empire State Building. **(**In the second version of the movie, he climbed the World Trade Center**.)** There he was attacked by airplanes.

Using the Period in Technical Expressions

The period is used in web addresses, where it is referred to as a "dot." (This usage comes from the early days of the Internet, where "dot" was the way UNIX software developers referred to the period.) In this usage, the period is generally used to separate the different parts of the addresses, as in **www.pearsonlongman.com**

The same usage applies to email addresses, where the period separates the domain from the service provider: **writers@pearsonlongman.com**

Phrases

Phrases are parts of sentences that come in several types, depending upon their function in the sentence.

Gerund

A gerund is a verb form that is spelled in the same way as the present participle, with an -ing ending, but a gerund is used as a noun, not an adjective, in a sentence.

> The problems of *parenting* were discussed at the symposium.
>
> The school taught *reading* and *writing* but little else.

A gerund phrase consists of a gerund and all its modifiers and complements. A gerund phrase acts as a noun in a sentence. In the following sentences, the gerund phrases are in **color** with the gerunds in ***color italics:***

> For Margaret, *remaining* in California became impossible once the high-tech sector began to shrink.
>
> *Running* five miles a day keeps a person in good condition.

Use gerund phrases sparingly especially at the beginning of a sentence. These phrases can become exceedingly long, and they lack any agency, so readers often do not know who is acting and what is really happening.

> *Running five miles a day in good weather conditions and with the proper clothing and gear* can keep a person in good condition.

Infinitive

An infinitive phrase consists of the present infinitive of the present perfect infinitive form of the verb and all its modifiers and complements. It acts as a noun, an adjective, or an adverb. In the following sentences, the infinitive phrases are in **color** with the infinitives in *color italics:*

> *To know* him is *to love* him. (*nouns*)
>
> Hard work is one way *to gain* successes in business. (*adjective*)
>
> She is proud *to have dedicated* her life to medicine. (*adverb*)

Keep infinitive phrases short and to the point. Long phrases can distract readers from the main point of the sentence.

Noun or Noun Substitute (Appositive)

An appositive is a word or phrase that provides additional information about the preceding noun or pronoun. Appositive phrases can provide useful definitional information in technical writing.

> The large screen television, *available in several configurations*, can be connected to a Mac or PC.

Use a pronoun that is part of a compound clause in the same case as the noun to which this clause refers.

> NOT: The partners—Willie, Travis, and *me*—plan to open a bicycle repair shop in July.
>
> BUT: The partners—Willie, Travis, and *I*—plan to open a bicycle repair shop in July.

(continued)

NOT: Only two people, the manager and *him,* knew the combination of the safe.

BUT: Only two people, the manager and *he,* knew the combination of the safe.

Participial

A participial phrase consists of a present or past participle and all its modifiers and complements. It acts as an adjective in a sentence. In the following sentences, the participial phrases are in **color** with the participles in ***color italics:***

Throughout his life, Whitman adhered to the beliefs *summarized* in the preface to his book.

A man *curled* in the fetal position with his arm *covering* his head is the subject of one of Rodin's most moving sculptures.

Prepositional

A prepositional phrase consists of a preposition, the object of the preposition, and all the words modifying this object. In the following sentences, the prepositional phrases are in **color,** and the prepositions and their objects in ***color italics:***

In many *cultures* whale meat has been an essential source *of protein.*

Some *of these* cultural *groups* resent efforts *by conservationists* to protect the whale, since these efforts would restrict the groups' ability to obtain food and would conflict *with their traditions*.

Keep prepositional phrases near the elements that they modify. Otherwise, the meaning can be misconstrued.

Prepositions

Prepositions are connecting words. Prepositions are important in technical writing because they show the ways in which ideas and concepts connect to each other. Each of the following prepositions shows a different relationship between the noun *stump* and the actions expressed by the verbs:

> The rabbit jumped *over* the stump, ran *around* the stump, sat *on* the stump, hid *behind* the stump, and crouched *near* the stump.

The group of words beginning with a preposition and ending with a noun or pronoun (called its *object*) is a *prepositional phrase*. Prepositional phrases function in a sentence as adjectives or adverbs.

Prepositions are among the most familiar and frequently used words in the language because they orient things and actions in space and time. The following is a list of common prepositions:

	Common Prepositions	
about	concerning	past
above	despite	save (meaning
across	down	"except")
after	during	since
against	except	through
along	for	throughout
among	from	till
around	in	to
at	inside	toward(s)
before	into	under
behind	like	underneath
below	near	until
beneath	of	unto
beside	off	up
between	on	upon
beyond	onto	with
but (meaning	out	within
"except")	over	without

A compound preposition is made up of more than one word. The following are some commonly used compound prepositions:

ahead of	in front of	out of
as for	on top of	together with
in back of		

Prepositions appear only in and at the beginning of prepositional phrases. In the following sentence, the prepositions are printed in **color** and the prepositional phrases in *italics*.

The term "transitory isotope" is used *by Dr. Helen Leroy in the introductory course.*

Note: The *to* in the infinitive form of a verb (such as "to describe") is not a preposition, it is part of the verb form.

Pronouns

The term pronoun covers a wide variety of words, many of which function in very different ways. Some pronouns change form, but most of them act as structure-class words, providing important information about the structural or grammatical relationships in a sentence.

A pronoun is a word that stands for or takes the place of one or more nouns. When a pronoun refers to a specific noun, that noun is called the antecendent of the pronoun. In the following sentences, the arrows indicate the *italicized* antecedents of the pronouns in **color** type:

Because *vitamins* can have toxic side effects, *they* should be administered with care.

A pronoun may also have another pronoun as an antecedent.

Most of the old records are scratched. *They* cannot be replaced.

Each of the mothers thought *her* child should receive the award.

Certain pronouns may lack a specific antecedent.

> *Who* can understand the demands made upon a child prodigy?
>
> *Everyone* knew that *something* was wrong.

There are seven categories of pronouns: *personal, demonstrative, indefinite, interrogative, relative, intensive,* and *reflexive.*

Personal Pronouns

Personal pronouns take the place of a noun that names a person or a thing. Like nouns, personal pronouns have numbers, genders, and cases. This means that they can be singular or plural; that they can be masculine, feminine, or neuter; and that they can function in the subjective, the objective, or the possessive case. In addition, personal pronouns are divided into three "persons": first-person pronouns refer to the person(s) speaking or writing, second-person pronouns refer to the person(s) being spoken or written *to,* and third-person pronouns refer to the person(s) or thing(s) being spoken or written *about.* The following is a list of all the personal pronouns:

	Singular	**Plural**
FIRST PERSON:	I, me, my, mine	we, us, our, ours
SECOND PERSON:	you, your, yours	you, your, yours
THIRD PERSON:	he, him, his, she, her, hers, it, its	they, them, their, theirs

Demonstrative Pronouns

Demonstrative pronouns point to someone or something. The demonstrative pronouns are *this* and *that* and their plural forms *these* and *those.*

Demonstrative pronouns are usually used in place of a specific noun or noun phrase.

> The sandwiches I ate yesterday were stale, but *these* are fresh.
>
> The design team named the program DirectionsFirst, because *that* was its chief feature (the ability to get driving directions quickly).

In addition, demonstrative pronouns are sometimes used to refer to a whole idea.

> Should we say farewell to the electoral college? *That* is a good question.

If you use a demonstrative pronoun in this way, be sure that the idea it refers to is clearly stated and not just vaguely suggested. (See also **Vague Reference.**)

Indefinite Pronouns

Indefinite pronouns do not take the place of a particular noun, although sometimes they have an implied antecedent. Indefinite pronouns carry the idea of "all," "some," "any," or "none." Some common indefinite pronouns are listed below:

everyone	somebody	anyone	no one
everything	many	anything	nobody

Some indefinite pronouns are plural, some are singular, and some can be either singular or plural

> *Everything* is going according to plan. (*singular*)
>
> *Many were* certain that the war, which officially started on July 28, 1914, would be over before autumn. (*plural*)
>
> *Some* of the material *was* useful. (*singular*)
>
> *Some* of the legislators *were* afraid to oppose the bill publicly. (*plural*)

Interrogative Pronouns

Interrogative pronouns are used to ask a question.

who	whom	whose	what	which

Who, whom, and *whose* refer to people. *What* and *which* refer to things.

> *What* were the effects of the Industrial Revolution on Europe during the first decade of the twentieth century?
>
> *Who* is Barbara McClintock, and for *what* is she best known?
>
> *Which* of the economic depressions have been most damaging?

Relative Pronouns

Relative pronouns are used to form adjective clauses and noun clauses.

who	whom	that	whoever	whichever
whose	which	what	whomever	whatever

Who, whom, whoever, and *whomever* refer to people. *Which, what, that, whichever,* and *whatever* refer to things. *Whose* usually refers to people but can also refer to things.

> The Society of Women Engineers is an organization *whose* mission involves encouraging women to seek careers in engineering.
>
> The food was given away to *whoever* wanted it.

Intensive Pronouns

Intensive pronouns are used to emphasize their antecedents. They are formed by adding *-self* or *-selves* to the end of a personal pronoun.

> The detectives *themselves* did not know the solution.
>
> The producer wasn't sure *herself* why the show was a success.

Reflexive Pronouns

Reflexive pronouns are used to refer to the subject of the clause or verbal phrase in which they appear. They have the same form as intensive pronouns.

During her illness Marjorie did not seem like *herself*.

If you have young children in the house, take precautions to prevent them from electrocuting *themselves* accidentally.

This plant can fertilize *itself*.

Proofreading Technical Documents

Important Items to Check for in Technical Documents

It is always important to proofread any document. When a team has collaborated on a document, it is especially important for someone, usually someone who was not part of the team and can bring a fresh eye, to proofread the document.

Technical documents requires special attention to items such as the following:

- Mathematical formulas
- Equations
- Numbers
- Technical language
- Acronyms
- Captions
- Headings
- Diagrams and other visuals

One good strategy is to proofread for specific things; in other words, to make several passes through the document, once for numbers, one for equations, once for captions, and so on. If the document is large, a team of proofreaders can help.

In publishing, proofreading is the second phase of looking over a manuscript. First comes copyediting, where the manuscript is reviewed for content, formatting, consistency, parallelism, and grammatical errors. See also **Copyediting.**

Copyediting and Proofreading Marks

In the pre-electronic document days, documents were marked up using special symbols that directed the typesetter to insert, delete, move, transpose, and perform other tasks on the text. The main methods of reviewing material today are tools such as track changes (in Microsoft Word) or comment boxes and other marking tools in Adobe Acrobat. Most mainstream electronic systems do not use the traditional proofreading symbols.

Writers still work with hard copy (paper) and use proofreading symbols, so it is good to become familiar with them.

Copyediting and Proofreading Marks

Instruction	Marginal Mark	In-text Mark	Final Copy
delete	ℓ	She submitted the the report.	She submitted the report
close up	⌒	She sub mitted the report.	She submitted the report.
delete and close up	ℓ⌣	she submited the report.	She submit the report.
insert space	#	She submitted␟he report.	She submitted the report.
make lowercase	(lc)	She submitted the Report.	She submitted the report.
wrong font	(wf)	She submitted the report.	She submitted the report.
set in italic	(ital)	She submitted the report.	She *submitted* the report.
set in roman	(rom)	She submitted the report.	She submitted the report.
set in bold	(bf)	She submitted the report.	**She submitted the report.**
insert period	⊙	She submitted the report	She submitted the report.
transpose	(tr)	Sgh the report submitted.	She submitted the report.
let stand	(stet)	She submitted the report.	She submitted the report.
hyphen	=	She submitted the year end report.	She submitted the year-end report.
comma	⌄	Yes she submitted the report.	Yes, she submitted the report.

Mark	Symbol	Example	Corrected
apostrophe		He read Lindas report.	He read Linda's report.
quotation marks		She said, I submitted the report.	She said, "I submitted the report."
parentheses		She Linda submitted the report.	She (Linda) submitted the report.
brackets		She Linda submitted the report.	She [Linda] submitted the report.
set in capitals	(cap)	she submitted the report.	She submitted the REPORT.
set in small capitals	(sm cap)	She submitted the report.	She submitted the REPORT.
subscript	(sub)	H$_2$0 is water.	H$_2$0 is water.
superscript	(sup)	a2 + b2 = c2	$a^2 + b^2 = c^2$
insert something	∧	She ^submitted the report.	She submitted the report.
spell out	(sp)	She submitted a 2nd report.	She submitted a second report.
start new paragraph	¶	¶ She submitted the report.	She submitted the report.
em dash		She submitted the report finally.	She submitted the report— finally.
en dash		She submitted pages 1 10 of the report.	She submitted pages 1–10 of the report.
colon		She submitted two reports monthly and quarterly.	She submitted two reports: monthly and quarterly.
semicolon		She submitted the report however, it was incomplete.	She submitted the report; however, it was incomplete.
set as question mark	?	Did she submit the report	Did she submit the report?
set as exclamation point	!	She submitted the report	She submitted the report!

Question Mark

Use a question mark at the end of a sentence that asks a direct question.

> Who invented the safety pin**?**
>
> Have you registered to vote**?**

Use a question mark at the end of an interrogative element that is part of another sentence.

> "Will he support our proposal**?**" she wondered.
>
> "Will he actively support women's rights**?**" she wondered.
>
> The telegram said that Malcolm is alive—can it be true**?**—and will be returned to the United States on Friday.

Usage varies somewhat on capitalization of questions following introductory elements. The more formal the question, the greater the tendency to use a capital letter. The less formal, the greater the tendency to use a lowercase letter.

> FORMAL: The book raises the question: What role should the United States play in the Middle East**?**
>
> INFORMAL: I wondered, should I bring my umbrella**?**

Use a question mark, usually in parentheses, to express doubt or uncertainty about a date, a name, or a word.

> Pong, an early computer game (developed in the 1970s**?**), remains quite popular.
>
> A dialect of Germanic, called Angleish (**?**), is the basis of modern-day English.

Quotation Marks

Quotation marks enclose quoted material and certain kinds of titles. They are always used in pairs.

For Direct Quotations

Use quotation marks to enclose a direct quotation—the exact words of a speaker or writer.

> When our company president said to the staff, "We want your best ideas," the staff replied, "We are happy to provide them."
>
> In *How to Write a User Manual in No Time*, author Jay Kraug tells readers that "with enough imagination, any user manual can become interesting and easy to use."

For Block Quotations

When quoting a prose passage of considerable length, you omit quotations marks but use the block quotation form. This form usually involves indenting the entire paragraph. Different documentation styles (APA, MLA: see Part 6) have different rules about the use of block quotations. For instance, APA style stipulates that quotations longer than 40 words use block quotation format.

For Quotations Within Quotations

Use single quotation marks to enclose quoted material contained within a quotation.

> In "Silence," Marianne Moore wrote: "My father used to say, 'Superior people never make long visits.'"
>
> Jensen looked up from his research and declared, "I've found the answer. It was Henry Clay who said, 'I would rather be right than President.'"

For Titles of Short Works

Use quotation marks to enclose the quoted titles of short stories, short poems, one-act plays, essays, articles, subdivisions of books, episodes of a television series, songs, short musical compositions, and dissertations.

> In his poem "Son of Frankenstein," Edward Field reveals the loneliness of the Frankenstein monster.

Use underlining for the titles of longer works.

With Other Punctuation Marks

Place a period or comma *inside* a closing quotation mark.

> "I don't want to talk grammar," Eliza Doolittle says in *Pygmalion*. "I want to talk like a lady."
>
> "After all," said Hillary, "We can always change the formatting."

Place a semicolon or a colon *outside* a closing quotation mark.

> The editor wrote that the research paper demonstrated the mechanical engineer's "talent with experimental design"; this comment, we felt, was a bit overstated.

Place a question mark or an exclamation point *inside* a closing quotation mark if the quotation itself is a question or exclamation.

> The song I was trying to recall is "Will You Love Me in December?"
>
> Upon reaching the summit of Mount Everest, Sherpa Tensing declared, "We've done the bugger!"

Place a question mark or an exclamation point *outside* the closing quotation mark if the sentence is a question or exclamation but the quotation itself is not.

> Who first said, "Big Brother is watching you"**?**
>
> What a scene she caused by saying, "I don't want to"**!**

If both the sentence and the quotation are questions or exclamations, use only one question mark or exclamation point, and place it *inside* the closing quotation mark.

> Why did he cause a scene by asking, "Who is that woman**?**"

Slash

Use a slash (/) to indicate that a pair of options exist.

> He prefers diet soda**/**low calorie yogurt.

Use a slash as part of a web site address (URL).

> Look here for more information: http:**//**www.nasa.gov**/**news**/**index.html.

Semicolon

The semicolon links independent clauses (clauses that can stand as complete sentences on their own) or separates items in a series. The semicolon indicates a pause longer than that taken for a comma but not as long as that taken for a period.

To Link Clauses

Use a semicolon between two closely related independent clauses that are not joined by a coordinating conjunction.

> Years ago caviar was an inexpensive food often given away free at taverns; today it is one of the most expensive foods in the world.
>
> Truffles are the food of the rich; turnips are the food of the poor.
>
> Truffles are the food of the rich; turnips, the poor.

The comma in the preceding sentence indicates that a part of the second clause has been left out. Use a semicolon between two independent clauses joined by a conjunctive adverb or by a transitional phrase.

> Columbus sent cacao beans back to Spain; however, the Spanish were not particularly impressed.
>
> Cortez brought hack more cacao beans; moreover, he brought back the knowledge of how to prepare them.
>
> In some cultures insects are considered delicacies; for example, the ancient Romans thought the cicada a delightful morsel.

Use a semicolon between two independent clauses joined with a coordinating conjunction if one or both of these clauses contain internal commas or if the clauses are particularly complex.

Between Items in a Series

Use semicolons to separate items in a series if the individual items are long or contain commas.

> In the language of flowers, each flower represents a particular attribute: belladonna, which is a deadly poison, silence; citron, which produces a sour, inedible fruit, ill-natured beauty; blue periwinkle, which is small and delicate, early friendship.
>
> The guide grouped wildflowers into many families, four of which were the cattail family, Typhaceae; the arrowhead family, Alismataceae; the yellow-eyed grass family, Xyridaceae; and the lizard-tail family, Saururaceae.

Sentences

Sentences are the backbone of written communication. A sentence is composed of one or more clauses and typically contains at least a subject and a verb.

Structure and Types

English has six basic patterns for sentences. Each pattern contains a subject and a verb. The pattern may also contain objects or complements that complete the meaning of the verb. The subject of the sentence is a noun or nounlike word group that answers the question "Who?" or "What?" about the verb. The subject is the part of the sentence that performs the main action.

> *Captain Cook* searched for a northwest passage to China.

In other sentences, the subject receives the action of the verb.

> The *deer* was wounded by the hunters.

Sometimes, the words *there* and *here* appear at the beginning of the sentence. These words are never the subject but simply serve to postpone the appearance of the subject, which is usually in the middle of the sentence. These type sentences can often be wordy and could be edited to be more precise and provide more information about who was involved.

> Here are four ways to increase productivity at this plant.

This sentence could be rewritten as follows.

> We can increase productivity at this plant in four ways.

Completeness

Avoid sentence fragments, that is, incomplete sentences. Although in creative writing sentence fragments are often used for emphasis, they are to be avoided in technical writing.

> NOT: Maddy invented a device for removing pet hair from clothing. A great thing for pet owners.
>
> BUT: Maddy invented a device for removing pet hair from clothing, which was a great thing for pet owners.

Make sure the subject goes with the verb.

> NOT: My opinion of his latest movie is poorly directed and ineptly filmed.
>
> BUT: My opinion of his latest movie is that it is is poorly directed and ineptly filmed.

Run-on

Sentences that should be split into two separate units are called run-on sentences. Technically, a run-on sentence is one where two independent clauses are placed together without any conjunction (and, or) or punctuation (comma, semi-colon). The following is a run-on sentence:

> The baseball team was testing its public address system it was bothering the neighbors

This sentence could become one of the following, depending on the writer's purpose:

> The baseball team was testing its public address system, and it was bothering the neighbors.
>
> The baseball team was testing its public address system; it was bothering the neighbors.
>
> The baseball team was testing its public address system. It was bothering the neighbors.

Style

Formal and Informal Style

See **Word Choice**.

Technical and Scientific Style

Writing styles change throughout the ages, and technical and scientific style is no different. For many years, engineers and scientists were encouraged to write in passive voice (see **use Passive Voice Sparingly**) to keep the data and experiment in the foreground and the scientist in the background.

> The experiment was conducted and the data analyzed.

This rule has changed. Science and engineering journals now encourage a more direct style when possible.

Other features of technical and scientific style include the following:

Brevity: keep ideas brief and to the point (see **Conciseness** on avoiding wordiness).

Concision: express ideas in a concise manner (see **Avoid Unnecessary Words and Phrases**).

Titles

Titles are important elements in reports, studies, books, journal articles, published papers, brochures, magazine articles, and other documents. Titles give readers a high-level snapshot of the content, scope, and approach to the document. Publishers and editors will often have guidelines for the length and wording of a title. In general, titles should be brief but informative. Long titles are hard to comprehend; vague titles will not provide readers with enough information to understand what the document is about.

Often a colon is added to provide space for a sub-title. This approach is used quite commonly in published papers and doctoral dissertations.

For titles of people (Dr., Ms., Mr.), see **Abbreviations and Acronyms.**

Tracking Changes and Version Control

In most business and professional settings, teams collaborate on documents. Many of these teams are virtual; that is, they are not located in the same physical setting. Team members can be spread out across a building, a state, a country, or the world. These documents need to be accessible electronically, and the team needs to keep track of the changes and who made them.

The most basic way of collaborating on a document is to use the "track changes" feature in a program such as Microsoft Word. Small teams, including groups of students working a team paper, can use these tools to insert changes and comments and share drafts. Hand-marked documents can be difficult to read, with all the boxes, cross-outs, and other features. In addition, with files often circulating via email, it becomes impossible to know who has the most recent version.

For that reason, most large organizations use tools that allow for what is called version control. Text or other material is checked in and out of a database, and all changes are stored. The most recent version can be compiled easily.

Transitions

Even if all the sentences in a paragraph relate to a controlling idea and follow an organized method of development, it is still helpful to the reader's understanding to link sentences with transitional devices that provide smooth passage from one idea to the next. These devices include not only *transitional words and expressions* but also *pronouns, repetition of key words and phrases,* and *parallel grammatical structure.*

Transitional Words and Expressions

Transitional words and expressions show the relationship of one term to another term, one sentence to another sentence, one idea to another idea, and even one paragraph to another paragraph. They serve as signposts that direct the reader through the passage.

The following is a list of some common transitional words and expressions and the relationships the might indicate:

Addition
again, also, and, besides, equally important, finally, first (second, third, and so on), furthermore, in addition, last, likewise, moreover, next, too

Similarity
in a similar fashion, likewise, moreover, similarly, so

Contrast
although, but, even so, for all that, however, in contrast, nevertheless, nor, on the contrary, on the other hand, or, still, yet

Time
afterward, at the same time, before, earlier, finally, in the past, later, meanwhile, next, now, previously, simultaneously, soon, subsequently

In the following paragraph, the transitional words and expressions are printed in **color:**

There are many possible new developments in cardiac care over the coming century, *too*. Implantable devices such as pacemakers are the most widely recognized, *but* new devices will be far smaller and easier to monitor. *In addition*, preventative care will help decrease heart disease. *Yet* new treatments for cardiac disease will always be needed.

Pronouns

Pronouns link sentences by referring the reader to their antecedents (see **Ambiguous Reference**). In the following excerpt, notice how the pronouns in the second clause link to the first clause:

> *There* is usually a coffee shop at every software company in the world. *Some* are brand-named shops, and *most* of *them* serve the usual variety of beverages.

Verbs

A verb expresses action—physical or mental—or a state of being. In the following sentence, the verbs are printed in color:

> Stephen King *writes* scary stories. (*physical action*)
>
> Many people *value* hard work and timeliness. (*mental action*)
>
> The young woman *is* a famous singer. (*state of being*)

Verb Forms

A verb has four basic forms, called *principal parts:* the present infinitive, the past tense, the past participle, and the present participle.

Present Infinitive	Past Tense	Past Participle	Present Participle
compute	computed	computed	computing
analyze	analyzed	analyzed	analyzing
fall	fell	fallen	falling
bring	brought	brought	bringing

Kinds of Verbs

Action Verbs

Action verbs are important for technical writing, because they can express with clarity and efficiency what is supposed to be done or what has been done. An action verb is called *transitive* when it takes an object (a noun or noun substitute that completes the idea expressed by the verb).

> Marie and Pierre Curie successfully *isolated* radium.
>
> Professor Higgins *introduced* Eliza Doolittle to society.

When the subject of a transitive verb performs the action received by the object, the verb is said to be in the active voice. The verbs in the preceding sentences are transitive active.

A transitive verb is in the passive voice when the active-voice object becomes the subject of that verb; the active-voice subject then appears as the actor only in a phrase with an introductory preposition such as *by*.

> *Radium was* successfully *isolated* by Marie and Pierre Curie.
>
> *Ronda Gayersk; was introduced* to the class by Professor Kennedy.

Only transitive verbs can be described as having voice. An active verb that has no object is intransitive.

> Sometimes even our old cat Max *runs*.
>
> Consider the corn in the field and how it *grows*.

State-of-Being Verbs

State-of-being verbs may be *linking* or *nonlinking*. They are, however, always intransitive because they never take objects. Instead, they take subject complements.

A linking verb expresses a state of being or condition and connects its subject to a word that describes or identifies that subject. The most common linking verb is *be*, including its forms: *am, is, are, was, were, will be, have been,* and so on.

> Radiotherapy *is* the treatment of disease with radiation.
>
> Jakob Nielsen *was* a popular speaker at the conference.
>
> Americans *were* aghast at the sinking of the *Lusitania*.

Other common linking verbs are *appear, become, feel, grow, look, remain, seem, smell, sound,* and *taste.*

> Houdini *became* world-famous for his daring escapes.
>
> The true fate of Amelia Earhart *remains* a mystery.

A linking verb connects the subject to a subject complement. A subject complement may be either a noun, which identifies the subject, or an adjective, which describes the subject.

> *Napoleon* was a brilliant *general*. (*noun subject complement*)
>
> After his defeat at Waterloo, *Napoleon* was *disconsolate*. (*adjective subject complement*)

A **nonlinking** state-of-being verb is not followed and completed by a subject complement but may be followed by an adverb modifier of time or place.

> Yesterday we *were* upstream from their camp.
>
> The next performance *is* at eight o'clock.

Auxiliary or Helping Verbs

Forms of the verbs *be* and *have* can help show important shifts in meaning in various forms of the main verb that signal *tense* and *mood*. When they serve this helping function they are called auxiliary or helping verbs.

> MAIN VERB: She *goes*.
>
> AUXILIARY VERBS: She *is* going; she *has* gone; they *have been* going.

Modals are used to form questions, to help express a negative, to emphasize, to show future time, and to express such conditions as possibility, certainty, or obligation. The words *do, does, did; can, could; may, might, must; will, shall; would, should;* and *ought to* are modals. A verb phrase may include both auxiliaries and modals. In each of the following sentences, the verb phrase is printed in *italics* and the modal in **color:**

> For a democracy to work, its citizens *must participate.*
>
> The election *may be decided* on the basis of personality.

Sometimes an auxiliary or modal is separated from the main part of the verb.

> *Do* you *know* the full name of the Imagist poet H. D.?
>
> The price of gold *has* not *been falling.*

Word Choice

Whenever you write, one of the most basic decisions you have to make is what level of formality and tone to use. Formal English, as its name suggests, adheres to the conventions of standard English and uses a professional, sometimes academic, tone. Most of the writing you do in college or in a profession—term papers, formal essays, theses, reports, instructions—should be in formal English. A formal tone conveys professionalism, and it also makes English easier for non-native speakers to understand. In addition, using standard words and sentence structures makes it easier for material to be translated into other languages. Informal English takes a more relaxed attitude to the conventions of standard English and may include more contractions, colloquialisms, jargon, and sometimes even slang. It is appropriate for informal writing situations—journal and diary entries, informal essays, and creative writing in which you try to capture the sound of everyday speech. Informal English is also the style of most email and instant messaging communication, but remember to always consider audience and purpose: an email message to your friend should be written differently from an email to your professor, manager, or supervisor.

INFORMAL:	The delegates *were savvy about the fact* that the document they were signing *wasn't* perfect.
FORMAL:	The delegates *understood* that the document they were signing *was not* perfect.
INFORMAL:	The candidate *slammed* her opponent for often *changing his tune* on the issues.
FORMAL:	The candidate *criticized* her opponent for often *changing his views* on the issues.
INFORMAL:	Stickley furniture may not be *real smooth*, but *it's pricey* and *fresh*.
FORMAL:	Stickley furniture may not be *very comfortable*, but *it is expensive* and *fashionable*.

Slang

Slang is extremely informal language. It consists of colorful words, phrases, and expressions added to the language to give it an exciting or ebullient flavor. Slang is usually figurative, highly exaggerated, and generally not appropriate for technical writing. After a time, slang can makes its way into the mainstream; for instance, the word *cool*, used to describe something that is approved of, exciting, or good, started as slang in the 1950s.

Colloquialisms

Colloquial language is the conversational and everyday language of educated people. Colloquialisms are the words and expressions that characterize this language. Though not as informal as slang, colloquial language is generally still too casual to be considered appropriate for professional or technical writing. Like jargon or slang, colloquialisms can cause problems for non-native speakers and for translations.

COLLOQUIAL:	The meeting will begin at 7:00 P.M. *on the dot.*
FORMAL:	The meeting will begin promptly at 7:00 P.M.

Commonly Confused Words

Accept, Except

Accept means to receive. *Except* means "with the exclusion of."

> She will *accept* the award on Friday.
>
> We all went to the ceremony *except* Lola.

Affect, Effect

Affect is a verb meaning "to influence." *Effect* is a noun meaning "the result of some influence."

> How can I *affect* the outcome of this project?
>
> What is the *effect* of this project on our bottom line?

Avoid using the noun *impact* as a verb in place of the verb *affect*.

> NOT: How can I *impact* the outcome of this project?
>
> BUT: How can I *affect* the outcome of this project?

Awhile, a while

Awhile is an adverb meaning "for a short time." *A while* is a noun phrase meaning "a period of time."

> Take your shoes off and stay *awhile*.
>
> After taking a long hike, Samantha rested *a while* and then kept going.

A lot, Alot

Always write *a lot* as two words, not as *alot*. In general, avoid using *a lot* or *lots* in technical writing. Instead, use language that is more precise.

> NOT: The meteorologists predict that we will get a lot of snow on Monday.
>
> BUT: The meteorologists predict that we will get between 10 and 12 inches of snow on Monday.

Amount, Number

Amount refers to mass or quantity. It is followed by the preposition *of* and a singular noun. *Number* refers to things that can be counted. It is followed by *of* and a plural noun.

> The *amount* of time he spent completing the job was far greater than the reward he derived from it.
>
> The *number* of domestic animals that have contracted rabies is alarming.

Bad, Badly

Use the adjective *bad* before nouns and after linking verbs. Use the adverb *badly* to modify verbs or adjectives.

> Several *bad* strawberries were hidden under the good ones.
>
> The students felt *bad* about the loss.
>
> The computer program performed *badly*.
>
> The book was *badly* written.

Because, Since

Both *because* and *since* indicate a relationship between two thoughts. *Because* indicates a strong causal relationship, that is, where one part caused the other part.

> *Because* we measured so carefully, our experimental results were quite precise.

Since is used in a similar manner but is a less effective way to indicate causality. Use since to indicate a time relationship.

> She has experienced leg weakness *since* 2005.

Between, Among

Use *among* with three or more people or objects. Use *between* with only two.

> According to the will, the funds were to be divided equally *between* the two children.
>
> He decided to leave his entire estate to his eldest child, rather than divide it *among* his six children.

Bi-, Semi-

Both *bi-* and *semi-* are prefixes that indicate time intervals. *Bi-* is used to indicate two of something or an event that happens twice.

> Bifocals (glasses with two lenses)
>
> Biweekly (two times per week)

Semi- is used to indicate half.

> Semiannually (two times per year)

Cite, Site

Cite is a verb meaning to quote or otherwise use material as evidence. (In its noun form, it is the familiar word *citation*.)

> In his report, he *cited* several scientific studies.
>
> During the meeting, she *cited* data from the latest research study.

Site is a noun meaning a place or location.

> We are breaking ground at the *site* of the new library.
>
> The new company web *site* is very attractive.

Compare to, Compare with

Use *compare to* when referring to the similarities between essentially unlike things. Use *compare with* when referring to the similarities and differences between things of the same type.

> Jarrad Gray *compares* the shape of e. coli bacteria *to* small rods.
>
> The professor *compared* study by Professor James Brown with one by Professor Marty Collegial.

Data, Criteria, Phenomena

These three words are plural and usually take plural verbs. Yet the word data is often used in the singular as well.

> The data suggest . . . (plural)
>
> The data suggests . . . (singular)

The decision on whether to treat data as singular or plural should be based on the specific meaning of the sentence or document as well as the field of study. In addition, certain technical and scientific journals have editors and style guides that provide guidance on this issue.

Due to, Because of

In formal writing, avoid using *due to* in place of *because of*.

> NOT: The shipment was delayed *due to* the bad weather.
>
> BUT: The shipment was delayed *because of* the bad weather.
>
> OR: The delay in the shipment *was due to* the bad weather.

e.g.

The Latin abbreviation *e.g.* means *exempli gratia*, or "for example." It is often used in scientific or technical writing as a shorthand way of indicating an example.

> Birds of prey (*e.g.*, hawks, falcons) have excellent eyesight from great distances.

Always use a comma after *e.g.*

This sentence can also be written as follows, but notice how the full *for example* phrase changes the flow:

> Birds of prey (*for example*, hawks and falcons) have excellent eyesight from great distances.

Etc.

Etc. is an abbreviation for the Latin phrase *et cetera*, meaning "and the rest." It is used to indicate "and other things" or "and so forth." Using etc. can be useful as a way of keeping a sentence short, but only if what is implied by "other things" is obvious to readers.

> She was running out of household staples: bread, milk, eggs, etc.

In most cases, technical writers avoid etc. because it is not precise or specific. Using etc. in the sentence below would only cause readers to become frustrated.

> To assemble the trailer hitch kit, you will need only a few basic tools: a crescent wrench, a regular screwdriver, and a hammer.

Even in the previous case, avoiding etc. will create a more precise sentence.

> She was running out of household staples: bread, milk, eggs, and other necessities.

Avoid using the phrase *and etc.*, which is redundant.

Fewer, Less

Use *fewer* to refer to things that can be counted. Use *less* to refer to a collective quantity that cannot be counted.

> This year *fewer* commuters are driving their cars to work.
>
> In general, smaller cars use *less* fuel than larger cars.

Former, Latter

Former means "the first mentioned of two." When three or more are mentioned, refer to the first mentioned as first. *Latter* means "the second mentioned of two." When three or more are mentioned, refer to the last mentioned as *last*.

> Anita and Jayne are athletes: the *former* is a gymnast, the *latter* a tennis player.
>
> The judges sampled four pies—apple, plum, rhubarb, and apricot—and gave the prize to the *last*.

He/She

Use the words *he* or *she* when they accurately represent the sex of the person you are writing about.

> Meredith Rocteau is a well-known biologist, and she is widely recognized around the world.

When you are writing about all people, use language that is not gender-specific and that is inclusive.

> NOT: When mankind is finally at peace, there will be no wars.
>
> BUT: When all people are at peace, there will be no wars.

See also **Inclusive Language.**

i.e.

The Latin abbreviation i.e. means *id est*, meaning "that is."

Insure, ensure

Insure and ensure are not used interchangeably in formal writing. Ensure means "to make certain that something will take place." Insure means "to guarantee or to protect against something."

> The end user must *ensure* that the product is registered.
>
> We can *insure* the automobile for up to $7,000 in damages.
>
> He needed to know that his health coverage *insured* him for prior injuries.

Its, It's

Its is the possessive case of the pronoun *it*. *It's* is a contraction of *it is* or *it has*. Avoid the contraction *it's* in formal writing.

> The cat was cleaning *its* paws.
>
> *It's* too late to submit an application.

Lay, lie

The verb *lie* means "recline." The verb *lay* means "put" or "place." Do not confuse the principal parts of these verbs.

Present Infinitive	Past Tense	Past Participle	Present Participle
lie	lay	lain	lying
lay	laid	laid	laying

The problem most people have with these verbs is using a form of *lay* when they mean *lie*. *Lie* is intransitive; it does not take an object. *Lay* is transitive; it does take an object.

> *Lie* on the floor. (*no object*)
>
> *Lay* the *book* on the table. (*object*)

> Not: Why is Millie *laying* on the couch in the nurse's office?
>
> But: Why is Millie *lying* on the couch in the nurse's office?
>
> Not: Peter *laid* in the sun too long yesterday.
>
> But: Peter *lay* in the sun too long yesterday.
>
> Not: She had just *laid* down when her cell phone rang.
>
> But: She had just *lain* down when her cell phone rang.

Maybe, May be

May be is a verb phrase. *Maybe* is an adverb meaning "perhaps."

> His findings *may be* accurate.
>
> *Maybe* they will find a solution.

Media, Medium

Media is the plural form of medium.

> With all the mass *media* of today, no one should be uninformed.
>
> DVD is the *medium* of choice for recording video.

People, Persons

Use *people* to refer to a large group collectively. Use *persons* to emphasize the individuals within the group.

The committee is investigating ways in which *people* avoid paying their full taxes.

The group thought it was near agreement when several *persons* raised objections.

Raise, Rise

The verb *rise* means "go up" or "get into a standing position." The verb *raise* means "lift." Do not confuse the principal parts of these verbs.

Present Infinitive	Past Tense	Past Participle	Present Participle
rise	rose	risen	rising
raise	raised	raised	raising

Rise is intransitive; it does not take an object. *Raise* is transitive; it does take an object.

Without yeast, the bread will not *rise*. (*no object*)

After they won the game, they *raised* the school *banner*. (*object*)

NOT: Every morning they *rise* the blinds before leaving for work.

BUT: Every morning they *raise* the blinds before leaving for work.

Sit, Set

The verb *sit* means "be seated." The verb *set* usually means "place" or "put in a certain position." Do not confuse the principal parts of these verbs.

Present Infinitive	Past Tense	Past Participle	Present Participle
sit	sat	sat	sitting
set	set	set	setting

The problem most people have with these verbs is using a form of *set* when they mean *sit*. *Sit* is intransitive; it does not take an object. *Set* is usually transitive; it does take an object.

> *Sit* in the chair by the fireplace. (*no object*)
>
> Please *set* the *table* for me. (*object*)

Note: *Set* is sometimes intransitive: *The sun sets*.

> NOT: Some people *set* in front of the television far too much.
>
> BUT: Some people *sit* in front of the television far too much.
>
> NOT: He *set* up until two in the morning, waiting for his daughter to come home from her date.
>
> BUT: He *sat* up until two in the morning, waiting for his daughter to come home from her date.

Sort of, Kind of

In formal writing, avoid using *sort of* and *kind of* as adverbs. Use the more professional words *rather* and *somewhat*.

> NOT: His description was *kind of* sketchy.
>
> BUT: His description was *rather* sketchy.
>
> NOT: She left *sort of* abruptly.
>
> BUT: She left *somewhat* abruptly.

That, Which

Use *that* to introduce a restrictive clause. Use *which* to introduce either a restrictive or a nonrestrictive clause. (In order to maintain a clearer distinction between *which* and *that*, some writers use *which* to introduce only a nonrestrictive clause.)

> Is this the manuscript *that* he submitted yesterday?
>
> This contract, *which* is no longer valid, called for a 30 percent royalty.

Their, There, They're

Their is a possessive pronoun meaning "belonging to them."

> The scientists' plant pathology lab is *their* best research facility.

There is an adverb meaning "that place or position."

> She went to graduate school and stayed *there* for ten years.

They're is a contraction for "they are."

> *They're* going to stay with us for the night.

To, Two, Too

To is a preposition (as in *to the store*) or sign of the infinitive form of a verb (as in *to be*).

Two is the number (2).

Too is a qualifier (as in *too much, too often*) and an adverb meaning "also."

Utilize, Use

Utilize is a verb meaning to put to use or make practical use of.

> Most plants *utilize* sunlight quite effectively to make energy.

Use is a verb meaning to deploy something.

> The team wanted to *use* the project management software.

Do not substitute *utilize* for *use*. Such substitution sounds stuffy and pretentious.

Which, Who

Use *which* to refer to objects. Use *who* to refer to people. (*That* usually refers to objects but at times may be used to refer to people.)

> The book, *which* was written by Nat Hentoff, is called *Jazz Is*.
>
> Nat Hentoff, *who* wrote *Jazz Is*, contributes articles to many magazines.

Who's, Whose

Who's is a contraction for *who is* or *who has*.

> *Who's* outside right now?
>
> *Who's* eaten all of the pumpkin pie?

In technical writing it is almost always better to spell out the two words rather than use the contraction *who's*.

Whose means belonging to or associated with a person.

> *Whose* lab notebook was left in the hallway?
>
> The doctor needed to know *whose* blood samples had been tested.

Your, You're

Your is a possessive form of "you."

> It is *your* car that the others want to drive.

You're is a contraction for "you are."

> *You're* not trying hard enough.

Documentation

Documentation

As a technical communicator, you need to document the sources of the information and data you quote or refer to in any research report. Proper documentation shows readers that the document is based on credible research sources, that you have followed the accepted form and conventions for your profession and discipline, and that you are ethical in your use of others' words and ideas. Sloppy or incomplete documentation erodes your credibility with your readers, and leaves you open to accusations of plagiarism or other breaches of ethical and professional conduct.

Document the source for any ideas, words, interpretations, or visuals that you have quoted or paraphrased from other sources.

Always Document

- **Quotations** or other uses of another author's exact wording
- **Paraphrases** of another author's words and ideas, even if you adapt them in your own words
- **Summaries** of other authors' ideas, findings, or conclusions
- **Visual illustrations**, including photographs, tables, charts, graphs, and illustrations, even if you edit or adapt them for your own purposes

You do not need to document *common knowledge*. You can assess whether something is common knowledge by considering whether it appears as an established fact in multiple sources. You would not need to document that Juneau is the capital of Alaska, for example, or that the freezing point of water is 32 degrees Fahrenheit. For more information, see the entries on **Citing Sources** and **Plagiarism** (Part 4).

When working with a large number of references, especially ones that might need to be formatted in different styles (e.g., in IEEE style for a journal article and in APA style for a dissertation), many writers find it convenient to work with bibliography software, such as Endnote (www.endnote.com), ProCite (www.procite.com), or Refworks (www.refworks.com). Some of these programs run on your personal computer (Endnote and ProCite), while others (RefWorks) work in conjunction with your company or university library. These programs allow you to enter the bibliographic information, such as author names, book or article title, publication date, page numbers, and so on, into a database. Each entry is like a card in the database. When you are ready to create your bibliography, you just select the citation style and let the program generate the reference list. Keep in mind that these programs can't help if you don't include all of the needed information. Some of these programs work in conjunction with word processing software to coordinate in-text citations with the references. Most libraries are familiar with these programs and a research librarian can help you determine what will work for you. Faculty and other researchers who have worked on large bibliographic projects will also know about these programs and can offer advice if you are a student.

Documentation Styles

Styles guides provide information regarding how to format documents, how to cite and reference sources, and how to use certain words, punctuation, and other elements of style and language. There are several style guides in publication. Many of them were developed for use in specific disciplines, though some are used across disciplines.

Be sure to structure your documents according to the appropriate style guide for your profession and your audience. If you are writing an article for submission to a particular publication, check the author's guidelines for information about which style guide you should follow. Following is a table of some of the most used style guides and the fields in which they are used.

Popular Style Guides

Style Guide	Commonly Known As	Publisher	Field
The Associated Press Stylebook	AP	Associated Press	Journalism
Publication Manual of the American Psychological Association	APA	American Psychological Association	Psychology and other behavioral and social sciences
The ACS Style Guide: A Manual for Authors and Editors	ACS	American Chemical Society	Chemistry and other sciences
Scientific Style and Format: The CSE Manual for Authors, Editors, and Publishers	CSE	Council of Science Editors	Natural sciences
MLA Style Manual and Guide to Scholarly Publishing	MLA	Modern Language Association of America	Literature and humanities
IEEE Standards Style Manual	IEEE	Institute of Electrical and Electronics Engineers	Engineering and other technical fields

Popular Style Guides *(continued)*

Style Guide	Commonly Known As	Publisher	Field
The Chicago Manual of Style	Chicago, CMS	University of Chicago Press	Publishing, sciences, art, history, humanities
The Bluebook: A Uniform System of Citation	Bluebook	Harvard Law Review Association	Law
Microsoft Manual of Style for Technical Publications	Microsoft	Microsoft	Computer and other information technology fields

In addition to using published style guides, many organizations use an in-house style guide that reflects their own style and is applied to their published documents. For example, an organization might set a style for telephone numbers, stating that a period is to be used instead of a hyphen: 555.555.5555.

Not all organizations develop an in-house style guide. Instead, they adopt all or part of one or more style guides for their own practices. There are many style guides beyond those listed in the table above, including the following:

- *AIP Style Manual* (published by the American Institute of Physics)
- *The Oxford Guide to Style: The Style Bible for All Writers, Editors, and Publishers*
- *The United States Government Printing Office Style Manual*
- *Webster's Standard American Style Manual*

Global and Digital Style Considerations

In the global exchange of ideas, you must be aware of style guides used in the countries of your chosen publications. For example, publications in Australia and the United Kingdom commonly use the Harvard Style. As noted earlier, always research which style guide is appropriate for the publications you might be writing for. Style can vary considerably from guide to guide, particularly in terms of references, as the following examples show.

Because many publications are available in both print and electronic versions, the examples on the following pages illustrate both as appropriate. "E" icons in the margins indicate electronic sources. Note that as new forms of digital media appear, style guides often do not keep pace with these changes. If no example exists in the style guide or a recent addendum, model the citation on the closest electronic publication available. For instance, web page citation style is often a good model for other electronic forms, such as blogs and wikis. Always include enough information so that another researcher can locate the original source.

Reference Examples

Each of the following examples references the same article. Note the variations from style to style—the use of author first names or initials; the use of commas, periods, and other punctuation; the placement of the year.

ACS Style

Cummings, D. I.; Arnott, R. W. C.; Hart, B. S. *Geology.* 2006, *34,* 249.
(Though not visible here, references in ACS style do not have a hanging indent.)

APA Style

Cummings, D. I., Arnott, R. W. C., & Hart, B. S. (2006). Tidal signatures in a shelf-margin delta. *Geology, (34)*4, 249–252.

Chicago Style

NOTES-BIBLIOGRAPHY STYLE (ART, HISTORY, HUMANITIES)

Cummings, Donald I., R. William C. Arnott, and Bruce S. Hart. "Tidal Signatures in a Shelf-Margin Delta." *Geology* 34 (2006): 249–252.

AUTHOR-DATE STYLE (SCIENCES)

Cummings, D.I., R.W.C. Arnott, and Bruce S. Hart. 2006. Tidal signatures in a shelf-margin delta. *Geology* 34(4): 249–252.

Harvard Style

Cummings, D. I., Arnott, R. W. C., & Hart, B. S. 2006. 'Tidal signatures in a shelf-margin delta', *Geology,* vol. 34, no. 4, pp. 249–252.

MLA Handbook for Writers of Research Papers

Cummings, Donald I., R. William C. Arnott, and Bruce S. Hart. "Tidal Signatures in a Shelf-Margin Delta." *Geology* 34.4 (2006): 249–52.

The rest of Part 6 provides details on four of the most popular documentation styles in technical communication: APA, CSE, IEEE, and MLA. Note that not all style guides contain sample references for newer media such as podcasts or for informal publications such as brochures or even magazines. In these cases, we have followed the guidelines in the particular style guide and adapted using the most similar type of citation. For instance, in the case of podcasts, the most similar would be broadcast audio.

APA Documentation

APA documentation style is used in most fields within the social sciences, including psychology, sociology, and communication. APA style uses in-text citations in parentheses that includes the author's name, year of publication, and page number for direct quotations.

In-text Citation

> Workers in today's world need a new set of skills, based not on the linear logic of previous decades, but on the "inventive, empathic, big-picture" skills of the "conceptual age" (Pink, 2005, p. 2).

The in-text reference in parentheses refers readers to a complete reference list of sources at the back of the paper or report.

References List

> Pink, D. H. (2005). *A whole new mind: Moving from the information age to the conceptual age*. New York: Riverhead.

Note: the author's name appears as last name, initials. Do not include the author's first name in APA style. Also, the year of publication appears in parentheses, followed by a period. Only the first word of the title and subtitle are capitalized.

Digital Object Identifier

Many scientific and technical articles have a Digital Object Identifier (DOI), which is more permanent and stable than a web site address (URL). Like an ISBN number, the DOI is uniquely associated with one publication and will locate the publication accurately no matter where it resides. If a DOI is available (you will usually see it on the first page of the published article), use the DOI rather than the web site address in your reference. Below is an example of a citation using a DOI. Note that although this article was retrieved from a database (OvidSP), the DOI makes the listing of the database or URL unnecessary.

> Brendgen, M., Vitaro, F. (2008). Peer rejection and physical health problems in early adolescence. *Journal of Developmental & Behavioral Pediatrics, 29*(3), 183–190. doi: 10.1097/DBP.0b013e318168be15

The APA publishes online supplements to its style guide and other information, available at http://apastyle.apa.org/pubmanual.html.

APA Sample References

Article, essay, or chapter in an edited collection

> Shneiderman, B. (2003). Direct manipulation: A step beyond programming languages. In N. Wardrip-Fruin & N. Montfort (Eds.), *The new media reader* (pp. 485-498). Cambridge, MA: MIT Press.

(E) Herring, S. C., Kouper, I., Scheidt, L. A., & Wright, E. L. (2004). Women and children last: The discursive construction of weblogs. In L. J. Gurak, S. Antonijevic, L. Johnson, C. Ratliff, & J. Reyman (Eds.), *Into the blogosphere: Rhetoric, community, and culture of weblogs*. Retrieved from http://blog.lib.umn.edu/blogosphere/women_and_children.html

Article in a scholarly journal with continuous pagination

Messmer, M. (1990). When bodies are weapons: Masculinity and violence in sport. *International Review for the Sociology of Sport, 25*, 203–220.

Article in a scholarly journal paginated by issue

Holtug, N. (1997). Altering humans: The case for and against human gene therapy. *Cambridge Quarterly of Healthcare Ethics, 6*(2), 157–160.

Article in a print-based scholarly journal that is accessed online

(E) Holtug, N. (1997). Altering humans: The case for and against human gene therapy. *Cambridge Quarterly of Healthcare Ethics, 6*(2), 157–160. doi.10.1017/S0963180100903098

Article in a popular magazine published monthly

Neimark, J. (1991, May). Out of bounds: The truth about athletes and rape. *Mademoiselle,* 196–199.

Article in a popular magazine published weekly or daily

Donahue, D., & Red, S. (1986, May 12). A Hyannis hitching. *People,* 53–56, 59.

(E) Saletan, W. (2007, Dec. 27). Personal space invaders: the top science-and-tech privacy threats of 2007. *Slate,* Dec. 31, 2007. Retrieved from http://www.slate.com/

Blog or wiki posting

Lessig, L. (2007, Dec. 20). Sunlight: Help on a distributed research project. *Lessig 2.0.* Retrieved from http://www.lessig.org/blog/

Book by a single author

Pink, D. H. (2005). *A whole new mind: Moving from the information age to the conceptual age*. New York: Riverhead.

Logie, J. (2006). *Peers, pirates, & persuasion: Rhetoric in the peer-to-peer*
(E) *debates*. West Lafayette, IN: Parlor Press. Retrieved from http://www.parlorpress.com/pdf/PeersPiratesPersuasion-Logie.pdf

Book in a second or subsequent edition

Schneiderman, B. (2004). *Designing the user interface: Strategies for effective human-computer interaction* (4th ed.). New York: Addison Wesley.

Book by an unknown author

Lonely planet bluelist. (3rd. ed.). (2008). Melbourne, Australia: Lonely Planet.

Book with a translator

Aristotle. (1987). *The poetics of Aristotle.* (S. Halliwell, Trans.) Chapel Hill: University of North Carolina Press.

Book by multiple authors

Salen, K., & Zimmerman, E. (2004). *Rules of play: Game design fundamentals.* Cambridge, MA: MIT Press.

Note: For works with more than six authors, use et al. ("and others") following the name and initials of the six authors to indicate that there are more than six authors.

Book by an organization or corporation

National Research Council. (1993). *The social impact of AIDS in the United States.* Washington, DC: National Academy Press.

Book with an editor

Gallegos, B. (Ed.). (1994). *English: Our official language?* New York: Wilson.

Brochure

National Institute of Outdoor Canine Activities (2005). *How to go dogsledding and love it* [Brochure]. New York.

Computer software

Bayarsky, N. L. (2001). Measurement analysis restructure tool (MART) [Software]. New Haven, CT: Active Next Analysis.

Conference proceedings

Liu, C., & Wu, J. (2007). Scalable routing in delay tolerant networks. In *MobiHoc'07: Proceedings of the Eighth ACM International Symposium on Mobile Ad Hoc Networking and Computing* (pp. 51–60). Montreal, Quebec, Canada, September 9–14, 2007. New York: ACM.

Note: You do not need to reference common software programs such as Microsoft Word or PowerPoint. You should reference more specialized software programs that will be unfamiliar to your reader.

Email

The APA Publication Manual, 5th edition, suggests that personal email communication is similar to other personal communications (letters, memos) and is therefore not something your reader can retrieve by looking at a reference. The APA manual recommends, therefore, that you cite personal communications within the text but not provide a reference.

Some forms of email communication may be public and available for retrieval; many organizations post important business emails on a company web site. In this case, you would cite the source within your paper and also provide a reference.

(E) Gurak, L. J. (2006, May 23). Announcing the new writing initiative. Email message posted to University email list and archived at http://www.messages.university.edu/mail-archives/newwritmsg.html

Encyclopedia or dictionary

Smart, P. J. (Ed). (2001). *The dictionary of new media and digital references*. Albany: SUNY Press.

Film, videotape, or DVD

Discovery Docs (Producer), & Herzog, W. (Writer/Director). (2005). *Grizzly man* [motion picture]. United States: Lions Gate Films.

(E) National Aeronautics and Space Administration (NASA). (2007, December 12). *Holiday message from the space station* [video file]. Retrieved from http://www.nasa.gov/multimedia/videogallery/

Government document

U.S. Department of Commerce. (2007). *Population profile of the United States* (Publication No. P-20, No. 808). Washington, DC: U.S. Government Printing Office.

(E) U.S. Food and Drug Administration. (Nov. 21, 2007). Use your microwave safely [fact sheet]. Washington, DC. Retrieved from http://www.fda.gov /consumer/updates/microwave112107.pdf

Interview

The APA Publication Manual, 5th edition, suggests that since interviews that you as an author conduct are not "recoverable data" (something that your reader can retrieve by looking at a reference) that you cite the interview within the text but not provide a reference.

If, however, the interview is available in an accessible format, such as video on a news web site or on YouTube, follow the instructions for citing a video.

Lecture or public speech

The APA Publication Manual, 5th edition, does not contain information on how to cite a lecture or speech, probably because a live lecture or speech is traditionally not considered "recoverable data." The advice, therefore, would be that you cite the speech within the text but not provide a reference.

If, however, the lecture or speech is available in an accessible format, such as video on a news web site or on YouTube, follow the instructions for citing a video.

Letter to the editor

Silva, J. T. (2003, August 5). The health care crisis [Letter to the editor]. *Houston-Post,* p. A14.

Ⓔ Barr, S. (2007, Dec. 31). Let grand jury take its course [Letter to the editor]. *Hartford Courant.* Retrieved from http://www.courant.com/news/opinion/letters/hc-digedlets1231.art0dec31,0,4717680.story

Newspaper article

Eskenazi, G. (1990, June 3). The male athlete and sexual assault. *New York Times,* Section 8, p. B6.

Ⓔ Crowley, C. F. (2007, Dec. 31). RPI scientist fights malaria at the source. *Times Union.* Retrived from http://timesunion.com/

Newspaper editorial

Help students—not the tax revolt [Editorial]. (2001, July 31). *Los Angeles Times,* p. B6.

Ⓔ Design observer: Writings on design and culture. (2007). Retrieved from http://www.designobserver.com/

Online article based on a print source

Kohl, J. R. (1999). Improving translatability and readability with syntactic cues. *Technical Communication Online, 46,* 149–166. Retrieved from http://www.ingentaconnect.com

Online article not based on a print source

Reid, A. (2003, Spring). New media's long history and global future: The *Uniplanet* project. *Kairos, 8*(1). Retrieved from http://english.ttu.edu/Kairos/8.1/binder2.html?coverweb/reid/reid.swf

Online article retrieved from a database

Alley, R. A. (2003, January). Ergonomic influences on worker satisfaction. *Industrial Psychology, 5*(11), 93–107.

Podcast

National Aeronautics and Space Administration (NASA). (2007, October 22).
To catch a galactic thief. Podcast retrieved from http://www.
nasa.gov/multimedia/podcasting/jpl-spitzer-20071022pod.html

Report or other publication where the author is a company or organization

Worldwatch Institute (2008). State of the world 2008. Washington, DC:
Worldwatch Institute.

E Pew Internet & American Life (2007). Teens and social media. Washington,
DC: Pew Internet & American Life Project. Retrieved from http://
www.pewinternet.org/PPF/r/230/report_display.asp

Sound recording

Ward, E. (2008). *In '60s San Francisco, love was the song*. Philadelphia, PA:
Fresh Air, produced by WHYY. Retrieved from http://www.npr.org
/templates/story/story.php?storyId=17898424

Unpublished material (dissertation, thesis, paper)

Grandon, J. L. (2008). *Creating surveys for research in public health*: *A review
of the literature*. Unpublished manuscript, University of Nebraska.

E Grandon, J. L. (2008). *Creating surveys for research in public health*:
A review of the literature. Unpublished manuscript, University of
Nebraska. Retrieved from http://www.grandonjl.net/papers99

Web site

If you need to cite a web site in its entirety (as opposed to a paper, report, or other
document that is contained on a web site), your reference should include the
author (if available), the title of the site, a date of publication, and the URL (web
address) for the home page. Use a date of retrieval only if the web site is likely to
change often. Do not add a period at the end of a URL. If a URL extends to more
than one line, break the line just before the slash and do not add a hyphen.

Your reference should appear as follows:

American Wind Energy Association. (2007). *Take action: contact congress*.
Retrieved from the American Wind Energy Association Web site:
http://www.awea.org/

Sample Pages in APA Style

Career Achievement Motivation in Women

Eileen B. Biagi

University of Notre Dame

Psychology

Dr. W. Bartlett

May 2008

1 Title page format.

2 Page header. The header provides a brief summary of the paper title as well as page numbers.

3   Abstract

Historically, women and men have differed in their career orientations. Farmer (1987) attempts to explain these differences with her model of career and achievement motivation. She proposes that interaction among personal, environmental, and background factors determines an individual's level of achievement motivation. Specific results of her study are presented and compared with other research. Support exists for her assertion that gender differences in this area are evident and that achievement motivation will vary with age and social changes. Other studies suggest that this motivation construct differs between females and males. However, these studies conflict with Farmer's in that they detect other types of gender differences. As yet, a comprehensive model that explains achievement motivation for women and men does not exist. But the models we do have for women's career achievement motivation suggest that they can serve as bases for changing women's positions in the job market and their attitudes about achievement in the career world.

3 Abstract. The abstract offers a short but comprehensive summary of the contents of the paper. It must be precise and informative, summarizing findings/results. Follow the assignment or publication guidelines for abstract length.

Career Achievement Motivation in Women

5 By identifying the factors that influence career achievement motivation in women, we can recognize, understand, and perhaps change the current position of women in the job market and women's attitudes about achievement in the career world. Do men and women differ in their career orientations? And, if they do, why do these differences occur and what factors contribute to the disparities? "Psychologists have generally viewed the need for achievement [achievement motivation] as a learned motive, with parents playing a major role in shaping their children's later strivings" (Benner, 1985, p. 12). However, this single motive
6 cannot easily explain the historical differences between and changes within women's and men's career orientations.

Farmer (1987) proposes one such model for explaining gender differences
7 in achievement motivation, basing the model on social learning theory (Bandura, 1987) and on a sociocultural perspective (Maehr, 1974, 1984). The model incorporates the idea of reciprocal determinism, that is, through interaction, the behavioral, cognitive, and environmental factors influence one another.
8 Farmer (1987) further presents three constructs that both interact among themselves and influence achievement motivation: personal, environmental, and background. Academic self-esteem, expressiveness, independence, ability attributes, and intrinsic values comprise the personal factors. Environmental elements include support from parents and teachers and support for working women, while the components of the background

---

4 Page numbering. Number pages with Arabic numerals in the upper right corner, starting with 1 on the title page and continuing throughout the paper.

5 Introduction and thesis. The author opens this paper by presenting her thesis and the particular problem under study. The thesis statement

usually appears near the beginning of the paper but it is not always the very first sentence.

6 In-text documentation. When the author's name is not given within the text itself, use a parenthetical reference including the name and page number. Use commas to separate elements.

7 Date reference. APA style requires the date, in parentheses, after the author's name.

8 Developing the analysis. As is typical for the social sciences, this paper begins with a review of previously published research on the topic.

9 References

Bandura, A. (1978). The self system in reciprocal determinism. *American Psychologist, 33,* 344–358.

10 Benner, D. G. (1985). Need for achievement. In D. G. Benner (Ed.), *Baker Encyclopedia of Psychology.* Grand Rapids: Baker Book House.

Farmer, H. (1985). Model of career and achievement motivation for women and men. *Journal of Counseling Psychology, 32* (3), 663–676.

Farmer, H. (1987). A multivariate model for explaining gender differences in career and achievement motivation. *Educational Researcher 16,* 5–9.

Ireson, C. (1978). Girls' socialization for work. In A. Stromberg & S. Harkess (Eds.), *Women Working* (pp. 128–139). Palo Alto: Mayfield.

Kaufman, D. R., & Richardson, B. L. (1982). *Achievement and Women.* New York: Free Press.

12 Lapsley, D. K., & Quintana, S. M. (1985). (Review of *Adolescent Sex Roles and Social Change*). *Sex Roles, 12,* 252–255.

Lueptow, L. B. (1984). *Adolescent Sex Roles and Social Change.* New York: **13** Columbia University Press.

14 Maehr, M. (1974). Culture and achievement motivation. *American Psychologist, 29,* 887–896.

Maehr, M. (1984). On doing well in science: Why Johnny no longer excels: Why Sarah never did. In S. Paris, G. Olson, & H. Stephenson (Eds.), *Learning and Motivation In the Classroom* (pp. 179–210). Hillsdale, NJ: Lawrence Erlbaum.

11 Rooney, G. (1983). Distinguishing characteristics of the life roles of worker, student, and homemaker for young adults. *Journal of Vocational Behavior, 22,* 324–342. Retrieved from InfoTrac: Expanded Academic Database ASAP.

---

9 Format of references. APA style refers to this page as "references," centering this heading on the page and including in alphabetical order all works cited in the paper.

10 Edited book. Use the abbreviation "Eds." for "editors" or "Ed." for "editor" when referring to an edited book.

11 Journal from a full-text database. The name of the source indicates that this is an electronic database.

12 Review of book. This reference is for a book review, not the book itself.

13 Publisher's name. The name of a university press is spelled out in full form, but the name of commercial publishers is not.

14 Journal article. The volume number, italicized, follows the name of the journal and a comma.

CSE Documentation

The Council of Scientific Editors, or CSE, uses a documentation style for writers who are publishing in the physical and life sciences. For online information about CSE style, see http://www.councilscienceeditors.org

As the CSE manual indicates, writers in the sciences use three systems for documenting sources: citation-sequence (similar to IEEE style, using numbered references), name-year (similar to APA style, using author-date), and citation-name (a hybrid of the first two). The 7th edition of *Scientific Style and Format: The CSE Manual for Authors, Editors, and Publishers* recommends using the citation-name system.

In the **citation-name system**, a complete list of sources referenced is presented at the end of the report or document, arranged in alphabetical order by author's last name (similar to MLA or APA). But each item in the alphabetized reference list is numbered, and these numbers are used in-text to direct readers to the references being cited.

In-text reference

Workers in today's world need a new set of skills, based not on the linear logic of previous decades, but on the "inventive, empathic, big-picture" skills of the "conceptual age."[6]

References list

1. Adams M. ～～～
2. Crouch P. ～～～
3. Gurel LD ～～～
4. Linus MM. ～～～
5. Norbert W. ～～～
6. **Pink DH. A whole new mind: moving from the information age to the conceptual age. New York (NY): Riverhead; 2005.**
7. Querty W. ～～～

Notes:

- CSE does not make recommendations about using italics or underlining to show titles; rather, writers should follow the guidelines for the particular journal or publication.
- Capitalize only the first word in a title; do not capitalize the first word of the subtitle.

CSE Sample References

Article, essay, or chapter in an edited collection

Lauwers E, Gielen G. High-level power estimation of analog front-end
blocks. In: Wambacq P, Giele G, Gerrits J, editors. Low-power design
techniques and CAD tools for analog and RF integrated circuits. New
York: Springer; 2001. p. 83–106.

Article in a scholarly journal with continuous pagination

Messmer M. When bodies are weapons: masculinity and violence in sport.
Int Rev Soc Sport. 1990; 25:203–220.

Article in a scholarly journal paginated by issue

Holtug N. Altering humans: the case for and against human gene therapy.
Cambridge Q Healthcare Ethics. 1997; 6(2):157–160.

Article in a popular magazine published monthly

Peterson L, Pai VS. Experience-driven experimental systems research.
Communications of the ACM. 2007. Nov, 38–44.

Article in a popular magazine published weekly or daily

Gawande A. 2007. The checklist. The New Yorker. Dec 10, 86.

E   Wapner, J. Group therapy: how poorly designed trials for cancer drugs are
hurting patients. Slate [Internet]. 2008 10 Jan. [cited 22 Jan 2008].
Available from: http://www.slate.com/id/2181789.

Article in a print-based scholarly journal that is accessed online

Witt ED. Mechanisms of alcohol abuse and alcoholism in adolescents: a case
for developing animal models. Behav and Neural Biology 1994;
62:168–177.

Blog or wiki posting

Bradley D. Nano pico femto satellites. Sciencebase [Internet]; 2008 Jan 11
[cited 2008 Jan 21]. Available from: http://www.
sciencebase.com/science-blog.

Book by a single author

Pink DH. A whole new mind: moving from the information age to the
conceptual age. New York (NY): Riverhead; 2005.

Book in a second or subsequent edition

Dubin D. Rapid interpretation of EKG's. 6th ed. Tampa (FL): Cover Publishing Company; 2000.

Book by an unknown author

Lonely planet bluelist. 3rd ed. Melbourne, Australia: Lonely Planet Publications; 2008.

Book with a translator

Hesiod. Works and days: a translation and commentary for the social sciences. Tandy DW, Neale, WC, translators. Berkeley (CA): University of California Press; 1997.

Book by multiple authors

Salen K, Zimmerman E. Rules of play: game design fundamentals. Cambridge (MA): MIT Press; 2004.

Note: For books with more than ten authors, list the first ten and then use "et al." or "and others."

Book by an organization or corporation

National Research Council (US). The social impact of AIDS in the United States. Washington (DC): National Academy Press (US); 1993.

Book with an editor

Gallegos B, editor. English: our official language? New York (NY): Wilson; 1994.

Brochure

National Consumers League. Food + drug interactions [Internet]. Washington (DC): National Consumers League; 21 Oct 2004 [cited 21 Jan 2008]. Available from: http://www.nclnet.org/publications/foodanddrugint.pdf.

Computer software

Bayarsky NL. Measurement analysis restructure tool (MART). New Haven (CT): Active Next Analysis; 2001.

Note: You do not need to reference common software programs such as Microsoft Word or PowerPoint. You should reference more specialized software programs that will be unfamiliar to your reader.

Conference proceedings

Flank S. Cross-language multimedia information retrieval. In: Proceedings of the sixth conference on applied natural language processing. San Francisco (CA): Morgan Kaufmann Publishers Inc.; c2000. p. 13–20.

Dictionary

Stedman's medical dictionary for the health professions and nursing. 6th ed. Hershey (PA): Wolters Kluwer Health; 2008.

Email

Gurak LJ. Announcing the new writing initiative [Internet]. Minneapolis (MN): University of Minnesota; 2006 [modified 2006 May 23; cited 2008 Jan 5]. Available from: http://www.messages.university. edu/mail-archives/newwritmsg.html.

Note: The CSE style guide, 7th edition, recommends that you put references to personal communications (letters, email, conversations) within the text of your document and do not cite these as sources. There are, however, forms of email communication that may be public and available for retrieval; many organizations post important business emails on a company web site. In this case, you would cite the source within your paper and also provide a reference, following the format where the email is archived (web site, blog, other).

Encyclopedia

Encyclopedia of Internet technologies and applications. Freire M, Pereira M, editors. Norwood (MA): IGI Publishing; 2008.

Film, videotape, or DVD

National Aeronautics and Space Administration (NASA) [Internet]. [Quicktime movie]. Journey to a comet. 3 Jul 2005 [cited 2008 1 Feb]. Available at: http://www.nasa.gov/mission_pages/ deepimpact/multimedia/07-03-05-movie.html.

Government document

Department of Commerce (US). Population profile of the United States (Publication No. P-20, No. 808). Washington (DC): Government Printing Office (US); 2007.

Interview

If you conduct an interview yourself, reference it within your paper but do not include it in your list of references. If the interview is available in an accessible format, such as video on a news web site or on YouTube, follow the instructions for citing a video.

Lecture or public speech

The CSE style guide, 7th edition, does not contain information on how to cite a lecture or speech, perhaps because such information is not recoverable by readers (e.g., it can't be listened to or looked up). If a lecture or speech is available in an accessible format, such as video on a news web site or on YouTube, follow the instructions for citing a video.

In the sciences, a paper may be presented orally at a conference. Normally, this paper will then be published in the conference proceedings. To cite the presentation, cite the paper in the proceedings.

Letter to the editor

> Barr S. Let grand jury take its course. Hartford Courant [Internet]. 31 Dec 2007. Available from: http://www.courant.com/news/opinion/letters/hc-digedlets1231.art0dec31,0,4717680.story.

Newspaper article

> Angier N. Tiny specks of misery, both vile and useful. New York Times. 2008 Jan 8; Sect. D:1 (col 1).

> **E** Boudreau J. Yahoo unveils newest cell phone home page. Mercury News.com [Internet]. 2008 Jan 7; [cited 2008 Jan 12]. Available from: http://www.mercurynews.com/ci_7903908?source=rss&nclick_check=1#.

Online article based on a print source

> Kohl JR. Improving translatability and readability with syntactic cues. Technical Comm [Internet]; 1999 [cited 2007 Jan 4]; 46: 149–166. Available from: http://www.ingentaconnect.com/content/stc/tc/1999/00000046/00000002/art00004.

Online article not based on a print source

> Ellison NB, Steinfield C, Lampe C. The benefits of Facebook "friends": social capital and college students' use of online social network sites. Journal of Computer-Mediated Communication [Internet]. 2007 [cited 15 Jan 2008] 1(4):article 1. Available from: http://jcmc.indiana.edu/vol12/issue4/ellison.html.

Online article retrieved from a database

> Fraser HJ, Collings MP, McCoustra MR, Williams DA. Thermal desorption of water ice in the interstellar medium. Monthly notices of the Royal Astronomical Society [database Aerospace & High Techn]; 2001 Nov; 327(4): 1165–1172. Oxford: Blackwell Publishing Ltd.

Podcast

National Aeronautics and Space Administration (NASA) [Internet] [Podcast]. To catch a galactic thief. 2007 Oct 22 [cited 2008 Jan 21]. Available at: http://www.nasa.gov/multimedia/ podcasting/jpl-spitzer-20071022pod.html.

Report or other publication where the author is a company or organization

National Academy of Sciences. Science, evolution, and creationism. Washington (DC): National Academies Press. 2008.

E Pew Internet & American Life. Teens and social media [Internet]. Washington (DC): Pew Internet & American Life project; 2007 [cited 12 Jan 2008]. Available at: http://www.pewinternet. org/PPF/r/230/report_display.asp.

Sound recording

Covey SR. The 7 habits of highly effective people [CD-ROM]. 15th edition. New York: Free Press; 2004. 13 CD set.

Unpublished material (dissertation, thesis)

Grandon JL. Creating surveys for research in public health: a review of the literature. University of Nebraska. Unpublished manuscript; 2008.

Web site

American Wind Energy Association. Take action: contact congress [Internet]; 2007 [cited 2007 Jan 10]. Available from: http:// www.awea.org/.

Sample Pages in CSE Style (Citation-name)

Heterogeneity and Anisotropy in Karst Aquifers:
How they Effect Groundwater Flow and
Contaminant Transport

Carolyn Moores

University of Idaho

Hydrogeology 509

12/9/03

Final Draft

1 Title page format. An informative title and subtitle with specific terminology from the study.

2 Title page format. Course name or other identifying information and the date appear at the bottom of the title page.

Abstract **3**

Groundwater and surface water form an integral link in the hydrologic cycle. No other aquifer system exposes this relationship more than a karst aquifer system. Efficient recharge and diverse flow paths blend surface and aquifer water chemistry and flow characteristics together in one system. To study a karst aquifer system is to study the terrain, rainfall, surface water, groundwater, and perhaps the surface water again, as springs discharge from the aquifer. The unpredictable flow paths due to widespread chemical dissolution in karst aquifers give them different hydraulic properties than other types of aquifers. The preferential dissolution along fractures, bedding planes, and highly irregular regolith and basement boundaries (known as "epikarst") present unique modeling problems and require special consideration and careful study to predict flow paths of water and potential contaminants.

4 ## Formation of Karst

Carbonate rocks originate as sedimentary deposits in marine environments. Post depositional changes include compaction, cementation, and dolomitization as the material lithifies. Perhaps the most important process that affects the hydraulic properties of the carbonate aquifer is the dissolution that occurs after lithification. Classic karst terrain evolves as the dissolution of bedrock, usually limestone or dolomite, produces distinguishing geomorphic features in the landscape such as sinkholes, caves, depressions, disrupted surface water drainages, and springs. Limestone and dolomite tend to weather quickly because of their chemistry and the slight acidity of rainwater.

5 $CaCO_3 + H_2O + CO_2 = Ca^{2+} + 2HCO_3^-$ *limestone*

$MgCO_3 + H_2O + CO_2 = Mg^{2+} + 2HCO_3^-$ *dolomite*

rock + water + carbon dioxide = calcium or + bicarbonate ion magnesium

---

3 Abstract. The CSE style guide recommends an abstract, noting that most journals in science and medicine require one.

4 Introduction and background. Technical study begins with an introduction, theory, or background section.

5 Equation. Equations are unnumbered and each set on a separate line. Note that equations, which are key to this paper, are introduced early in the discussion.

6 Studies have shown that the amount of calcite dissolution is limited in closed systems only by the amount of CO_2 present[1]. However due to high CO_2 concentrations in most soils from respiration and decay of organic matter, most systems are considered open and much more calcite will be dissolved. The weathering yields groundwater high in carbonates and ionic strength. As the solution cracks grow in size, increases in surface area exposure and subsequent increases in flow rates further accelerate the dissolution process.

Karst tends to form in areas of high rainfall, high driving head, and, of course, where soluble rocks are at or near the surface. In the eastern United States, where rainfalls are typically higher than 40 inches per year karst is more commonly found, although karst aquifers are used all around the country. Karstification, or the development of the karst, is higher when the difference between the top of the aquifer and regional base level is greater, creating a greater driving hydraulic head. Several examples of karst terrain exist where carbonate rocks outcrop **7** directly at the surface, such as the Edwards aquifer in south-central Texas[1]. Other examples of karst find the carbonate rocks overlain by other materials. This situation is referred to as "mantled" karst. One example of mantled karst is Lake Barco in Florida where the lake itself sits on the unconsolidated surficial deposits slumped into a sinkhole in the underlying highly soluble limestone[4].

The wide range in size and shape of dissolution features makes karst aquifers nearly impossible to characterize without empirical testing procedures[9]. The cracks can range from small linear dissolution of fractures millimeters or centimeters in width to the widening of joints tens of meters up to the circular conduits hundreds of meters in width that we refer to as caves. Sometimes the caverns can fill with unconsolidated sediment as sinkholes collapse and fill with

---

6 Use a superscript number at the end of the sentence. This number corresponds to the numbered list of references.

7 Second reference to a source. The same reference to the paper by Brown is again referenced with the superscript numeral "1."

8 References

1. Brown DS, Patton JT. Recharge to and discharge from the Edwards Aquifer in San Antonio area, Texas. USGS Open File Report 96–181; 1995.

9 2. Flocks JG, Kindinger JL, Davis JB, Swarzenski, P. Geophysical investigations of upward migrating saline water form the Lower to Upper Floridan aquifer, Central Indian River Region, Florida. Karst Interest Group Proceedings. USGS Water Resources Investigations Report 01-4011:135–140; 2001.

3. Hubbell JM, Bishop CW, Johnson GS, Lucas JG. Numerical groundwater flow modeling of the Snake River Plain aquifer using the super-position technique. Ground Water. 1997; 35(1):59–66.

10 4. Katz BG, Lee TM, Plummer LN, Busenberg E. Chemical evolution of ground water near a sinkhole lake, northern Florida. Water Resour. Res. 1995; 31(6):1549–1564.

5. Loper, D. Steps toward better models of transport in karstic aquifers. Karst Interest Group Proceedings. USGS Water Resources Investigations Report 01–4011. 2002; p. 56–57.

6. Martin, JB, Dean, RW. Exchange of water between conduits and matrix in the Floridan aquifer. Chem. Geology. 2001; 179(1-4): 145–165.

7. Plummer, N. Tracing and dating young groundwater. 1999. USGS Fact Sheet 134–99.

8. Sepulveda, N. Comparisons among ground-water flow models and analysis of discrepancies in simulated transmissivities of the Upper Floridan aquifer in ground-water flow model overlap areas. Karst Interest Group Proceedings. 2001; USGS Water Resources Investigations Report 01-4011 p. 58–67.

8 Format of references. In CSE styles the title "references" indicates that this page contains only those works that were actually cited in the paper. Some papers also provide a list of additional references, as a resource to the reader.

9 Listing a work by multiple authors. CSE style requires you to include names of all authors up to the 10th author. If there are eleven or more authors, use the phrase "and others" after listing the first eleven.

10 Journal article. The volume number and issue number in parentheses follow the journal article name and the abbreviated title of the journal.

9. Spangler, LE. Use of dye tracing to determine conduit flow paths within source-protection areas of a karst spring and wells in the Bear River Range, Northern Utah. Karst Interest Group Proceedings. 2002; USGS Water Resources Investigations Report 02-4174 p. 75–80.

10. Taylor, CJ, Greene EA. Quantitative approaches in characterizing karst aquifers. In: Kuniansky, E, Kuniansky, E, editors. USGS Karst Interest Group Proceedings. 2001; Water Resources Investigations Report 01-4011 p. 164–166.

11. **11.** USGS Aquifer Information Pages—Aquifer Basics [Internet]. United States Geological Survey (US); [updated 2006 Apr 1; cited 2007 October 5]. Available from: http://capp.water.usgs. gov/aquiferbasics/sandcarb.html.

12. Wolfe WJ, Haugh CJ. Preliminary conceptual models of the occurrence, fate, and transport of chlorinated solvents in the karst regions of Tennessee. USGS Water-Investigations Report. 1997; 97-4097.

11. Government web site. This resource page comes from the US Geological Survey (USGS) and is referenced with a publication date and an access date (both in square brackets) and the words "Available from," a colon, and the URL (web address) for the site.

IEEE Documentation

IEEE is the Institute of Electrical and Electronics Engineers, which, as noted on their web site (www.ieee.org/web/aboutus), is an organization that includes professionals from many technical fields, not just electrical engineering. This organization publishes numerous journals and uses its own citation style. The IEEE citation style is also used in other technical publications.

IEEE documentation style uses a **footnote-type system**, enclosing numbers in square brackets. Bibliographic sources go in an appendix called "Annex," and specifically, after the annex used for normative references or standards. B stands for the informative bibliography. All normal in-text citations include "B" and the number in sequence of a source's *first* use in the text.

In-text Citation

> Norman [B1] has argued that technologies should be intuitive and easy to understand for the end user. For instance, door handles should be placed in ways that make it obvious which end of the handle to press.

The references list in Annex B would then contain the cited source, in alphanumeric sequence. Titles of books and articles can use title case, with all major words capitalized, or simply capitalize the first word of the title and the subtitle. All items are separated by commas (see sample entries below). Entries are in block format and begin with the corresponding [B1]. The first and last page numbers of the reference being used may be listed at the end, if appropriate.

Bibliography

[B1] Norman, D., *The Design of Everyday Things*, New York: Basic Books, 2002, pp. 5-6.

[B2] Love, A., *Mathematical Theory of Elasticity 2*, Cambridge, UK: Cambridge University Press, 1934.

The entire *IEEE Standards Style Manual* is available online at http://standards. ieee.org/guides/style/2007_Style_Manual.pdf. Note that the IEEE Style Guide defers to the Chicago Manual of Style on issues not covered by the IEEE. See http://www.chicagomanualofstyle.org/home.html for more information.

IEEE Sample References

Article, essay, or chapter in an edited collection

[B1] Kostelnick, C. "Melting-pot ideology, modernist aesthetics, and the emergence of graphical conventions: The statistical atlases of the United States, 1874–1925," in *Defining Visual Rhetorics*, C. A. Hill and M. Helmers, Eds. Mahwah, NJ: Erlbaum, 2004, pp. 215–242.

Article in a scholarly journal with continuous pagination

[B2] Song, D. L., Conrad, M. E., Sorenson, K. S., and Alvarez-Cohen, L., "Stable carbon isotope fractionation during enhanced situ bioremediation of TCE," *Environ. Sci. Technol.*, vol. 36, pp. 2262–2268, 2002.

Article in a scholarly journal paginated by issue

[B3] Johannesen, C., Funderburg, T., and Welhan, J., "Basalt surface morphology as a constraint in stochastic simulation of high conductivity zones in the Snake River Plain aquifer, Idaho," *Geol. Soc. America*, vol. 31, no. 4, pp. A-12, 1998.

Article in a magazine published monthly

[B4] Anderson, G. T., Tunstel, E. W., and Wilson, E. W., "Robot System to Search for Signs of Life on Mars," *IEEE Aerospace and Electronic Systems Magazine*, vol. 22, no. 12, pp. 23–30, 2007.

Article in a popular magazine published weekly

[B5] M. Gladwell, "Open Secrets: Enron, Intelligence, and the Perils of Too Much Information," *New Yorker,* retrieved Jan. 8, 2008 from http://newyorker.com/reporting/2007/01/08/070108fa_fact_gladwell.

Blog or wiki posting

[B6] Mikell Taylor, "CES: robots for girls?" *Automaton: IEEE Spectrum's Blog on Robots and Other Silicon-Brained Contraptions*, retrieved Jan. 10, 2008 from http://blogs.spectrum.ieee.org/automaton/2008/01/10/ces_robots_for_girls.html

Book by a single author

[B7] Love, A., *Mathematical Theory of Elasticity 2*, Cambridge, UK: Cambridge University Press, 1934.

Book in a second or subsequent edition

[B8] Paul, C. R., *Analysis of Multiconductor Transmission Lines*, 2nd ed., New York: Wiley-IEEE, 2007.

Book by an unknown author

[B9] *Lonely Planet Bluelist*. 3rd ed., Melbourne, Australia: Lonely Planet, 2008.

Book with a translator

[B10] Heisenberg, W., *The Physical Principles of the Quantum Theory*, 2nd ed., Eckhart E., and Hoyt, F. C., trans., New York: Dover, 1949.

Book by multiple authors

[B11] Salen, K. and Zimmerman E., *Rules of Play: Game Design Fundamentals*. Cambridge, MA: MIT Press, 2004.

Note: For books with more than ten authors, list the first ten and then use "and others."

Book by an organization or corporation

[B12] Pfizer Global Pharmaceuticals, *The Faces of Public Health*, Surrey, UK: Pfizer Global Pharmaceuticals, 2004.

Book with an editor

[B13] *Rapid Prototyping of Application Specific Signal Processors*, Gadient, A. J., Frank, G. A., and Richards, M. A., editors, Dordrecnt, the Nether lands: Kluwer Academic Publishers, 1997.

Brochure

[B14] NIH Publication No. 03-5329, "Harmful interactions: Mixing alcohol with medicines," National Institute on Alcohol Abuse and Alcoholism, 2007.

Computer software

[B15] *STORYPACE* [Computer Software], Watertown, MA: Eastgate Systems, Inc., retrieved December 20, 2007 from http://www.eastgate.com/Storyspace.html.

Note: You do not need to reference common software programs such as Microsoft Word or PowerPoint. You should reference more specialized software programs that will be unfamiliar to your reader.

Conference proceedings

[B16] Subhash, G., and Zhang H. H., "Thermodynamic and Mechanical Behavior of Hafnium/Zirconium-Based Bulk Metallic Glasses," *Proceedings of the International Conference on Mechanical Behavior of Materials (ICM-9)*, Geneva, Switzerland, May 25–29, 2003.

Dictionary

[B17] Microsoft Corporation, *Microsoft Computer Dictionary*, Fifth Edition, Redmond, WA: Microsoft Press, 2002.

Email

Note: The 2007 IEEE manual does not discuss references to personal communications (letters, email, conversations) within the text of your document. There are, however, forms of email communication that may be public and available for retrieval; many organizations post important business emails on a company web site. In this case, you would cite the source within your paper and also provide a reference, following the format where the email is archived (web site, blog, other). The reference is then:

[B18] Gurak, L. J., "Announcing the New Writing Initiative," Minneapolis, MN: University of Minnesota, May 23, 2006, retrieved Jan. 10, 2008 from http://www.messages.university. edu/mail-archives/newwritmsg.html.

Encyclopedia

[B19] *Encyclopedia of Internet Technologies and Applications*, Freire M., and Pereira M., editors, Norwood, MA: IGI, 2008.

Encyclopedia Entry

[B20] Knox, B. M. W., "Books and Readers in the Greek World: From the Beginnings to Alexandria," *Cambridge History of Classical Literature*, vol. 1, eds. Easterling, P. E., and Knox, B. M. W., Cambridge: Cambridge UP, 1985, 1–16.

Interview

If you conduct an interview yourself, reference it within your paper but do not include it in your bibliography. If the interview is available in an accessible format, such as video on a news web site or on YouTube, follow the instructions for citing a video.

Film, videotape, or DVD

[B21] Pollak, R., Producer, *One Woman, One Vote*. [Videorecording]. Washington, DC: PBS Video, 1995.

Government document

E [B22] U.S. Environmental Protection Agency (EPA), *Weather makes a difference: 8-hour ozone trends for 1997–2006*, 2007, retrieved Jan. 15, 2008 from http://www.epa.gov/airtrends/weather/region09.pdf.

[B23] Department of Commerce (US), Population profile of the United States (Publication No. P-20, No. 808), Washington DC: Government Printing Office (US), 2007.

Newspaper article

[B24] Turner, J., "Disorder Kills Without Warning," *The Toronto Star*, pp. F1–F2, June 26, 1998.

⑤ [B25] Boudreau, J., "Yahoo Unveils Newest Cell Phone Home Page," MercuryNews.com [Internet]. Jan. 7, 2008, retrieved Jan. 12, 2008 from http://www.mercurynews.com/ci_7903908?source=rss&nclick_check=1#.

Lecture or public speech

[B26] Richards, M. A., "Space-Time Adaptive Processing Architecture," panel session at the Mass. Inst. of Technology Lincoln Laboratory Adaptive Sensor Array Processing (ASAP) Workshop, Lexington, Massachusetts, May 16, 1994.

Letter to the editor

[B27] Silva, J. T., "The Health Care Crisis" [letter to the editor], *Houston-Post,* p. A14, Aug. 5, 2003.

Newspaper editorial

⑤ [B28] Brooks, D. "Middle-Class Capitalists," *New York Times.* Jan. 11, 2008, retrieved Jan. 13, 2008 from http://www.nytimes.com/2008/01/11/opinion/11brooks.html.

[B29] "Help students—not the tax revolt" [editorial], *Los Angeles Times,* p. B6, July 31, 2001.

Podcast

[B30] Institute of Electrical and Electronics Engineers (IEEE), "Cars, Eco-footprints, Tech Toys, and Electric Cars," IEEE Spectrum Radio [podcast], December 2007, retrieved Jan. 15, 2008 from http://IEEE.org/radio?id=2318.

Online article based on a print source

[B31] Cabilio P., Farrell P., "A Computer-based Lab Supplement to Courses in Introductory Statistics," *The American Statistician* 2001; 55; 228–232, retrieved Jan. 15, 2008 from http://www.jstor.org/journals/astata.html.

Online article not based on a print source

[B32] Braaten, D., Haas, N. Li, X., Rowley, R. J., Hulbutta, K., Kostelnick, J., and Meisel, J., "Sea Level Rise and GIS Data," 2007, retrieved Jan. 10, 2008 from http://cresis.ku.edu/research/data/sea_level_rise/index.html.

Online article retrieved from a database

[B33] Ghosh, D., Chowdhury, A. R., and Saha, P., "On the Various Kinds of Synchronization in Delayed Duffing-Van der Pol System," *Communications in Nonlinear Science & Numerical Simulation*, vol. 13, no. 4, pp. 790–803, July 2008, retrieved Jan. 10, 2008 from *Academic Search Complete*.

Report or other publication where the author is a company or organization

[B34] Michigan Department of Natural Resources, *A Biological Survey of Seeley Drain*, Oakland County, Michigan, May 10, 1989, pp. 1–11.

Sound recording (music, other audio)

[B35] Covey S. R., *The 7 habits of highly effective people* [CD-ROM], 15th edition, New York: Free Press, 2004.

Unpublished material (dissertation, thesis)

[B36] Shah, J. J., "Microfluidic Devices for Forensic DNA Analysis." Ph.D. diss., George Mason University, 2008.

Web site

If you need to cite a web site in its entirety (as opposed to a paper, report, or other document that is contained on a web site), your reference should include the author or organization, the title of the site, a date of publication, date of retrieval, and the URL (web address) for the home page. If a URL extends to more than one line, break it at a slash and do not add a hyphen.

[B37] Lynch, P., Horton, S., *Web style guide,* 2nd edition (online version), 2002, retrieved Jan. 14, 2008 from http://www.webstyleguide.com/site/index.html.

Sample Pages in IEEE Style

 Microbial Remediation

Prepared by:

Carolyn Moores

University of Idaho

Chemical Geology

May 1, 2005

 Title page format. The title page for an IEEE style paper includes the title, the words "Prepared by" followed by the name and affiliation of the author or authors at the left margin. The bottom of the page includes the course name and date.

2 Abstract

In order to develop a successful microbially mediated remediation strategy for the TCE contamination plume at the TAN site of the INEEL, research needs to be done to accurately characterize methanotrophs and symbiant microbial communities to enhance natural attenuation of TCE. Cometabolism of the TCE by methanotrophs offers a promising solution to the problem of the growing plume because methanotrophs live in the aquifer and have been demonstrated to fortuitously degrade TCE through cometabolism. By better understanding the methanotrophic community, exactly what species live in the aquifer, how they are distributed in the media, and what other microbial communities naturally assist in the TCE degradation by methanotrophic communities, strategies can be developed to successfully increase the TCE degradation, such as injection of the proper chemicals and nutrients to increase or stimulate the metabolism of the methanotrophs and other organsisms which may contribute to their cometabolism by proving vital nutrients and H+ ions **3** for the dechlorination reaction.

4 Geologic Setting

The Idaho National Laboratory (INL) is an 890 mi^2 site operated by the Department of Energy located on the Eastern Snake River Plain (ESRP) in southeastern Idaho. The ESRP is a northeast-trending crescent-shaped depression extending from near the western boundary of Yellowstone National Park to the Thousand Springs area north of the Snake River. The ESRP is structurally bounded on the northwest by faulting and on the southeast by down warping and faulting **5** [B1].[1] The volcanic basin is filled with 1,000 to 2,000 feet or more of Pliocene and younger basaltic rocks overlying older rhyolytic ash-flow tuffs. The many flow groups that comprise the subsurface are composed of discrete flows that erupted **7** about the same period of time. Median flow thickness ranges from 7 to 25 feet [B2]. Periods of erosion and deposition between eruptions created sedimentary

6 [1] The numbers in brackets correspond to those of the bibliography in Annex B.

---

2 Abstract. Paper begins with an optional abstract.

3 Format. Document is single spaced, with headings centered.

4 Subheadings. Subheads can be helpful, but are optional.

5 In-text citation. First citation in text uses letter of the Annex (B for informative bibliography), the number 1, and a required footnote at the end of the sentence.

6 Footnote for citations. Footnote required to explain the first citation.

7 Citation sequence. Citations are numbered in order of their *first* appearance in the text.

interbeds which occasionally punctuate the basalt flow groups. The thickness of the interbeds ranges from feet to tens of feet.

Hydrologic Setting

The Snake River Plain Aquifer (SRPA) is a large, freshwater aquifer hosted in the fractured basalts of the ESRP. It is technically an unconfined aquifer, but it is often treated in numerical modeling as a semi-confined or confined aquifer. The depth to the water table and the large relative thickness of the aquifer mean changes in the water table are negligible relative to the overall thickness, allowing transmissivity to be handled with linear modeling equations. Around TAN area on the INL, the water table is approximately 200 to 250 feet deep. Transmissivities in the aquifer span orders of magnitude due to the heterogeneity of the aquifer. Preferential flow paths created by the fractured basalt combined with variable locations of sedimentary interbeds made of weathered volcanic rock to clay and wind-blown sand create a mosaic of hydraulic conductivities and transmissivities. The temperature and pH of the aquifer are relatively stable, but can be altered by the presence of contaminants. The pH is generally neutral and the temperature has been recorded at 12–13°C [B3].

Groundwater Contamination

The INL has been in operation for 50 years. During that period of time, research in harnessing nuclear energy and expanding its practical applications has been the primary objective of the laboratory. In reaching its goals, the INL has produced waste products both radiological and chemical in nature. Disposing of these wastes

2

8 Running header. Header includes a shortened title and the date.

9 Acronyms. Acronyms are written out in first use and then defined in parentheses.

Annex B

(informative)

Bibliography

[B1] Garabedian, S. P., "Hydrology and digital simulation of the regional aquifer system, ESRP, Idaho," Professional Paper 1408-F, ISBN 0607715154, 1992.

[B2] Johannesen, C., Funderburg, T., and Welhan, J., "Basalt surface morphology as a constraint in stochastic simulation of high conductivity zones in the Snake River Plain aquifer, Idaho," *Geol. Soc. America*, vol. 31[4], pp. A-12, 1998.

[B3] Macbeth, T. W., Cummings, D. E., Spring, S., and Petzke, L. M., "Molecular characterization of a dechlorinating community resulting from in situ biostimulation in a TCE contaminated deep, fractured basalt aquifer and comparison to a derivative laboratory culture," *Appl. Environ. Microbiol.*, vol. 70, pp. 7329–7341, 2004.

[B4] Newby, D. T., Reed, D. W., Petzke, L. M., Igoe, A. L., Delwiche, M. E., Roberto, F. F., McKinley, J. P., Whiticar, M. J., and Colwell, F. S., "Diversity of methanotroph communities in a basalt aquifer," *FEMS Microbiol. Ecology*, vol. 48, pp. 333–344, 2004.

[B5] Lehman, R. M., Roberto, F. F., Earley, D., Bruhn, D. F., Brink, S. E., O'Connell, S. P., Delwiche, M. E., and Colwell, F. S., "Attached and unattached bacterial communities in a 120-meter corehole in an acidic, crystalline rock aquifer," *Appl. Environ. Microbiol.*, vol. 67, pp. 2095–2106, 2001.

[B6] Tobin, K. J., Onstott, T. C., DeFlaun, M. F., Colwell, F. S., and Fredrickson, J., "In situ imaging of microorganisms in geologic material," *Journal of Microbiol. Methods*, vol. 37, pp. 201–213, 1999.

---

10 Annex title. Annex B stands for informative bibliographic sources.

11 Type of Annex. Informative label appears next, in parentheses, not bolded.

12 Bibliography title. Sources are listed in a bibliography. Both headings are in large boldface font.

13 Order of reference. Entries listed in order of first appearance in the document.

14 Format of references. All entries are single-spaced and in block format.

15 Listing a work by multiple authors. Multiple authors are listed up to ten names.

16 Journal titles. Many known journals use accepted title abbreviations.

MLA Documentation

Modern Language Association (MLA) documentation style is used primarily in literature and the humanities. While not a common format in technical writing, MLA is the preferred documentation style in English departments and foreign languages, which prefer MLA style for college and university research papers.

MLA style uses in-text name and page number citations for all quotations, paraphrases, and summaries of source material. The parentheses include only the author's name and a page number with no comma between them. If the source has no named author, a shortened title is used instead. MLA also encourages "signal phrases" in the sentence that include an author's name and sometimes the title of the work. In that case, you include only the page number in parentheses.

> Hunter explains how the discriminatory compensation practice for "typically female jobs" in the market "is carried over to the firm" (16).
> The discriminatory compensation practice for "typically female jobs" in the market "is carried over to the firm" (Hunter 16).

Block quotations are used for long quotations of five or more lines. Indent the text five spaces on both sides and place the in-text citation after the period of the last sentence.

> If, when dollar values are assigned to the compensable factors, market values for typically female jobs are used as benchmarks for female jobs in the establishment, then discrimination in the market place is carried over to the firm. (Hunter 16)

The list of references is called the Works Cited, which appears double-spaced at the end of the regular document. Entries must include the first (if available) and last name, the title of the work, the edition or editors, if applicable, the publisher and the year. Titles of books and articles all use title case, with all major words capitalized. All items are separated by periods, except when colons are used for journals and before page numbers (see sample entries below). Entries use a hanging indent of five spaces after the first line.

> Hunter, Frances C. Equal Pay for Comparable Worth. New York: Praeger, 1986.603

■ The *MLA Syle Manual and Guide to Scholarly Publishing*, Third Edition (2008), includes several significant changes to documentation and formatting guidelines. These include italicizing, rather than underlining, all titles; adding both issue and volume number to all journal entries; and adding the medium of publication (Print, Web, Audio, CD, Film, etc.) to every works-cited entry, as well as a simplified style for all online citations. The MLA recommends that undergraduate students continue to use the current documentation guidelines, as reflected below, until the new seventh edition of the *MLA Handbook for Writers of Research Papers* is published, probably in spring 2009. You may want to ask your instructor which guidelines to follow.

- Place content notes on a Notes page directly before the Works Cited.
- Capitalize all the key words in a title and the subtitle.

MLA Sample Works Cited Entries

Article, essay, or chapter in an edited collection

> Wysocki, Anne Frances. "Seriously Visible." Eloquent Images: Word and Image in the Age of New Media. Eds. Mary E. Hocks and Michelle R. Kendrick. Cambridge, MA: MIT, 2003. 37–59.

Article in a scholarly journal with continuous pagination

> Knadler, Stephen. "E-Racing Difference in E-Space: Black Female Subjectivity and the Web-Based Portfolio." Computers and Composition 18.3 (2001): 235–55.

Article in a scholarly journal paginated by issue

> Douglas, J. Yellowlees. "What Hypertexts Can Do That Print Narratives Cannot." Reader 28 (Fall 1992): 1–22.

Article in a popular magazine published monthly

> Denise Wilson and Ella Kliger. "Learning From Katrina." IEEE Spectrum. Jan. 2008. 15 Jan. 2008 <http://IEEE.org/magazineindex>.

Article in a popular magazine published weekly or daily

> Gladwell, Malcolm. "Open Secrets: Enron, Intelligence, and the Perils of Too Much Information." New Yorker. Jan. 8 2007. 15 Jan 2008 <http://newyorker.com/reporting/2007/01/08/070108fa_fact_gladwell>.

Article in a print-based scholarly journal that is accessed online

> Enoch, Jessica. "A Woman's Place Is in the School: Rhetorics of Gendered Space in Nineteenth-Century America." College English. 70: 3 (Jan. 2008). 15 Jan 2008 <http://www.ncte.org/pubs/journals/ce/contents/125132. htm>.

Blog or wiki posting

Taylor, Mikell. "CES: robots for girls?" Automaton: IEEE Spectrum's Blog on Robots and Other Silicon-Brained Contraptions (blog). Jan 10 2008. 15 Jan 2008 <http://blogs.spectrum.ieee.org/automaton/2008/01/10/ces_robots_for_girls.html>.

Book by a single author

Schriver, Karen A. Dynamics in Document Design: Creating Text for Readers. New York: John Wiley, 1997.

Book in a second or subsequent edition

Banham, Reyner. Theory and Design in the First Machine Age. 2nd ed. Cambridge, MA: MIT P, 1999.

Book by an unknown author

Lonely Planet Bluelist. 3rd ed. Melbourne, Australia: Lonely Planet, 2008.

Book with a translator

Gu, Batoong, Changfu Chang, and Shunzhu Wang, Eds. and Trans. Contemporary Western Rhetoric: Critical Methods and Paradigms. Beijing, China: China Social Sciences Academy Press, 1998.

Book by multiple authors

Wysocki, Anne Frances, Johndan Johnson-Eilola, Cynthia Y. Selfe, and Geoffrey Sirc. Writing New Media: Theory and Application for Expanding the Teaching of Composition. Logan, UT: Utah State UP, 2004.

Book review

Razee, Alan. "Review of Eloquent Images: Word and Image in the Age of New Media." Resource Center for Cyberculture Studies (RACC) Book Reviews. Dec 1 2007. 15 Jan 2008 <http://rccs.usfca.edu/booklist.asp>.

Book by an organization or corporation

Microsoft Corporation. Microsoft Manual of Style for Technical Publications, Third Edition. Redmond, WA: Microsoft Press, 2003.

Book with editors

Fleckenstein, Kristie S., Linda T. Calendrillo, and Demetrice A. Worley, eds. Language and Image in the Reading-Writing Classroom: Teaching Vision. Mahwah, NJ: Lawrence Erlbaum Associates, 2002.

Brochure

> National Institute of Outdoor Canine Activities. How to Go Dogsledding
> and Love It (brochure). New York, 2005.

Computer software

> STORYSPACE. Computer software. Watertown, MA: Eastgate Systems, Inc.
> N.d. 15 Jan 2008 <http://www.eastgate.com/Storyspace.html>.

Note: You do not need to reference common software programs such as Microsoft Word or PowerPoint. You should reference more specialized software programs that will be unfamiliar to your reader.

Conference proceedings

> Marshall, Catherine C. and Gene Golovchinsky. "Saving Private Hypertext:
> Requirements and Pragmatic Dimensions for Preservation."
> Proceedings of the Fifteenth ACM Conference on Hypertext. New
> York: Association for Computing Machinery, 2004. 130–38.

Dictionary

> Microsoft Corporation. Microsoft Computer Dictionary, 5th ed. Redmond,
> WA: Microsoft Press, 2002.

Edited Collection

> MAIN REFERENCE
> Snyder, Ilana, ed. Page to Screen: Taking Literacy into the Electronic Era.
> London: Routledge, 1998.

> SUBSEQUENT REFERENCE
> Johnson-Eilola, Johndan. "Living on the Surface: Learning in the Age of
> Global Communication Networks." Snyder 185–210.

Email

The *MLA Handbook for Writers of Research Papers*, 5th edition, suggests that email communication, similar to other personal communications (letters, memos), is not something your reader can retrieve by looking at a reference. The *MLA Handbook* recommends that if you cite an email within the text, either provide a copy in the appendix or list it on the Works Cited as a personal communication.

> Hocks, Mary E. "Visual Rhetoric Sources" (email). Jan. 10, 2008.

There are forms of email communication that may be public and available for retrieval; many organizations post important business emails on a company web site.

In this case, you would cite the source within your paper and also provide a reference.

> Gurak, L. J. "Announcing the new writing initiative" (email). May 23, 2006. 15 Jan 2008 <http://www.messages.university.edu/mail-archives/newwritmsg.html>.

Encyclopedia

> Encyclopedia of Internet Technologies and Applications. M. Freire, M. Pereira, Eds. Norwood, MA: IGI, 2008.

Encyclopedia Entry

> Hocks, Mary E. "Reception Theory." Encyclopedia of English Studies and the Language Arts, Vol. II. Gen. Ed. Alan Purves. Urbana, IL: National Council of Teachers of English, 1994. 1019–1020.

Interview

If you conduct an interview yourself, you may cite it within your paper and include it in an appendix or reference it as a personal communication in your works cited. If the interview is available in an accessible format, such as video on a news web site or on YouTube, follow the instructions for citing a video.

> Hocks, Mary E. Personal Interview. 15 Jan. 2008.

Film, videotape, or DVD

> Calendar Girls (DVD). Burbank, CA: Touchstone Pictures, 2003.

Government document

E   U.S. Environmental Protection Agency (EPA). Weather Makes a Difference: 8-hour Ozone Trends for 1997–2006. 2007. 15 Jan. 2008 <http://www.epa.gov/airtrends/weather/region09.pdf>.

> U.S. Department of Commerce. Population profile of the United States (Publication No. P-20, No. 808). Washington, DC: Government Printing Office, 2007.

Lecture or public speech

> Gerrard, Lisa. "Gender, Culture, and the Internet." Presented at the Conference on College Composition and Communication. March 1997.

Letter to the editor

> Silva, J. T., "The health care crisis" (Letter to the editor). August 5, 2003. Houston-Post, p. A14.

Newspaper article

> Leyden, Peter. "The Changing Workscape." Special Report, Part III. Minneapolis Star Tribune 18 June 1995: 2T–6T.

(E) Boudreau, J. "Yahoo unveils newest cell phone home page." MercuryNews.com. Jan 7, 2008. 12 Jan. 2008 <http://www. mercurynews.com/>.

Newspaper editorial

(E) Brooks, David. "Middle-Class Capitalists." New York Times. Jan. 11, 2008. 13 Jan. 2008 <http://www.nytimes.com/2008/ 01/11/ opinion/11brooks.html>.

"Help students—not the tax revolt." (Editorial). Los Angeles Times. 31 July 2001: B6.

Online article based on a print source

Cabilio P. and P. Farrell. "A computer-based lab supplement to courses in introductory statistics." The American Statistician 55 (2001): 228–232. Online <http://www.jstor.org./journals/astata.html>.

Online article not available in print

Comstock, Michelle and Mary E. Hocks. "Re-framing Voice, Music, and the Cultural Soundscape: Toward a Rhetoric of Sound in Composition Studies." Computers and Composition Online. 5:2 (Fall 2006). 15 Jan. 2008 <http://www.bgsu.edu/cconline/sound/>.

Online article retrieved from a database

Pareck, Rupal. "The Importance of Web-based Software." Accounting Today. Supplement on Firm of the Future. June 18, 2007: 24A. 15 Jan. 2008 <www.Lexisnexis.com>.

Podcast

Institute of Electrical and Electronics Engineers (IEEE). "Cars, Eco-footprints, Tech Toys, and Electric Cars" (podcast) IEEE Spectrum Radio. Dec 2007. 15 Jan. 2008 <http://IEEE.org/radio?id=2318>.

Report or other publication where the author is a company *or* organization

> Worldwatch Institute. State of the World 2008. Washington, DC: Worldwatch Institute, 2008.

Pew Internet & American Life Project. "The Internet and Education:

Findings of the Pew Internet & American Life Project." 2001.

15 Jan. 2008 <http://www.pewinternet.org/PPF/r/39/report_

display.asp>.

Special Issue of Journal

Webb Peterson, Patricia, and Wilhelmina Savenye. Distance Education.
Spec. Issue of Computers and Composition 18.4 (2001).

Sound recording (music, other audio)

Covey, S. R. The 7 Habits of Highly Effective People (CD-ROM). Fifteenth
edition. New York: Free Press, 2004. 13 CD set.

Unpublished material (dissertation, thesis)

Heineman, David Scott, Ph.D. The Digital Rhetorics of Hacktivism:
Anti-institutional Politics in Cyberspace. Iowa City: University of
Iowa, 2007. 241 pages.

Web site

If you need to cite a web site in its entirety (as opposed to a paper, report, or other
document that is contained on a web site), your reference should include the title
of the site, a date of publication (or n.d.), a date of retrieval, and the URL (web
address) for the home page. Put the URL in angle brackets (< >). If a URL extends to more than one line, break it at a slash and do not add a hyphen.

Lynch, Patrick and S. Horton. Web style guide: 2nd edition (online
version). 2002. 10 Jan. 2008 <http://www.webstyleguide.
com/site/index.html>.

Sample Pages in MLA Style

① John Michael Jansen

Professor M, Hall

English 102

6 November 2000

② Comparable Worth in Salaries

Industry needs to adopt an evaluative system that is blind to gender in comparing all jobs in order to achieve comparable worth in salaries. When jobs are worth the same to an organization, those who staff the jobs should be paid the same salary. However, opponents counter that the market system properly determines wages and salaries, and that businesses and professions must base wage and salary decisions on the free market. How do we solve the dilemma, especially to provide equity to women and minorities who have suffered the most discrimination in comparable worth jobs? The solution appears to lie in adopting fair job evaluation systems that award equal pay to women and men who perform the same jobs. This would reconcile the concerns of minority workers, including women, and those of employers and taxpayers as well as labor economists, so long as all share the economic and cultural costs of such programs. As a start, this can be accomplished when all parties concerned involve themselves in implementing and funding such procedures within their own companies.

Women in the 1980s earned on the average about 30 to 35 percent less than men earned (Aaron and Loughy 4). It appears to be common knowledge that **③** within a corporation male janitors can earn more money than female secretaries **④** and that highway repair flagmen earn more than the company's female nurse. The inequities extend beyond one corporation. Truck drivers, usually male, earn more than teachers, usually female. Problems exist even within the professions. Paul Recer points out that "women physicians in 1988 earned only 62.8 percent of the

---

① Identification and title page format. A separate page is not required by MLA guidelines. Name and other identifying information is placed in a block at the left margin at the top of the page.

② Introduction and thesis. The thesis statement appears in the first sentence in the form of an

argumentative proposition that identifies a problem within industry. The thesis statement usually appears near the beginning of the paper but it is not always the very first sentence.

③ In-text documentation. Because names of authors are not incorporated into the text discussion, they appear in a parenthetical reference

along with the page number on which the information is found.

④ Common knowledge. Views expressed by the author about janitors and flagmen do not come from any particular source and need no citation common knowledge. Thus, they do not need to be documented.

⑥ pay received by male doctors [...] a decline from 1982, when female physicians earned 63.2 percent of the pay of male doctors" (A4).

Although men and women freely make career decisions in the same way, our culture places restrictions on women's choices. Some of the differences between men's and women's earnings are due to decisions women make about how many ⑦ hours to work each week, what occupations to choose, and when to enter or to leave the work force. On the other hand, custom and tradition also influence wages in different jobs and occupations, a view that acknowledges that women improve their relative earnings by entering jobs traditionally held by men. Women, who enter these same jobs and work just as many hours, earn, in many cases, less salary than their male counterparts. Recer also cites a study of the American Medical Women's Association that reveals that female physicians "are classed in the four lowest-paid specialities: general practice, pediatrics, psychiatry, and internal medicine" (A4).

⑧ However, the narrower the definition of industries and occupations, the greater the apparent segregation of women and men who hold particular jobs becomes. According to the U.S. Department of Labor classifications, 37 percent of all women work in industries in which at least 66 percent of the employees are women (Norwood 4). Furthermore, it seems to be common knowledge that the larger the proportion of female workers in an industry, the lower the average hourly wage they receive.

Can comparable worth be used to set wage standards? To do so requires some form of job evaluation, a method of ranking jobs according to their value to the employer. Its purpose, Meeker says, is to provide a system of comparing ⑨ "dissimilar work [...] to determine appropriate wage levels" (674). The result of ⑩ the evaluation process becomes a hierarchy of job values, a model reflecting the

⑤ Paper format. All pages of the paper, excluding the title page, are numbered in the upper right corner.

⑥ Use of evidence. The author cites published evidence and examples to support his thesis. Note that the page number of the newspaper in parentheses also includes the section of the newspaper (A4).

⑦ Developing the argument. After presenting some examples,

the author shows how society places restrictions on the occupational choices that most women make. Note the use of the transitional phrase *on the other hand*.

⑧ Transition. The author accomplishes a smooth transition between paragraphs by using the conjunctive adverb *however* in the opening sentence of the next paragraph.

⑨ Direct quotation. Notice that a capital letter does not follow the opening quotation mark. The in-text citation includes only the page number, not source name, because Meeker (source) is mentioned directly in the text itself.

⑩ Ellipsis within a quotation. The three dots indicate that the author did not present the full sentence from the source.

(11) Works Cited

(12) Aaron, Henry J., and Cameron M. Lougy. The Comparable Worth Controversy. Washington: Brookings Institute, 1986.

Hutner, Frances C. Equal Pay for Comparable Worth. New York: Praeger, 1986.

McArthur, Leslie Zebrowitz. "Social Judgment Biases in Comparable Worth Analysis." Comparable Worth: New Directions for Research. Ed. Heidi I. Hartmann. Washington: National Academy, 1985. 53–70.

Meeker, Suzanne E. "Equal Pay, Comparable Work, and Job Evaluation. "Yale Law Journal 90 (1981): 674–92. InfoTrac: Expanded Academic ASAP. CD-ROM. Information Access. 1989.

(13) Norwood, Janet L. The Female-Male Earnings Gap: A Review of Employment Earnings Issues. U.S. Department of Labor. Bureau of Labor Statistics, Report 673. Washington: GPO, Sept. 1982. 1 Oct. 1996. <http://www.gpo.gov/lbrpt673.jnorwood.htm>.

Recer, Paul. "Study: Health Field Still Sex-Segregated." South Bend Tribune 8 Sept. 1991, metro ed.: A4.

Schwab, Donald P. "Job Evaluation Research and Research Needs." Comparable Worth: New Directions for Research. Ed. Heidi 1. Hartmann. Washington: National Academy, 1985. 37–52.

Treiman, Donald J. Job Evaluation Research: An Analytic Review. Washington: National Academy of Sciences, 1979.

---

(11) In MLA style, references page is called "works cited."

(12) Typical entry for a book.

(13) Entry for a government source from a website.

References

Belanger, S. E., Kendall, S. L., Matoush, T. L., & Wu, Y. D. (2005). *Business and technical communication: An annotated guide to sources, skills, and samples.* Westport, CT: Praeger.

Crescimanno, B. (2005, December). Sensible forms: A form usability checklist. *A List Apart: For People Who Make Web Sites.* No. 209. Retrieved May 26, 2008 from http://www.alistapart.com/articles/sensibleforms.

Fogg, B. J. (2002, May). Stanford guidelines for web credibility. A research summary from the Stanford Persuasive Technology Lab. Palo Alto, CA: Stanford University. Retrieved May 26, 2008 from http://www.webcredibility.org/guidelines.

Habig, A. (2006). Physics 3061: Introductory laboratory electronics (class syllabus). University of Minnesota. http://neutrino.d.umn.edu/phy3061/P3061SYL.html

Irby, K. (2003, April 2). *LA Times* photographer fired over altered image. *Poynter Online.* Retrieved May 26, 2008 from http://www.poynter.org/content/content_view.asp?id=28082.

Jarrett, C. (2004, December). Hooray, I'm doing the forms! *Intercom Online: The Magazine for the Society of Technical Communicators*, pp. 6–8. Retrieved May 26, 2008 from http://www.stc.org/intercom/.

Kohl, J. (1999). Improving translatability and readability with syntactic cues. *Technical Communication,* 2nd quarter, 149–166.

Lynch, P. and Horton, S. (2002). *Web style guide;* 2nd edition (online version). Retrieved May 26, 2008 from http://www.webstyleguide.com/site/index. html.

McCloud, S. 1994. *Understanding comics.* New York: HarperPerennial.

National Institute on Aging and the National Library of Medicine. (2002). Making your web site senior friendly: A checklist. Retrieved May 26, 2008 from http://www.nlm.nih.gov/pubs/checklist.pdf.

Pew Internet & American Life Project. (2001). The Internet and education: Findings of the Pew Internet & American Life Project. Retrieved May 26, 2008 from http://www.pewinternet.org/ PPF/r/39/report_display.asp.

Pew Internet & American Life Project. (2006). Online health search 2006. Retrieved May 26, 2008 from http://www.pewinternet.org/PPF/r/190/report_display.asp.

Price, J. and Price, L. (2002). *Hot text: Web writing that works.* Indianapolis, IN: New Riders.

Tufte, E. R. (1983). *The visual display of quantitative information.* Chesire, CT: Graphics Press.

Tufte, E. R. (1997). *Visual explanations: Images and quantities, evidence and narrative.* Chesire, CT: Graphics Press.

Tufte, E. (2005). PowerPoint does rocket science: Assessing the quality and credibility of technical reports. Retrieved May 26, 2008 from http://www.edwardtufte.com/bboard/q-and-a-fetch-msg?msg_id=0001yB.

University of Minnesota (2000). QuickStudy: Library research guide. Evaluating sources. http://tutorial.lib.umn.edu/infomachine.asp?moduleID=9&lessonID=42&pageID=348

U.S. Environmental Protection Agency (EPA). (2007). Weather makes a difference: 8-hour ozone trends for 1997–2006. Retrieved May 26, 2008 from http://www.epa.gov/airtrends/weather/region09.pdf.

U.S. Office of Health and Human Services. (1993). Office for Protection from Research Risks. Tips on informed consent. Retrieved May 26, 2008 from http://www.hhs.gov/ohrp/humansubjects/guidance/ictips.htm.

W3C. (2004). Web content accessibility guidelines 2.0. Retrieved May 26, 2008 from http://www.w3.org/TR/2004/WD-WCAG20-20040311/.

Weber, J. H. (1999). Choosing and using help topics. *Technical Editors' Eyrie: Resources for Technical Editors.* Retrieved May 26, 2008 from http://www.jeanweber.com/howto/planhlp2.htm.

Wikipedia. (2007). Researching with Wikipedia. Retrieved October 18, 2007 from http://en.wikipedia.org/wiki/Wikipedia:Researching_with_Wikipedia.

Wysocki, A. F., Sorby, S. and Baartmans, B. (2002). *Introduction to 3D spatial visualization: an active approach.* CD-ROM with workbook. Clifton Park, NY: Thomson Delmar.

Credits

Text Credits

Pages 10, 16, 25–26: Reprinted by permission of Steve Gough; **p. 13:** Reprinted by permission of Drupal Association, http://drupal.org; **p. 29:** Source: Liz Tasker; **pp. 31–32:** Microsoft Office Project® is a registered trademark of Microsoft Corporation. Screen shots reprinted with permission from Microsoft Corporation; **p. 36:** Source: Liz Tasker; **pp. 39–40:** Microsoft Word® is a registered trademark of Microsoft Corporation. Screen shots reprinted with permission from Microsoft Corporation; **p. 51:** Franklin Street, Fifth Floor, Boston, MA 02110-1301 USA. Everyone is permitted to copy and distribute verbatim copies of this license document, but changing it is not allowed; **p. 56:** Reprinted by permission of Society for Technical Communication, www.stc.org; **p. 57:** Copyright © 2000, 2001, 2002 Free Software Foundation, Inc., **pp. 60–61:** Courtesy California Contractors State License Board; **p. 64:** Microsoft Entourage® is a registered trademark of Microsoft Corporation. Screen shot reprinted with permission from Microsoft Corporation; **p. 76:** Courtesy of www.lexar.com; **p. 80:** Courtesy of *New Scientist* Magazine, www.sciencejobs.com; **p. 88:** Reprinted by permission of Virginia Tech Career Services; **p. 92:** Microsoft Word® is a registered trademark of Microsoft Corporation. Screen shot reprinted with permission from Microsoft Corporation; **p. 94:** Copyright 2007 - Monster, Inc. All rights reserved. Reprinted with permission; **p. 97:** Courtesy of www.bookfactory.com; **p. 117:** Reproduced with permission of Medtronic, Inc, www.medtronic.com; **p.126:** *SnagIt* Newsletter, copyright 2007 by TechSmith Corporation, www.techsmith.com. Reprinted by permission; **pp. 130–131, 134:** Microsoft Word® is a registered trademark of Microsoft Corporation. Screen shots reprinted with permission from Microsoft Corporation; **p. 140:** Wag-A-Lot Press Release, May 11, 2005. Reprinted by permission of Craig B. Koch; **pp. 143–44, 158–167, 218–219:** Reprinted with permission from Steve Gough; **pp.189–204:** From *Investigation of the Low-Temperature Fracture Properties of Three MnRoad Asphalt Mixtures*, 2006–15 Final Report, www.lrrb.org. Reprinted by permission of the Minnesota Department of Transportation; **p. 221:** Copyright SurveyMonkey.com. Reprinted by permission of Ryan Finley, author/owner; **pp. 230–231:** Courtesy of The Toro Company; **pp. 235–241:** Courtesy Plastaket Manufacturing Co., Inc.; **p. 242:** Courtesy TechSmith Corporation; **p. 246:** Courtesy University of Minnesota; **p. 248:** Courtesy American Society for Nutrition, © 2007; **p. 249:** Reprinted by permission of All Media Guide LLC, www.allmusic.com. **p. 276:** Courtesy Rehabilitation International; **p. 283:** Courtesy of Geophysical Fluid Dynamics Laboratory, National Oceanic and Atmospheric Administration; 284, 285 Courtesy California Department of Fish and Game; **p. 286:** From Surface Water Management Plan (2005). Reprinted by permission of City of Shoreview, MN, www.ci.shoreview.mn; **pp. 287, 288:** Courtesy of Minnesota Pollution Control Agency; **p. 298:** VTC Tutorial for Adobe InDesign is the sole property of Virtual Training Company (www.vtc.com). All rights reserved. Reprinted by permission; **p. 304:** Courtesy of the City of Roswell, Georgia; **p. 306:** Courtesy Mark Dewell, Heritage Railway Association Webmaster; **p. 314:** Microsoft Word® is a registered trademark of Microsoft Corporation. Screen shot reprinted with permission from Microsoft Corporation; **p. 326:** Reprinted by permission of Center for Assistive Technology and Environmental Access (CATEA), www.catea.org; **p. 327:** Courtesy of Pliny McCovey, Jr; **p. 330:** The screen capture taken from pbs.org contains copyrighted material of the Public Broadcasting Service, WGBH Educational Foundation and Twin Cities Public Television. Reprinted with permission; **p. 336:** Courtesy of The Lawrence Berkeley National Lab; **p. 337:** Adobe product screen shot reprinted with permission from Adobe Systems Incorporated; **p. 340:** Google™ is a trademark of Google, Inc. Reprinted by permission; **p. 343:** From *CIO* magazine webpage. Reprinted by permission of CIO magazine, www.cio.com; **pp. 345–346:** Microsoft Office Project® is a registered trademark of Microsoft Corporation. Screen shots reprinted with permission from Microsoft Corporation; **p. 348:** Microsoft Excel® is a registered trademark of Microsoft Corporation. Screen shot reprinted with permission from Microsoft Corporation; **pp. 350, 368:** Microsoft Windows Movie Maker® is a registered trademark of Microsoft Corporation. Screen shot reprinted with permission from Microsoft Corporation.; **p. 355:** Courtesy of International Association for Food Protection; **p. 360:** Photo courtesy of Anchor Optics®; **p. 361:** © 2008 A.D.A.M., Inc. Reprinted by permission; **p. 363:** Source: NASA and Ann Field (STScI); **p. 367:** From *Introduction to 3D Spatial Visualization: An Active Approach 1st edition* by Sorby, 2003. Reprinted with permission of Delmar Learning, a section of Thomson Learning: www.thomsonrights.com. Fax 800 730–2215; **p. 372:** Source: NOAA, Great Lakes Environmental Research Laboratory; **p. 379:** Adobe product screen shot reprinted with permission from Adobe Systems Incorporated; **p. 389:** Courtesy University of Minnesota; **p. 391:** Reprinted by permission, www.RefWorks.com; **pp. 395, 398, 404:** © 2007, Regents of the University of Minnesota. All Rights Reserved. Reprinted by permission; **p. 403:** Reprinted with permission of the Society for Technical Communication; **p. 407:** Google™ is a trademark of Google, Inc. Reprinted by permission; **pp. 413–414:** © University of Pittsburgh, 2006. All rights reserved. Reprinted by permission; **pp. 416–417:** From "A Review of Wind-Resource-Assessment Technology" by Shikha Singh, T. S. Bhatti and D. P. Kothari, *Journal of Energy Engineering*, (2006), Vol. 132, No. 1. Copyright © 2006, ASCE. Reprinted with permission from ASCE. All rights reserved; **p. 422:** Source: U.S. Fish and Wildlife Service; **p. 430:** Reprinted by permission of American Wind Energy Association; **p. 435:** Copyright SurveyMonkey.com. Reprinted by permission of Ryan Finley, author/owner; **pp. 439–443:** Courtesy Danish Wind Industry Association, windpower.org; **p. 445:** Copyright © 2000, 2001, 2002 Free Software Foundation, Inc., 51 Franklin Street, Fifth Floor, Boston, MA 02110-1301 USA. Everyone is permitted to copy and distribute verbatim copies of this license document, but changing it is not allowed.

Photo Credits

Page 98: The Thomas Edison Papers, microfilm edition. Courtesy, Rutgers University; **p. 268:** Scott Adams, Inc./Dist. By UFS, Inc./United Media; **p. 269:** UFS, Inc./United Media; **pp. 334–335:** Idaho National Laboratory; **p. 365:** AV Digital Productions, Ltd. (AVDP); **p. 375:** Centers for Disease Control; **p. 405:** From: Benjamin Franklin: In His Own Words. Manuscript essay of letter to Jan Ingenhousz, 1777. Manuscript section/Library of Congress; **pp. 409–410:** NASA Jet Propulsion Laboratory; **p. 437:** Walter Geiersperger/Corbis.

Index

Curriculum vitae (c.v.), 78, 82. *See also*
    Resumes
Customization of data, 284

Dash (—), 468, 501
data, 558
Database publishing. *See* Content manage-
    ment systems
Databases, 398
    delivery format and, 14
    as secondary source of information, 406
Data display, 278–290
    emphasizing specifics and, 279
    highlighting trends/relationships and,
        280–282
    individual customization of data and,
        284–288
    qualitative data visuals and, 285–290
    quantitative data visuals and, 279–285
    reporting qualitative research results and,
        288–290
    visualizing complex numerical data
        and, 282
Dates, 517–519
    abbreviations in, 450
    APA documentation style and, 581
Decorative fonts, 316
Deductive argument, 27, 30
Definite article (the), 23, 459, 514
Definition, paragraph development
    and, 523
Delivery formats, 14, 118
Demographics
    of audiences, 5
    web site design and, 382
Demonstrative pronouns, 533
Dependent clause, comma and, 475
Descriptive documents, 15, 17
Design considerations, 14, 135
    ethics and, 18
    for blogs, 58
    for brochures, 62
    for email, 66
    for FAQs, 75
    for forms, 71
    for instructions, 77
    for job descriptions/job advertisements, 81
    for laboratory notebooks, 103
    for letters, 114
    for medical information, 118
    for memos, 123
    for newsletters, 132
    for press releases, 141
    for product descriptions, 145
    for proposals, 154

    for reports, 216
    for resumes, 93
    for specifications, 219
    for surveys, 222
    for user manuals, 243
    for web sites, 250
    for white papers, 260
    for wikis, 58
Designing
    callouts, 307
    captions, 300
    charts, 275
    documents, 291–318, 374
    for pocket technologies, 339–341
    graphs, 322
    headings, 303
    labels, 307
    lists, 310
    maps, 324
    storyboards, 350
    tables, 358
    tables of contents, 313
    visuals and text to work together, 372
    web pages, 381–386
despite, 531
Detailed lists, assembly manuals and, 232
Development, classical modes of, 17
Diagrams
    assembly manuals and, 233
    procedure manuals and, 231
Dictionaries, 399
    APA documentation style for, 576
    CSE documentation style for, 586
    IEEE documentation style for, 596
    MLA documentation style for, 607
Digidesign Pro Tools, 267
Digital cameras
    digital video (DV) cameras and, 364
    photograph file formats/quality
        and, 332
Digital communication, 57. *See also* Blogs;
    Email; Wikis
    audience considerations and, 57
    plagiarism and, 431
Digital documents, laboratory notebooks
    and, 98
Digital Object Identifier (DOI), 573
Digital phones, 339
Digital photographs, editing, 336
Digital research, strategies for, 400
Digital resumes, 81, 93–95
Digital video (DV) cameras, 364
Direct articles, translation/localization
    and, 23
Director (Adobe), 350

Fragments, 545
Frequently Asked Questions (FAQs), 73–75
from, 531
Front matter, in reports, 156
furthermore, 483, 548
Future perfect tense, 499
Future tense, 498

Gantt, Henry L., 274
Gantt charts, 31, 272, 274, 345
Gender-biased language, avoiding, 508
Geographical Information Systems (GIS),
   284, 326
Geographic proximity, context and, 29
Geology, field notebooks for, 100
Gerunds (*-ing*), 528
  headings and, 302
  user guides and, 226
GIF files, 333, 380
GIMP (GNU Image Manipulation Pro-
   gram), 337
GIS (Geographical Information Systems),
   284, 326
Global audiences. *See* International
   communication
Global positioning systems (GPS), 325,
   339
Glossaries, research reports and, 188
Gobbledygook, 466
Google
  Picasa and, 337
  as research tool, 399, 400
Google Earth, 327
Google Maps, 324
Government documents
  APA documentation style for, 576
  CSE documentation style for, 586
  IEEE documentation style for, 597
  MLA documentation style for, 608
Government web sites, 247
GPS (global positioning systems), 325, 339
Grammar, guidelines for, 447–566
Graphic Interchange Format (GIF) files,
   333, 380
Graphs, 319–323
  vs. charts, 271
  data display considerations and, 280–282
  ethics and, 18
  procedures/processes and, 45
  visual persuasion and, 374
  white papers and, 260
Grids, 298
Group projects, project management for, 31
grow, 552

Guidelines
  for animation, 369
  for audio, 267
  for blogs, 59
  for brochures, 62
  for callouts, 308
  for captions, 301
  for cartoons/comics, 270
  for charts, 275
  for clip art, 278
  for color use, 380
  for cover letters, 85
  for data display, 390
  for email, 67
  for FAQs, 75
  for field notebooks, 103
  for fonts, 318
  for forms, 72
  for grammar, 447–566
  for graphs, 323
  for headings, 303
  for instructions, 77
  for job descriptions/job advertisements, 82
  for job interviews, 87
  for labels, 308
  for laboratory notebooks, 103
  for letters, 115
  for lists, 311
  for literature reviews, 426
  for maps, 328
  for medical communication, 119
  for memos, 124
  for multimedia, 331
  for newsletters, 132
  for online help, 135
  for page layout, 299
  for photographs, 338
  for pocket technologies, writing for, 341
  for podcasts, 344
  for presentations, 139
  for press releases, 142
  for product descriptions, 145
  for project management visuals, 346
  for proposals, 155
  for punctuation, 447–566
  for reports, 217
  for resumes, 96
  for specifications, 220
  for spreadsheets, 349
  for storyboards, 353
  for surveys, 223, 435
  for symbols, 357
  for tables, 359
  for tables of contents, 315

Mute button, web page design and, 384
my, 533

Names
  abbreviations preceding, 449
  capitalization and, 462
Name-year system, CSE documentation
    style and, 583
National Institutes on Drug Abuse
    (NIDA), research reports and,
    184
Navigation. *See also* Links
  context-specific symbols and, 355
  page layout and, 291
  storyboards and, 352
  web site design and, 384
Navigational aids, 5
Navigation bars, 383–385
near, 531
nearby, 549
Needs assessments, 7–11
  sample of, 10
  template for, 8–10
neither... nor, 484, 486
Netscape Composer, 245
Neutral point of view (NPOV), 55, 58
nevertheless, 483, 548
Newsletters, 125–133
  electronic, 126
  print, 127
  templates for, 129
Newspaper articles
  APA documentation style for, 577
  CSE documentation style for, 587
  IEEE documentation style for, 597
  MLA documentation style for, 609
  as secondary source of information, 406
Newspaper editorials
  APA documentation style for, 577
  IEEE documentation style for, 598
  MLA documentation style for, 609
next, 548
NIDA (National Institutes on Drug Abuse),
    research reports and, 184
nobody, 534
Nominalizations, avoiding, 478
Nondefinite count nouns, 458
Nonlinear argument, 30
Nonlinking verbs, 551, 552
Nonrestrictive clause, comma and, 473
no one, 534
nor, 548
  comma and, 469

compound subject and, 486
  as coordinating conjunction, 483
"Note" section, product manuals and, 239
not only... but also, 484
Noun phrases, translation/localization
    and, 24
Nouns, 514–516
  as antecedent of pronouns, 532
  appositives and, 529
  avoiding strings of, 478
  collective, 487
  compound, 504
  forms/kinds of, 514
  nondefinite count, 458
  plural in form but singular in meaning,
    488
  plural of, 515
  possessive, 455–457
now, 548
now that, 484
NPOV (neutral point of view), 55, 58
number/amount, 556
Numbered headings, feasibility reports and,
    171
Numbered items
  procedure manuals and, 231
  product manuals and, 241
Numbered list of references, CSE documen-
    tation style for, 591
Numbered lists, 309
  in consulting reports, 161
  quick reference guides and, 242
Numbered sections, white papers and, 255
Numbered steps
  assembly manuals and, 234
  product manuals and, 238
  user guides and, 227
Numbers, 517–519
  abbreviations and, 450
  compound, 506
  hyphen and, 506
Numerical data, visualizing, 282
with this object in mind, 549

Objectivity, sources and, 404
Observations, as primary source of informa-
    tion, 406
Occupational Health and Safety Administra-
    tion (OSHA), 51
of, 531
off, 531
Older audiences, computer use by, 4
on, 531

Revision Symbols

Boldface numbers refer to pages in the handbook.

ab	abbreviation	**449**	shift	shift in voice or number	**491**
agr	agreement	**485**	sub	subordination	**483**
awk	awkward construction	**476**	t	verb tense error	**496**
cap	capitalization	**461**	trans	transition needed	**548**
coh	coherence	**520**	var	sentence variety	**544**
coord	coordination	**483**	vb	verb form error	**550**
cs	comma splice	**474**	w	wordy	**476**
d	diction, word choice	**553**	ww/wc	wrong word/word choice	**553**
dev	development needed	**521**	//	faulty parallelism	**493**
dm	dangling modifier	**452**	.?!	end punctuation	**525**
doc	check documentation	**567**	:	colon	**467**
frag	sentence fragment	**545**	'	apostrophe	**455**
fs	fused sentence	**545**	--	dash	**504**
hyph	hyphen	**504**	( )	parentheses	**527**
ital	italics	**512**	[ ]	brackets	**459**
lc	lowercase letter	**461**	...	ellipsis	**503**
mm	misplaced modifier	**452**	/	slash	**542**
no ¶	no paragraph needed	**519**	;	semicolon	**542**
num	number	**517**	" "	quotation marks	**469**
¶	paragraph	**519**	,	comma	**469**
¶ dev	paragraph pronoun needed	**521**	⌒	close up	
ref	unclear paragraph reference	**466**	∧	insert missing element	
sp	spelling error	**555**	ℓ	delete	
			∾	transpose order	